Mathematical Methods for Physicists and Engineers

Second Corrected Edition

Royal Eugene Collins

DOVER PUBLICATIONS, INC.
Mineola, New York

Bibliographical Note

This Dover edition, first published in 1999, is an unabridged and corrected republication of the first edition of *Mathematical Methods for Physicists and Engineers* published by Reinhold Book Corporation in 1968.

Library of Congress Cataloging-in-Publication Data

Collins, Royal Eugene.
    Mathematical methods for physicists and engineers / Royal Eugene Collins. — 2nd corr. ed.
        p.   cm.
    Includes bibliographical references and index.
    ISBN 0-486-40229-0 (pbk.)
    1. Mathematics. 2. Mathematical physics. 3. Engineering mathematics. I. Title.
QA37.2.C656   1999
510—dc21   98-28913
                                                                                      CIP

Manufactured in the United States of America
Dover Publications, Inc., 31 East 2nd Street, Mineola, N.Y. 11501

ACKNOWLEDGMENTS

In preparing this text I have drawn heavily on a great number and variety of texts on applied mathematics. Many are cited as desirable references for the student. I am indebted to all of these authors as my teachers. I am also indebted to my students over the years who, by their responses, have taught me how to teach this material in a more effective way each year.

I express my appreciation to Mrs. Gloria Parr and Mrs. Shirley Double for their patience with my poor handwriting in typing this manuscript, and to my son, Mr. Roy Collins, for his skillful preparation of the illustrations.

R.E.C.

INTRODUCTION

The physical sciences and engineering have become increasingly mathematical during the 20th century, and indeed the same trend is also observed in the biological sciences.

Advanced undergraduates, or beginning graduate students, in these fields find themselves in need of a certain *mathematical* tools in order to comprehend the physical principles, and their applications, encountered in their major courses. Here we emphasize *mathematical tools* rather than mathematics because the student of physics or engineering must *use* these elements of mathematics as *tools* to achieve the desired understanding of the physics rather than simply comprehend the mathematics itself. This distinction is important.

The mathematical tools most urgently needed, beyond a one-semester course in *ordinary* differential equations, as offered in the mathematics departments of most colleges and universities, are as follows: a knowledge of vector calculus, matrix algebra, and linear vector operations; the many and varied methods of solving linear boundary value problems, including a good familiarity with the more common special functions of mathematical physics; some knowledge of the calculus of variations, and variational and perturbation approximations applicable to boundary value problems and nonlinear differential equations; the ability to evaluate *complicated* integrals, or sum series; a facility for curve fitting and numerical approximation methods; a familiarity with the basic elements or probability and how these apply to physical problems; and, finally, some knowledge of integral equations.

This is a formidable list, and if the student were to try to *completely* master these many areas form the mathematician's point of view, it would be more than a lifetime task. However, the *essential* needs of the student can be met without such complete mastery.

Most universities attempt to meet this need with a one-year course in the mathematics department, usually called "Mathematics for Engineers and Physicists" and commonly taught in the student's junior year. However, rarely is even a small part of the above list actually treated in the course. As a result, a great amount of time must usually be devoted to the teaching of mathematics in *each* and *every* course in physics or engineering. This situation exists partially as a result of the form of most texts available for such a course and partially because of the method of presentation in the mathematics class.

The text presented here is intended to help remedy this situation in two ways: first, by including an adequate treatment of all the needed topics in *applied* mathematics; and second by presenting the material in the form of *programmed* instruction. The text has been designed to be entirely readable so that the material can be comprehended by advanced undergraduate, or beginning graduate, students without *any* assistance from the instructor. Consequently, class time can be devoted almost entirely to reviewing the problems executed by the student, thus dispensing with formal lectures on the mathematical methods.

It is recommended that the student be assigned a segment to read along with a corresponding set of problems to solve. These can then be reviewed and discussed in the following class session. Occasionally, the instructor may find it necessary to work out some of the more difficult problems or to enlarge on a particular topic with additional examples. Further, some of the text segments and problems sets are rather long and may require two or three class sessions to cover adequately. Nevertheless, it should be possible to cover all of this text in two semesters if this method of teaching is used. The author has found it to be most effective over a period of several years using essentially the material of this text as prepared notes issued to students.

Some of my mathematician friends may question the lack of rigor in this text, but they must remember the objective of the text and forgive me. The intent is to *eliminate* most of this *same* material from present physics and engineering courses by putting *all* of it into *one* course. The mathematician may wish to call the course something other than Mathematics; I prefer Mathematical Tools. This compaction of *applied* topics will also leave physics and engineering students free to include in their curricula additional courses in *pure* mathematics, subjects such as point set theory, topology, and others that are strongly recommended, especially for physics students. In addition, this method will allow students to concentrate on *physics,* or *engineering,* in their major courses. The end result is that courses in physics and engineering will be stronger, and students better educated.

R. E. Collins

CONTENTS

CHAPTER ONE

Elementary vector calculus; the vector field

Here we present a brief but essentially complete presentation of those elements of vector calculus essential for the physicist or engineer. A knowledge of the elementary *algebra* of vectors is assumed.

THE POSITION VECTOR AND THE LINE ELEMENT; COORDINATE TRANSFORMATIONS

We consider a point in space defined with reference to a set of *rectangular cartesian coordinates* by the three numbers, x_1, x_2, x_3, which are the coordinates of the point, or the orthogonal projections of the point, on the *three*† axes. The directed line from the origin to the point is the *position vector of the point*. We define the base set of unit vectors on these axes as $1_1, 1_2, 1_3$ where‡

(1.1)
$$\begin{cases} 1_i \cdot 1_j = \delta_{ij} = \begin{cases} 1, i = j \\ 0, i \neq j \end{cases} \\ 1_i \times 1_j = 1_k; \qquad i, j, k \text{ in cyclic order} \end{cases}$$

---

† In the next chapter we consider higher dimensional spaces.
‡ *Cyclic Order* means the permutations of order of indices as:

$$(i, j, k) \to (k, i, j) \to (j, k, i) \to (i, j, k)$$

Then we write,

$$(1.2) \qquad\qquad \mathbf{r} = \sum_{i=1}^{3} \mathbf{1}_i x_i$$

as the representation of the position vector on these axes.

Now the *differential displacement vector* is just given by,

$$(1.3) \qquad\qquad \mathbf{dr} = \sum_{i=1}^{3} \mathbf{1}_i \, dx_i$$

since these unit vectors, $\mathbf{1}_i$, $i = 1, 2, 3$, have the same magnitude *and direction* at all points of space, i.e., $\mathbf{dr}$ is constructed as,

$$(1.4) \qquad\qquad \mathbf{dr} = \lim_{\mathbf{r}' \to \mathbf{r}} (\mathbf{r}' - \mathbf{r})$$

or

$$(1.5) \qquad\qquad \mathbf{dr} = \lim_{\mathbf{r}' \to \mathbf{r}} \sum_{i=1}^{3} (\mathbf{1}_i' x_i' - \mathbf{1}_i x_i)$$

but, since for any two points we have $\mathbf{1}_i' = \mathbf{1}_i$, the unit vectors factor out and Eq. (1.3) results.

The magnitude of $\mathbf{dr}$ is the *line element*, ds. This can be constructed as

$$(1.6) \qquad\qquad ds = |\mathbf{dr}| = (\mathbf{dr} \cdot \mathbf{dr})^{\frac{1}{2}}$$

or

$$(1.7) \qquad\qquad (ds)^2 = \sum_{i=1}^{3} (dx_i)^2$$

by virtue of Eq. (1.1).

Now we do not always use *rectangular orthogonal* axes. Thus we now examine what happens to $\mathbf{dr}$ and ds under some *arbitrary change of variables*,

$$(1.8) \qquad\qquad x_i = x_i(q_1, q_2, q_3), \qquad \hat{i} = 1, 2, 3$$

Thus,

$$(1.9) \qquad\qquad dx_i = \sum_{j=1}^{3} \frac{\partial x_i}{\partial q_j} dq_j, \qquad i = 1, 2, 3$$

and these can be inserted into Eq. (1.3) for the dx's and the *order* of summation interchanged to give,

$$(1.10) \qquad\qquad \mathbf{dr} = \sum_{j=1}^{3} \left(\sum_{i=1}^{3} \frac{\partial x_i}{\partial q_j} \mathbf{1}_i \right) dq_j$$

In these equations the dq's may have *any* dimensions, or units, but the x's *must* all have the same dimensions in order to maintain dimensional homogeniety of our equations. We also note that the quantities in parenthesis in Eq. (1.10) are vectors in the rectangular space, but having components, $\partial x_i / \partial q_j$, which are also *not* dimensionally homogeneous. Thus now multiply and divide each term here by a factor, h_j, so that we have,

$$(1.11) \qquad\qquad \mathbf{dr} = \sum_{j=1}^{3} \left(\frac{1}{h_j} \sum_{i=1}^{3} \frac{\partial x_i}{\partial q_j} \mathbf{1}_i \right) h_j \, dq_j$$

and require that the h_j be chosen so that the factors $h_j\,dq_j$ all have the same dimensions as the x_i.

Furthermore we now note that the vectors forming each parenthesis above, denoted now as,

$$(1.12) \qquad \mathbf{1}'_j = \frac{1}{h_j} \sum_{i=1}^{3} \frac{\partial x_i}{\partial q_j} \mathbf{1}_i$$

are dimensionless and will be *unit vectors*, if we require,

$$(1.13) \qquad \mathbf{1}'_j \cdot \mathbf{1}'_j = \frac{1}{h_j^2} \sum_{i=1}^{3} \left(\frac{\partial x_i}{\partial q_j}\right)^2 = 1$$

for each j. This condition then determines the h_j in terms of the coordinate transformation by the relations,

$$(1.14) \qquad h_j^2 = \sum_{i=1}^{3} \left(\frac{\partial x_i}{\partial q_j}\right)^2, \qquad j = 1, 2, 3$$

so the h_j are fixed except for sign by these equations†.

We now have the expression for $\mathbf{dr}$ in the form,

$$(1.15) \qquad \mathbf{dr} = \sum_{j=1}^{3} \mathbf{1}'_j h_j\,dq_j$$

and here let us again look at the square of the line element, ds^2. We have,

$$(1.16) \qquad (ds)^2 = \mathbf{dr} \cdot \mathbf{dr} = \sum_{j=1}^{3} \sum_{k=1}^{3} (\mathbf{1}'_j \cdot \mathbf{1}'_k) h_j h_k\,dq_j\,dq_k$$

This is distinctly different from our Eq. (1.7) for the rectangular system but it reduces to a very *similar* form *if* we require,

$$(1.17) \qquad \mathbf{1}'_j \cdot \mathbf{1}'_k = 0, \qquad j \neq k$$

This states that these new unit vectors are mutually *orthogonal*, or perpendicular. Hence for *any orthogonal coordinate system* (i.e., Eq. (1.17) satisfied) we have,

$$(1.18) \qquad (ds)^2 = \sum_{j=1}^{3} h_j^2 (dq_j)^2$$

and it is just such systems of coordinates that we will nearly always employ.

Note that the *criterion of orthogonality*, Eq. (1.17), has the form,

$$(1.19) \qquad \sum_{i=1}^{3} \frac{\partial x_i}{\partial q_j} \frac{\partial x_i}{\partial q_k} = 0, \qquad j \neq k$$

if we make use of Eqs. (1.12) which define the $\mathbf{1}'_j$ vectors.

SCALE FACTORS; AREAS AND VOLUMES

The elements $h_j, j = 1, 2, 3$, introduced above are called the scale factors of the "curvilinear" coordinate system, q_j. Here we note how area elements, and volume elements, are expressed in terms of these h_j and the dq_j.

---

† We always take the h_j *positive* since the signs on the h_j factors cancel in Eq. (1.11) for $\mathbf{dr}$.

Consider Eq. (1.15) for $\mathbf{dr}$ and choose $dq_2 = dq_3 = 0$, then we have a vector displacement we will call $\mathbf{dr}_1$ given by

(1.20) $$\mathbf{dr}_1 = \mathbf{1}'_1 h_1 \, dq_1$$

This is in the direction of $\mathbf{1}'_1$, for $dq_1 > 0$, and is a *line element of length* $h_1 \, dq_1$, from the definition of $\mathbf{dr}$. Similarly if *at the same point in space* we let $dq_1 = dq_3 = 0$ we form another line element,

(1.21) $$\mathbf{dr}_2 = \mathbf{1}'_2 h_2 \, dq_2$$

in the direction of $\mathbf{1}'_2$, $dq_2 > 0$, and of *length, $h_2 \, dq_2$.*
   The vector product, or cross product, is,

(1.22) $$\mathbf{dr}_1 \times \mathbf{dr}_2 = (\mathbf{1}'_1 \times \mathbf{1}'_2) h_1 h_2 \, dq_1 \, dq_2 = \mathbf{dA}_3$$

which in magnitude is the *area* of the parallelogram indicated in Fig. 1-1, with

(1.23) $$|\mathbf{1}'_1 \times \mathbf{1}'_2| = \sin \theta$$

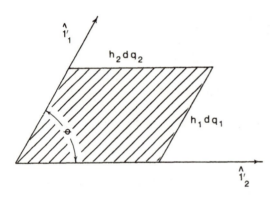

FIGURE 1.1.   Element of Area on the $q_1 q_2$ Plane.

Obviously if the coordinate system is *orthogonal*, $\sin \theta = 1$. Also we note that the direction of this *vector* element of area is in the direction of $\mathbf{1}'_3$, i.e., from our definitions above we have the cyclic rule

(1.24) $$\mathbf{1}'_i \times \mathbf{1}'_j = \mathbf{1}'_k \qquad (i, j, k \text{ in cyclic order})$$

just as for the original unit vectors.
   If we now form the scalar, or dot, product of a line element,

(1.25) $$\mathbf{dr}_3 = \mathbf{1}'_3 h_3 \, dq_3$$

with $\mathbf{dA}_3$ we obtain,

(1.26) $$d\tau = (\mathbf{1}'_1 \times \mathbf{1}'_2 \cdot \mathbf{1}'_3) h_1 h_2 h_3 \, dq_1 \, dq_2 \, dq_3$$

as the *volume* of the elemental parallelopiped. Obviously, for *orthogonal systems* this is just

(1.27) $$d\tau = h_1 h_2 h_3 \, dq_1 \, dq_2 \, dq_3$$

and we will restrict our consideration to such systems. Here $h_1 h_2 h_3$ is the Jacobian of the transformation from the x_i to the q_j.

INVERSE TRANSFORMATION

We call attention to the fact that we generally require that it be possible to carry out the inverse transformation,

$$(1.28) \qquad q_j = q_j(x_1, x_2, x_3), \qquad j = 1, 2, 3$$

so that,

$$(1.29) \qquad dq_j = \sum_{i=1}^{3} \frac{\partial q_j}{\partial x_i} dx_i, \qquad j = 1, 2, 3$$

Substitution of these expressions for the dq_j in Eq. (1.9) shows that the necessary conditions for this inverse to exist are,

$$(1.30) \qquad \sum_{j=1}^{3} \frac{\partial x_i}{\partial q_j} \frac{\partial q_j}{\partial x_l} = \delta_{il} = \begin{cases} 1, i = l \\ 0, i \neq l \end{cases}$$

Also, in order that we should be able to solve Eq. (1.9) for the dq_j as functions of the dx_i, it is required that the determinant of the coefficients should not vanish, i.e.,

$$(1.31) \qquad \mathrm{Det}\left(\frac{\partial x_i}{\partial q_j}\right) \neq 0$$

These ideas become more concrete when we consider an example.

Example (1)

We here examine the *spherical coordinate system* in terms of the general ideas outlined above. The equations of transformation from the rectangular system, now denoted as x, y, z, to the spherical system, r, θ, ϕ, are:

$$(1.32) \qquad \begin{cases} x = r \sin \theta \cos \phi \\ y = r \sin \theta \sin \phi \\ z = r \cos \theta \end{cases}$$

as the equivalent of Eq. (1.8).
The scale factors are

$$(1.33) \qquad h_r^2 = \left(\frac{\partial x}{\partial r}\right)^2 + \left(\frac{\partial y}{\partial r}\right)^2 + \left(\frac{\partial z}{\partial r}\right)^2$$

$$(1.34) \qquad h_\theta^2 = \left(\frac{\partial x}{\partial \theta}\right)^2 + \left(\frac{\partial y}{\partial \theta}\right)^2 + \left(\frac{\partial z}{\partial \theta}\right)^2$$

and

$$(1.35) \qquad h_\phi^2 = \left(\frac{\partial x}{\partial \phi}\right)^2 + \left(\frac{\partial y}{\partial \phi}\right)^2 + \left(\frac{\partial z}{\partial \phi}\right)^2$$

which reduce to,

$$(1.36) \qquad \begin{cases} h_r = 1 \\ h_\theta = r \\ h_\phi = r \sin \theta \end{cases}$$

if we take the *positive square roots*.

Next we verify that the r, θ, ϕ system is an orthogonal system, i.e.,

(1.37)
$$\frac{\partial x}{\partial r}\frac{\partial x}{\partial \theta} + \frac{\partial y}{\partial r}\frac{\partial y}{\partial \theta} + \frac{\partial z}{\partial r}\frac{\partial z}{\partial \theta} = 0$$

and similarly with (r, ϕ), (θ, ϕ) replacing (r, θ) here. This does indeed prove to be true for all these combinations as required by Eq. (1.14).

Having constructed the scale factors and shown that the r, θ, ϕ system is an *orthogonal* system we know from our discussion above that we can form a *unit* vector $\mathbf{1}_r$ in the direction of increasing r at *any* point in space, and similar vectors, $\mathbf{1}_\theta$ and $\mathbf{1}_\phi$, in the directions of increasing θ and ϕ, respectively, at this *same* point. These, $\mathbf{1}_r$, $\mathbf{1}_\theta$, $\mathbf{1}_\phi$ form a *local* orthogonal rectangular set of axes, i.e., as depicted in Fig. 1-2.

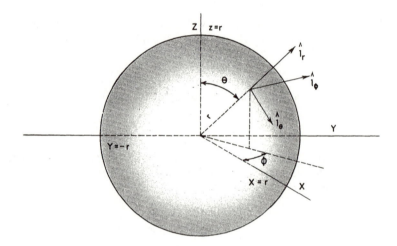

FIGURE 1.2.   Unit Vectors for the Spherical Coordinate System.

However, as can be seen from the equations defining these unit vectors, $\mathbf{1}_r$ for example,

(1.38)
$$\mathbf{1}_r = \mathbf{1}_x \sin \theta \cos \phi + \mathbf{1}_y \sin \theta \sin \phi + \mathbf{1}_z \cos \theta$$

formed as prescribed by Eq. (1.12), is *not a constant vector*. That is, although its *length* is always unity, its *direction* varies from point to point in space. For example we can *differentiate* $\mathbf{1}_r$ with respect to θ or ϕ and the result is *not zero*, whereas *all* derivatives of $\mathbf{1}_x$, $\mathbf{1}_y$, and $\mathbf{1}_z$ are zero.

From our general formulation we see that, for example,

(1.39)
$$\mathbf{1}_\theta \times \mathbf{1}_\phi r^2 \sin \theta \, d\theta \, d\phi = \mathbf{1}_r h_\theta h_\phi \, d\theta \, d\phi$$

is an element of area oriented in space such that the vector $\mathbf{r}$ from the origin is orthogonal to this area, i.e., this is the magnitude of the element of area on a spherical surface of radius r at the point r, θ, ϕ, and directed *outward* from the surface.

GRADIENT OF A SCALAR

We consider some continuous function, $F(q_j)$, of a set of *orthogonal* coordinates, $q_j, j = 1, 2, 3$, having continuous derivatives; at least continuous *first* derivatives. The *total* differential of this function is given by,

$$(1.40) \qquad dF = \frac{\partial F}{\partial q_1} dq_1 + \frac{\partial F}{\partial q_2} dq_2 + \frac{\partial F}{\partial q_3} dq_3$$

and represents the *difference* in the values of F at the two points $[(q_1 + dq_1), (q_2 + dq_2), (q_3 + dq_3)]$ and (q_1, q_2, q_3).

Note that if we multiply and divide the first term by h_1, the second by h_2, and the third by h_3 this appears as,

$$(1.41) \qquad dF = \left(\frac{1}{h_1}\frac{\partial F}{\partial q_1}\right)(h_1 \, dq_1) + \left(\frac{1}{h_2}\frac{\partial F}{\partial q_2}\right)(h_2 \, dq_2) + \left(\frac{1}{h_3}\frac{\partial F}{\partial q_3}\right)(h_3 \, dq_3)$$

and this has the *form* of the scalar, or dot product, of the displacement vector, **dr**, having components $h_i \, dq_i$, with another vector,

$$(1.42) \qquad \nabla F = \mathbf{1}'_1 \frac{1}{h_1}\frac{\partial F}{\partial q_1} + \mathbf{1}'_2 \frac{1}{h_2}\frac{\partial F}{\partial q_2} + \mathbf{1}'_3 \frac{1}{h_3}\frac{\partial F}{\partial q_3}$$

which we call the *gradient of F*, or just "grad F". Thus we see

$$(1.43) \qquad F(\mathbf{r} + \mathbf{dr}) - F(\mathbf{r}) = \nabla F \cdot \mathbf{dr} = dF$$

in *any orthogonal* coordinate system, with ∇F defined by Eq. (1.42).

Note that the *magnitude* of the gradient, $|\nabla F|$, is

$$(1.44) \qquad |\nabla F| = \left\{\sum_{j=1}^{3} \frac{1}{h_j^2}\left(\frac{\partial F}{\partial q_j}\right)^2\right\}^{\frac{1}{2}}$$

Hence we can write Eq. (1.43) above as,

$$(1.45) \qquad dF = |\nabla F| \, ds \cos \gamma$$

since $ds = |\mathbf{dr}|$, and we here define γ as the angle between the direction of ∇F and $\mathbf{dr}$. This follows from the definition of the scalar product. Since $|\nabla F|$, being just a function of the q_j, has a fixed value at any point in space, and we can choose some *fixed magnitude* for the displacement ds, we ask *in what direction* must we make the displacement in order to achieve *maximum change in F*, i.e., $dF = $ maximum. Quite obviously this occurs for $\gamma = 0$, or $\cos \gamma = 1$. Thus the gradient vector is always directed in the direction of the maximum space rate of change of F.

Also observe that $dF = 0$ corresponds to $\gamma = \pi/2$, but since $dF = 0$ defines the *surface of constant F*, we see that the vector ∇F is *always at right angles to the surfaces of constant F*. This is indicated in Fig. 1-3.

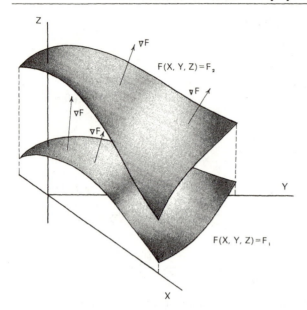

FIGURE 1.3. Surfaces of Constant $F(x, y, z)$ and Their Relation to the Gradient Vector, ∇F.

EXECUTE PROBLEM SET (1-2)

SURFACE INTEGRALS IN VECTOR FIELDS

Just above we saw that we could form a vector, ∇F, from some scalar function, $F(q_j)$ in a certain way. The components of this vector were *each* a function of the coordinates. We call such a vector a *vector field*. A vector field assigns, at each point of space, a vector, say $V(r)$, whose direction and magnitude will in general be different at each point, r, of this space.

Such vector fields may be formed in many ways, the gradient field is just a *very* special case. Thus we write for the general vector field,

$$(1.46) \qquad \mathbf{V} = \mathbf{1}'_1 V_1(\mathbf{r}) + \mathbf{1}'_2 V_2(\mathbf{r}) + \mathbf{1}'_3 V_3(\mathbf{r})$$

where the components, $V_i(\mathbf{r})$, are distinct functions of the coordinates which we here indicate in the general way as the position vector, $\mathbf{r}$.

Now we visualize some surface, S, in this space as indicated in Fig. 1-4. We *choose* a positive side for the surface, then subdivide it into infinitesimal elements of area, ΔS_i. At the center of each such element we erect a unit vector, $\mathbf{n}_i$, normal to the area and, in this way, have the vector element of area

$$(1.47) \qquad \Delta \mathbf{S}_i = \mathbf{n}_i \, \Delta S_i$$

such as for element number 1 in our picture.

Also at the center of each element we evaluate the vector field, i.e., $\mathbf{V}(\mathbf{r}_1)$ for the first element in our picture, then form the sum, I_N,

$$(1.48) \qquad I_N = \sum_{i=1}^{N} \mathbf{V}(\mathbf{r}_i) \cdot \Delta \mathbf{S}_i$$

over all such area elements making up S, here assumed to be N in number.

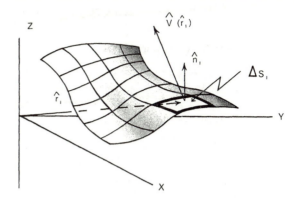

FIGURE 1.4. Subdivision of a Surface for Construction of the Surface Integral of a Vector Function.

In the formal limiting process we begin with "small" elements ΔS_i then show that as *all* $\Delta S_i \to 0$ and $N \to \infty$, in *any* way the above sum approaches a unique limit. In such fashion we define the *surface integral of* **V** *over S* as,

$$(1.49) \qquad\qquad I = \int_S \mathbf{V(r)} \cdot \mathbf{dS}$$

Example (2)

Compute the vertical force on a hemisphere, whose base is horizontal, when the external curved surface, above this base, is exposed to a uniform pressure P.

Since by definition of the pressure the force on an element of area **dS** on the surface, with *outward* normal, is $-P\,\mathbf{dS}$, we have as the *vertical component* of this force†

$$(1.50) \qquad\qquad dF_z = \mathbf{1}_z \cdot (-P\,\mathbf{dS}) = (-P\mathbf{1}_z) \cdot \mathbf{dS}$$

Then since in spherical coordinates

$$(1.51) \qquad\qquad \mathbf{dS} = \mathbf{1}_r r^2 \sin\theta\, d\theta\, d\phi$$

we have, using Eq. (1.38) for the representation of $\mathbf{1}_r$,

$$(1.52) \qquad\qquad dF_z = -Pr^2 \sin\theta \cos\theta\, d\theta\, d\phi$$

Integrating and observing that both P and r^2 are constant and can be brought outside the integral, we obtain

$$(1.53) \qquad F_z = \int_S (-P\mathbf{1}_z) \cdot \mathbf{dS} = -Pr^2 \int_0^{2\pi} \int_0^{\pi/2} \sin\theta \cos\theta\, d\theta\, d\phi$$

so that

$$(1.54) \qquad\qquad F_z = -\pi r^2 P$$

where the minus sign indicates the resultant is in the direction of the $-z$ axis.

† Note that here the vector **V** of Eq. (1.49) is $(-\mathbf{1}_z P)$.

DIVERGENCE OF A VECTOR

Suppose we have given a vector field, $V(r)$, and select a volume τ bounded by a surface S. Then we can form the integral of V over the *closed surface* S and divide the result by the inclosed volume τ. If $V(r)$ is a *continuous vector function* in the infinitesimal neighborhood of a point r then the limit

(1.55)
$$\text{Div } V(r) = \lim_{\tau \to 0} \frac{\oint_S V \cdot dS}{\tau}$$

will exist.† Here S encloses the point r in the limit, and we call the quantity *defined* here the *divergence* of V at r.

Before proceeding with further analytical enlargement on this definition it will be helpful to have some physical concept of the meaning of the divergence of a vector. This is best conveyed by thinking here of V as the velocity of a fluid flowing in space. Then $V \cdot dS$ represents the volume of fluid per unit time *flowing out* from the volume τ through the area dS. Thus the numerator on the right in Eq. (1.55) represents the total volume of fluid per unit time *leaving* τ. As we let $\tau \to 0$ we see $\text{Div}[V(r)]$ as the volume of fluid per unit time, *per volume of space*, leaving the neighborhood of the point r, i.e., as pictured in Fig. 1-5.

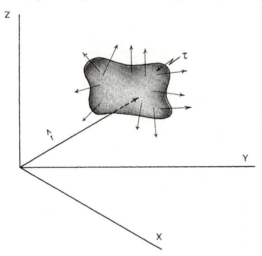

FIGURE 1.5.   Flow through the Surface of a Volume Element Inclosing a Point.

Note that our definition says nothing about the *shape* of the volume at any stage of the limiting process. We will not go through the process of a mathematical proof here, but we state that in the limit the shape does not matter. Physically one can see that in the case of an *incompressible* fluid, the amount per unit time passing through one surface will also pass through another surface inclosing the first. Parallel ideas are used in the mathematical proof.

Let us then *choose* our volume element in the form of a rectangular parallelopiped with sides $h_1 \Delta q_1$, $h_2 \Delta q_2$, $h_3 \Delta q_3$ surrounding the point $r = (q_1, q_2, q_3)$, as indicated in Fig. 1-6. Observe that the contribution to the surface integral of $V \cdot dS$ over S on the end shaded and

---

† We use the symbol $\oint$ to indicate an integration over a *closed* domain.

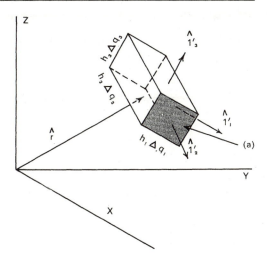

FIGURE 1.6. Volume Element in the Curvilinear Coordinates q_1, q_2, q_3.

indicated by (a) is approximately,

(1.56) $$\int_{(a)} \mathbf{V} \cdot \mathbf{dS} \approx (V_2 h_1 h_3)_{\bar{q}_1, q_2 + \frac{\Delta q_2}{2}, \bar{q}_3} \Delta q_1 \Delta q_3$$

where $\bar{q}_1$, $\bar{q}_3$ indicate values of these coordinates at some mean point on this end. (This is just an application of the mean value theorem of calculus.)

We can write similar expressions for the other surfaces, for example on the end opposite to (a) we have,

(1.57) $$\int_{(a')} \mathbf{V} \cdot \mathbf{dS} = -(V_2 h_1 h_3)_{\bar{q}_1, q_2 - \frac{\Delta q_2}{2}, \bar{q}_3} \Delta q_1 \Delta q_3$$

where the minus sign results because the area element vector is *outward* from the surface while $\mathbf{V}$ is inward so that the cosine of the angle between $\mathbf{V}$ and $\mathbf{dS}$ is (-1). Collecting all such integrals and adding to form the integrand of Eq. (1.55), then dividing by the volume, $\Delta \tau = h_1 h_2 h_3 \Delta q_1 \Delta q_2 \Delta q_3$, we have:

(1.58)
$$\frac{\oint_S \mathbf{V} \cdot \mathbf{dS}}{\tau} = \frac{(h_2 h_3 V_1)_{q_1 + \frac{\Delta q_1}{2}, \bar{q}_2, \bar{q}_3} - (h_2 h_3 V_1)_{q_1 - \frac{\Delta q_1}{2}, \bar{q}_2, \bar{q}_3}}{h_1 h_2 h_3 \Delta q_1}$$
$$+ \frac{(h_1 h_3 V_2)_{\bar{q}_1, q_2 + \frac{\Delta q_2}{2}, \bar{q}_3} - (h_1 h_3 V_2)_{\bar{q}_1, q_2 - \frac{\Delta q_2}{2}, \bar{q}_3}}{h_1 h_2 h_3 \Delta q_2}$$
$$+ \frac{(h_1 h_2 V_3)_{\bar{q}_1, \bar{q}_2, q_3 + \frac{\Delta q_3}{2}} - (h_1 h_2 V_3)_{\bar{q}_1, \bar{q}_2, q_3 - \frac{\Delta q_3}{2}}}{h_1 h_2 h_3 \Delta q_3}$$

Hence taking limits as all $\Delta q_j \to 0$ we obtain,

(1.59) $$\mathrm{Div}\ \mathbf{V} = \frac{1}{h_1 h_2 h_3} \left[\frac{\partial}{\partial q_1}(h_2 h_3 V_1) + \frac{\partial}{\partial q_2}(h_1 h_3 V_2) + \frac{\partial}{\partial q_3}(h_1 h_2 V_3) \right]$$

in *any* orthogonal curvilinear coordinate system.

Example (3)

The divergence of **V** in spherical coordinates can readily be written down since we have already shown that for $q_1, q_2, q_3 \rightarrow r, \theta, \phi$ we have the scale factors as given in Eq. (1.36). Thus

(1.60) $$\text{Div } \mathbf{V} = \frac{1}{r^2 \sin \theta} \left[\frac{\partial}{\partial r}(r^2 \sin \theta V_r) + \frac{\partial}{\partial \theta}(r \sin \theta V_\theta) + \frac{\partial}{\partial \phi}(r V_\phi) \right]$$

which simplifies to,

(1.61) $$\text{Div } \mathbf{V} = \frac{1}{r^2} \frac{\partial}{\partial r}(r^2 V_r) + \frac{1}{r \sin \theta} \frac{\partial}{\partial \theta}(\sin \theta V_\theta) + \frac{1}{r \sin \theta} \frac{\partial V_\phi}{\partial \phi}$$

for spherical coordinates. Here we have used the obvious notation,

(1.62) $$\mathbf{V} = \mathbf{1}_r V_r + \mathbf{1}_\theta V_\theta + \mathbf{1}_\phi V_\phi$$

for the vector **V**.

DIVERGENCE AS A VECTOR OPERATOR

In rectangular coordinates $h_x = h_y = h_z = 1$ and the divergence of **V** appears as

(1.63) $$\text{Div } \mathbf{V} = \frac{\partial V_x}{\partial x} + \frac{\partial V_y}{\partial y} + \frac{\partial V_z}{\partial z}$$

where V_x, V_y, V_z are the rectangular components of **V**. Also note that the gradient of a scalar F appears in this coordinate system as,

(1.64) $$\nabla F = \mathbf{1}_x \frac{\partial F}{\partial x} + \mathbf{1}_y \frac{\partial F}{\partial y} + \mathbf{1}_z \frac{\partial F}{\partial z}$$

Thus if we *define* the *vector operator* "del", or "nabla" as it has sometimes been called, as,

(1.65) $$\nabla = \mathbf{1}_x \frac{\partial}{\partial x} + \mathbf{1}_y \frac{\partial}{\partial y} + \mathbf{1}_z \frac{\partial}{\partial z}$$

then we can look upon Eq. (1.63) as the dot, or scalar, product of this operator and the vector **V**, thus,

(1.66) $$\text{Div } \mathbf{V} = \nabla \cdot \mathbf{V}$$

Here we must execute the dot operations, then the derivative operations.

Note that this *operator structure is not valid in all systems of coordinates*, only in rectangular coordinates. However it is often convenient to carry out various manipulations with "∇" as an operator until the last step; say for example the end result is symbolically ∇ · **V**. We interpret this as the divergence and to write it out, say in spherical coordinates, we must go back to Eq. (1.59) for the proper form.

LAPLACIAN OPERATOR

Frequently in physical problems we encounter the divergence of a vector **V** which itself is the gradient of a scalar function. Symbolically, in *operator form*, we write

(1.67) $$\text{Div}(\nabla F) = \nabla \cdot (\nabla F) = (\nabla \cdot \nabla)F = \nabla^2 F$$

where the *scalar operator*, ∇^2, represented here in *rectangular coordinates* as the dot product of the operator of Eq. (1.65) with itself,

$$(1.68) \qquad \nabla^2 = \frac{\partial^2}{\partial x^2} + \frac{\partial^2}{\partial y^2} + \frac{\partial^2}{\partial z^2}$$

is called the *Laplacian* operator.

If we use our previous definition of "Div" and "grad", Eqs. (1.42) and (1.59) we see the *general* form for $\nabla^2 F$ in any *orthogonal system* as,

$$(1.69) \qquad \nabla^2 F = \frac{1}{h_1 h_2 h_3} \left[\frac{\partial}{\partial q_1}\left(\frac{h_2 h_3}{h_1} \frac{\partial F}{\partial q_1} \right) + \frac{\partial}{\partial q_2}\left(\frac{h_1 h_3}{h_2} \frac{\partial F}{\partial q_2} \right) + \frac{\partial}{\partial q_3}\left(\frac{h_1 h_2}{h_3} \frac{\partial F}{\partial q_3} \right) \right]$$

for any function F. Deleting F we then have the general representation of the Laplacian *operator* in any orthogonal system.

EXECUTE PROBLEM SET (1-4)

DIVERGENCE THEOREM

Here we develop one of the most valuable theorems of vector calculus. It is based on the formal definition of the divergence of a vector field as given in Eq. (1.55).

Consider the surface integral of a vector function $\mathbf{V}(\mathbf{r})$ over a closed surface S, bounding a volume τ. It is *not* necessary that S be simply connected, but we will here put a restriction on the function $\mathbf{V}(\mathbf{r})$. This vector function must be continuous throughout τ and have first derivatives everywhere in τ. We depict the domain schematically in Fig. 1-7, where we have indicated the case of the surface, S, *not* being simply connected. Here S is made up of two

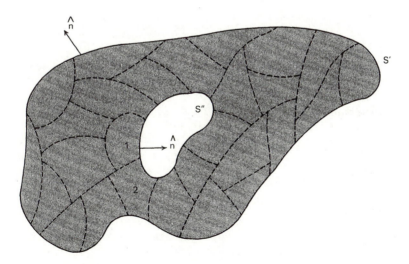

FIGURE 1.7. Subdivision of a Volume into Elements.

nonconnected surfaces S' and S'', so we are considering the integral,

(1.70)
$$I = \oint_S \mathbf{V} \cdot \mathbf{dS} = \oint_{S'} \mathbf{V} \cdot \mathbf{dS} + \oint_{S''} \mathbf{V} \cdot \mathbf{dS}$$

The positive direction of $\mathbf{dS}$ is *everywhere* outward from the volume τ, which is the shaded region, i.e., as indicated by the normal vector $\mathbf{n}$ in our picture.

Now we subdivide τ into volume elements, $\Delta\tau_i$, $i = 1, 2, \ldots N$, as indicated by the dotted lines in our picture, and consider the surface integrals of $\mathbf{V(r)}$ over the closed surfaces, ΔS_i, bounding these volume elements.

For example now consider the *sum* of two such integrals, say for the elements 1 and 2 in our picture. We have,

(1.71)
$$\oint_{\Delta S_1 + \Delta S_2} \mathbf{V} \cdot \mathbf{dS} = \oint_{\Delta S_1} \mathbf{V} \cdot \mathbf{dS}_1 + \oint_{\Delta S_2} \mathbf{V} \cdot \mathbf{dS}_2$$

Note that a *part* of ΔS_1 is on S' and that ΔS_2 has parts on S' and S'' also. But most significant is that along the *interface* between Volume Elements 1 and 2 we have $\mathbf{dS}_1$ directed *from* $\Delta\tau_1$, *into* $\Delta\tau_2$, while $\mathbf{dS}_2$, at the same point and of the same magnitude, is directed *from* $\Delta\tau_2$ into $\Delta\tau_1$. Thus for this particular element we see, since $\mathbf{V}$ *has but one value*, i.e., is continuous across the interface,

(1.72)
$$\mathbf{V} \cdot \mathbf{dS}_1 + \mathbf{V} \cdot \mathbf{dS}_2 = 0$$

on the common interface.

Thus we see that in the sum on the right in Eq. (1.70) the contributions on the common interface all add to zero. From this it is evident that,

(1.73)
$$\oint_S \mathbf{V} \cdot \mathbf{dS} = \sum_{i=1}^{N} \oint_{\Delta S_i} \mathbf{V} \cdot \mathbf{dS}_i$$

since in the sum the only parts which do not "cancel" each other are those surface elements bounding the whole region.

In this sum we multiply each term by $\Delta\tau_i/\Delta\tau_i = 1$ so that it appears as,

(1.74)
$$\oint_S \mathbf{V} \cdot \mathbf{dS} = \sum_{i=1}^{N} \left\{ \frac{\oint_{\Delta S_i} \mathbf{V} \cdot \mathbf{dS}_i}{\Delta\tau_i} \right\} \Delta\tau_i$$

where ΔS_i incloses the volume $\Delta\tau_i$. We then take the *limit* as all $\Delta\tau_i \to 0$, and see that, from the definition of the divergence, each bracket in the sum approaches $\nabla \cdot \mathbf{V}$ at a point in the infinitesimal volume, $\Delta\tau_i \to d\tau$. Also we see the limit of the sum as the *volume integral* over the whole region bounded by S, (S' and S''). Thus we have,

(1.75)
$$\oint_S \mathbf{V} \cdot \mathbf{dS} = \int_\tau \nabla \cdot \mathbf{V}\, d\tau$$

as the *divergence theorem*.

Example (4)

We illustrate the power and utility of the divergence theorem by applying it in the derivation of the heat flow, or diffusion,[†] equation.

† The *form* of these equations is the same.

The law of heat conduction states that the flow of heat is parallel but oppositely directed to the gradient of the temperature, T, and the quantity per unit area, per unit time, flowing in that direction is given by,

$$(1.76) \qquad \mathbf{J} = -k\nabla T$$

where k is the thermal conductivity of the medium and is essentially constant. We call $\mathbf{J}$ the heat flux density vector.

We also define the quantity of heat per unit volume of the medium as, $C\rho T$, where ρ is the mass per unit volume and C is the specific heat per unit mass of the medium. Actually this is approximate, but is sufficiently accurate in most situations.

Now in some arbitrary fixed region in the medium, of volume τ, the quantity of heat is,

$$(1.77) \qquad H = \int_\tau C\rho T \, d\tau$$

and this may be *increasing* at the rate

$$(1.78) \qquad \frac{dH}{dt} = \int_\tau C\rho \frac{\partial T}{\partial t} \, d\tau$$

where the time derivative d/dt can be taken inside as a *partial* derivative since τ is a *fixed* region of the coordinate space. Note that we assume C and ρ constant in time.

Since heat is a form of energy, and is conserved, this rate of increase must arise as a result of *flow* of heat *into* τ if there are *no sources in* τ. From above the *net* rate of flow of heat *into* τ is given by,

$$(1.79) \qquad -\oint_S \mathbf{J} \cdot \mathbf{dS} = \oint k\nabla T \cdot \mathbf{dS}$$

where S is the surface bounding τ. Thus we have,

$$(1.80) \qquad \oint_S k\nabla T \cdot \mathbf{dS} = \int_\tau C\rho \frac{\partial T}{\partial t} \, d\tau$$

for *any region* τ free of heat sources.

Here we apply the *divergence theorem* to the surface integral to obtain,

$$(1.81) \qquad \int_\tau \nabla \cdot (k\nabla T) \, d\tau = \int_\tau C\rho \frac{\partial T}{\partial t} \, d\tau$$

or,

$$(1.82) \qquad \int_\tau \left\{ k\nabla^2 T - C\rho \frac{\partial T}{\partial t} \right\} d\tau = 0$$

where we assume k to have the same value everywhere. Since this integral must vanish for *every* volume element τ, free of heat sources, we conclude that,

$$(1.83) \qquad k\nabla^2 T = C\rho(\partial T/\partial t)$$

in all regions free of heat sources. This is the heat-flow equation.

LINE INTEGRALS

Let there exist a vector field, $\mathbf{V}(\mathbf{r})$, and a space curve C. Such a curve can be prescribed as the intersection of two surfaces in three dimensions, say,

(1.84)
$$\begin{cases} f(q_1, q_2, q_3) = 0 \\ g(q_1, q_2, q_3) = 0 \end{cases}$$

where $f(q_j)$ and $g(q_j)$ are functions of the orthogonal coordinates, $q_j, j = 1, 2, 3$. Such a curve is depicted in Fig. 1-8. We have at each point of the curve C, specified by the position vector $\mathbf{r}$, a line element ds, and correspondingly a displacement $\mathbf{dr}$ directed along the curve. We take $\mathbf{dr}$ positive going from end point A of C to end point B.

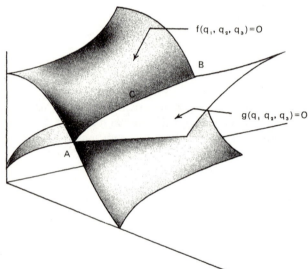

$f(q_1, q_2, q_3)=0$

B

$g(q_1, q_2, q_3)=0$

A

FIGURE 1.8.   A Space Curve as the Intersection of Two Surfaces.

At each such point on C the vector $\mathbf{V}$ has a value and we form the scalar, or dot product, $\mathbf{V} \cdot \mathbf{dr}$. Then we sum, or integrate, to form,

(1.85)
$$I = \int_A^B {}_{(C)} \mathbf{V}(\mathbf{r}) \cdot \mathbf{dr}$$

as the *line integral* of $\mathbf{V}(\mathbf{r})$ on C from A to B. In general the value of such an integral depends not only on the function, $\mathbf{V}(\mathbf{r})$, and the end points, but also on the curve C along with the integration is carried out.

Observe that from the physical point of view the line integral of $\mathbf{V}$ represents the *sum* of the *tangential components* of $\mathbf{V}$, V_t, multiplied by the magnitude, ds, of the corresponding elements of displacement for all the elements, ds, of the path, C.

Example (5)

Evaluate the line integral of

(1.86)
$$\mathbf{V} = y\mathbf{1}_x + x\mathbf{1}_y + x\mathbf{1}_z$$

on the curve C defined by,

(1.87)
$$\begin{cases} x + y + z - 2 = 0 \\ x^2 - y = 0 \end{cases}$$

between the points $A = (1, 1, 0)$ and $B = (0, 0, 2)$.

We have,

(1.88)
$$I = \int_B^A {}_{(C)}\mathbf{V} \cdot d\mathbf{r} = \int_{(1,1,0)}^{(0,0,2)} (y\,dx + x\,dy + x\,dz)$$

and we proceed to express y, dy and z, dz all in terms of x and dx by using Eq. (1.87). We obtain

(1.89)
$$I = \int_1^0 x^2\,dx - x\,dx = \tfrac{1}{6}$$

as the final result.

CURL OF A VECTOR

Given a vector field, $\mathbf{V(r)}$, we select a *plane closed curve*, C, inclosing an area, ΔS. We define a positive side of the surface by a unit normal vector, $\mathbf{n}$, and a positive sense for integration of $\mathbf{V}$ around the curve C as indicated in Fig. 1-9. Then we *define*

(1.90)
$$\mathbf{n} \cdot (\text{curl } \mathbf{V}) = \lim_{\Delta S \to 0} \frac{\oint_C \mathbf{V} \cdot d\mathbf{l}}{\Delta S}$$

where we use $d\mathbf{l}$ in lieu of $d\mathbf{r}$ on C, as the component of *the curl of* $\mathbf{V}$ parallel to $\mathbf{n}$, at the point $\mathbf{r}$. Here the limit $\Delta S \to 0$ is taken in such manner that the point $\mathbf{r}$ is always in the area element ΔS and $\mathbf{n}$ is the normal to ΔS at $\mathbf{r}$.

Strictly speaking the *shape* of the plane closed curve is immaterial and actually the area ΔS need not be restricted to be just the *plane* area bounded by C. We could for example let C be the perimeter of the circle forming the boundary of a spherical cap and ΔS be the area of

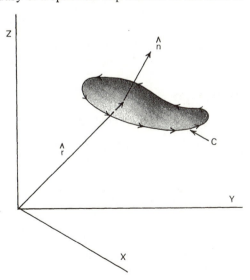

FIGURE 1.9.   An Area Element with Unit Normal Vector n at a Point r.

the curved cap. However *in the limit*, as $\Delta S \to 0$, we see that we approach a *plane* area ΔS even in this case.

Now we evaluate the components of the "vector-like" quantity, curl **V**, in an arbitrary curvilinear *orthogonal* coordinate system. Choose the normal **n** to be the $1'_i$ vector and let the curve C be the "*rectangle*" with sides $h_j\Delta q_j$, $h_k\Delta q_k$. (Note $i \neq j \neq k$, and later we will permute i, j, k in cyclic manner to obtain the other components.) Here we have the picture of this "rectangle," with $1'_i$ *out* of the plane of the paper.

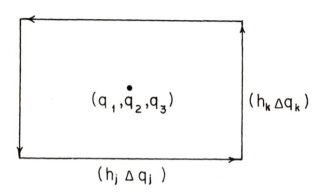

$$(q_1, \dot{q}_2, q_3) \qquad (h_k \Delta q_k)$$

$$(h_j \Delta q_j)$$

FIGURE 1.10.   Rectangular Path About a Point q_1, q_2, q_3.

On the right end of the rectangle the component of **V** parallel to the curve is V_k so we have $V_k h_k \Delta q_k$, *evaluated* at the *midpoint* of this line $q_i, q_j + (\Delta q_j/2), q_k$, as the contribution to the line integral around C. Similarly for the other sides of the "rectangle." Also the *area* of the rectangle is $h_j h_k \Delta q_j \Delta q_k$. Thus we examine,†

$$\left[\frac{(V_k h_k)_{q_i, q_j + \frac{\Delta q_j}{2}, q_k} - (V_k h_k)_{q_i, q_j - \frac{\Delta q_j}{2}, q_k}}{h_j h_k \Delta q_j \Delta q_k}\right]\Delta q_k + \left[\frac{(V_j h_j)_{q_i, q_j, q_k - \frac{\Delta q_k}{2}} - (V_j h_j)_{q_i, q_j, q_k + \frac{\Delta q_k}{2}}}{h_j h_k \Delta q_j \Delta q_k}\right]\Delta q_j$$

in the limit as Δq_j, Δq_k both go to zero. The result is:

(1.91)
$$(\text{curl } \mathbf{V})_i = \frac{1}{h_j h_k}\left[\frac{\partial(h_k V_k)}{\partial q_j} - \frac{\partial(h_j V_j)}{\partial q_k}\right]$$

as the q_i component of curl **V**. Simply by *cyclic permutation of the indices, $i \to j \to k \to i$,* we obtain the j and k components of curl **V**.

CURL V AS AN OPERATOR

If we apply Eq. (1.91) in rectangular coordinates we can construct the vector,

(1.92)
$$\text{curl } \mathbf{V} = \mathbf{1}_x\left(\frac{\partial V_z}{\partial y} - \frac{\partial V_y}{\partial z}\right) + \mathbf{1}_y\left(\frac{\partial V_x}{\partial z} - \frac{\partial V_z}{\partial x}\right) + \mathbf{1}_z\left(\frac{\partial V_y}{\partial x} - \frac{\partial V_x}{\partial y}\right)$$

and this can be constructed as the *cross* product, or vector product, of our del operator, ∇, defined in Eq. (1.65), with the vector **V**. That is,

(1.93)
$$\text{curl } \mathbf{V} = \nabla \times \mathbf{V}$$

† Note that we must take account of the fact that actually h_k, as well as V_k varies over the distance $q_j - \Delta q_j/2$ to $q_j + \Delta q_j/2$, etc.

where we execute the vector operations of the cross products of the unit vectors and then execute the differentiation operations.

Again we *caution* the student, this operational use of ∇ is generally only valid in *rectangular coordinates*. However we may always assume rectangular coordinates are used through any sequence of manipulations with the ∇ operator and then apply the proper *form* for the chosen coordinate systems at the end.

For example if the end result of several manipulations is $\nabla \times \mathbf{V}$ we *must* revert to Eq. (1.91) for the proper form. In *spherical* coordinates this would appear as:

(1.94)
$$\nabla \times \mathbf{V} = \mathbf{1}_r \left[\frac{1}{r \sin \theta} \frac{\partial}{\partial \theta} (\sin \theta V_\phi) - \frac{1}{r \sin \theta} \frac{\partial V_\theta}{\partial \phi} \right] + \mathbf{1}_\theta \left[\frac{1}{r \sin \theta} \frac{\partial V_r}{\partial \phi} - \frac{1}{r} \frac{\partial}{\partial r} (r V_\phi) \right]$$
$$+ \mathbf{1}_\phi \left[\frac{1}{r} \frac{\partial}{\partial r} (r V_\theta) - \frac{1}{r} \frac{\partial V_r}{\partial \theta} \right]$$

which is *not* the result of using ∇ as a vector operator crossed into $\mathbf{V}$.

EXECUTE PROBLEM SET (1-6)

STOKES' THEOREM

This theorem relates a line integral to a surface integral in a manner quite analogous to the relation of the surface and volume integrals seen in the divergence theorem.

Suppose we are given the line integral of a vector $\mathbf{V(r)}$ about a *closed space curve, C*,

(1.95)
$$I = \oint_C \mathbf{V} \cdot \mathbf{dr}$$

We then visualize C as bounding some surface S as indicated in Fig. 1-11. Then we subdivide S into area elements ΔS_i, $i = 1, 2, \ldots N$, each small enough to consider approximately plane, and consider the line integrals of V about each of the curves, C_i, bounding these area elements. We require $\mathbf{V}$ *to be a continuous vector function* having continuous first derivatives.

Examine now the two integrals around the adjoining elements 1 and 2. On their *common boundary* the integrands $\mathbf{V} \cdot \mathbf{dr}$ have the *same* magnitude at a given point, but if we integrate counter clockwise (with regard to $\mathbf{n}$, the normal) the *signs* are *opposite*. Thus for example

(1.96)
$$\oint_{C_1} \mathbf{V} \cdot \mathbf{dr} + \oint_{C_2} \mathbf{V} \cdot \mathbf{dr} = \oint_{C_1 + C_2} \mathbf{V} \cdot \mathbf{dr}$$

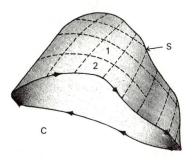

FIGURE 1.11. Subdivision of an Open Surface S Bounded by a Closed Curve C.

where by $C_1 + C_2$ we mean the outer curve bounding the *two* elements, i.e., the contribution from the two integrals on C_1 and C_2 on the *common* boundary "cancel." Generalizing then we see that,

(1.97)
$$\oint_C \mathbf{V} \cdot d\mathbf{r} = \sum_{i=1}^{N} \oint_{C_i} \mathbf{V} \cdot d\mathbf{r}$$

Then we multiply each term in this sum by $\Delta S_i / \Delta S_i = 1$ to obtain:

(1.98)
$$\oint_C \mathbf{V} \cdot d\mathbf{r} = \sum_{i=1}^{N} \left\{ \frac{\oint_{C_i} \mathbf{V} \cdot d\mathbf{r}}{\Delta S_i} \right\} \Delta S_i$$

The next step is now obvious, we take the limit as *all* $\Delta S_i \to 0$ and $N \to \infty$, noting that the bracket becomes $(\text{curl } V)_n$, the component normal to the surface, and hence,

(1.99)
$$\oint_C \mathbf{V} \cdot d\mathbf{r} = \int_S (\text{curl } \mathbf{V}) \cdot d\mathbf{S}$$

which is *Stokes' Theorem*. It is most important to note that the given curve C is the bounding curve of an arbitrarily large family of surfaces, S, i.e., the "shape" of S in Fig. 1-11 is quite arbitrary.

Example (6)

Prove that if $\mathbf{V} = \nabla F$, where F is some scalar function of the coordinates, then the line integral of $\mathbf{V}$ along a curve C connecting two points, A and B, is independent of the choice of curve C, i.e., the integral,

(1.100)
$$I = \int_A^B {}_{(C)}\mathbf{V} \cdot d\mathbf{r}$$

has the *same* value for all curves C' connecting A to B if $\mathbf{V} = \nabla F$.

We make the following construction. Let C and C' be two distinct curves connecting A and B as in Fig. 1-12 then note that

(1.101)
$$\int_B^A {}_{(C')}\mathbf{V} \cdot d\mathbf{r} = - \int_A^B {}_{(C')}\mathbf{V} \cdot d\mathbf{r}$$

is true for *any* line integral, i.e., $d\mathbf{r} \to -d\mathbf{r}$ so $B \to A$, $A \to B$.

Now consider the integral around the *closed* contour, A to B on C, then B to A on C'. Then by Stokes' Theorem,

(1.102)
$$\oint_{C+C'} \mathbf{V} \cdot d\mathbf{r} = \int_S (\text{curl } \mathbf{V}) \cdot d\mathbf{S}$$

where S is an area bounded by the two curves. Now in Eq. (1.92), if we substitute $\mathbf{V} = \nabla F$, we see that

(1.103)
$$\text{curl } \nabla F = \nabla \times \nabla F = 0$$

is an *identity* for *any* function F. Thus the right member above is zero. Also we can write the

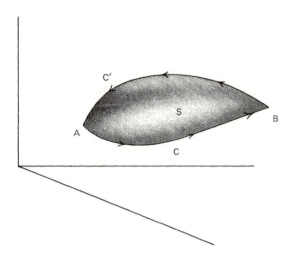

FIGURE 1.12.   Two Paths of Integration From A to B and the Enclosed Area.

left member as the sum of two integrals so,

(1.104)
$$\oint_{C+C'} \mathbf{V} \cdot \mathbf{dr} = \int_A^B {}_{(C)}\mathbf{V} \cdot \mathbf{dr} + \int_B^A {}_{(C')}\mathbf{V} \cdot \mathbf{dr}$$

Hence in view of Eq. (1.101)

(1.105)
$$\int_A^B {}_{(C)}\nabla F \cdot \mathbf{dr} = \int_A^B {}_{(C')}\nabla F \cdot \mathbf{dr}$$

which proves the two line integrals are the same for *any* curves C and C' *when* $\mathbf{V} = \nabla F$, and the integral is therefore *independent* of the path of integration.

EXECUTE PROBLEM SET (1-7)

Table 1-1 following the problems presents some useful identities which may also be used as exercises for the student.

PROBLEMS

Set (1-1)

(1) Show that the cylindrical coordinate system, r, θ, z defined by,

$$x = r \cos \theta$$

$$y = r \sin \theta$$

$$z = z$$

is an orthogonal system and determine the scale factors, h_r, h_θ, h_z.

(2) Use the formal definition in the text to construct the unit vectors, $\mathbf{1}_r$, $\mathbf{1}_\theta$, $\mathbf{1}_z$ appropriate to the cylindrical coordinate system.

(3) Write the expression for the line element in cylindrical coordinates.

(4) Prove that $\mathbf{1}_r \times \mathbf{1}_\theta = \mathbf{1}_z$ and the cyclic, right hand, rule holds for $r, \theta, z \to z, r, \theta, \to \theta, z, r$, for these *cylindrical* coordinates.

(5) Show that for the unit vector $\mathbf{1}_r$ and $\mathbf{1}_\theta$ in the *spherical* coordinate system (Fig. 1-2) the relation:

$$\frac{\partial \mathbf{1}_r}{\partial \theta} = \mathbf{1}_\theta$$

Derive similar relations for the other unit vectors in this coordinate system.

(6) Show that the distance between two points, r, θ, z and r', θ', z', in cylindrical coordinates is:

$$R = \sqrt{r^2 + r'^2 - 2rr'\cos(\theta - \theta') + (z - z')^2}$$

Set (1-2)

(1) Write out the expressions for gradient of F in cylindrical and spherical coordinates.
(2) For a *fixed* difference δF between the values of F on two *adjacent* surfaces, $F = $ constant, what is the *relation* between the normal distance between the surfaces and the *magnitude* of ∇F on one surface. (This will be a *first*-order estimate of the relation for *small* δF.)
(3) Use what you know about the relation of ∇F to surfaces of $F = $ constant to show that

$$ax + by + cz - h = 0$$

represents a *family* of *plane* surfaces oriented in a certain fashion in the x, y, z space, and everywhere parallel.
(4) Show that in an *orthogonal* coordinate system, q_1, q_2, q_3 the surfaces $q_i = $ constant have unit vectors *normal* to the surfaces given by

$$\mathbf{1}_{ni} = \frac{\nabla q_i}{|\nabla q_i|}$$

using ∇ in rectangular coordinates, and that at a point of intersection of the *three* surfaces $q_i = $ constant, $i = 1, 2, 3$, these form an orthogonal set of unit vectors, i.e., the three surfaces are mutually orthogonal at *any* point of intersection. How are these vectors, $\mathbf{1}_{ni}$, related to the $\mathbf{1}_i'$ defined in the text?

Set (1-3)

(1) Given the vector field,

$$\mathbf{V} = \mathbf{1}_x(x^2 + y^2) - \mathbf{1}_y\, y + \mathbf{1}_z yz$$

evaluate the surface integral of $\mathbf{V}$ over the outside of the half-cylinder whose axis is on the z axis and with the curved surface in the domain of $x > 0$, and in the length $0 < z < L$.
(2) Show that for any *open* surface S the integral

$$\int_S \mathbf{1}_n \cdot d\mathbf{S} = A_n$$

where A_n is the area of the projection of the surface S on a *plane* having $\mathbf{1}_n$ as unit normal vector.

(3) Show that if **h** is any *constant* vector and U is some function then

$$\int_S U\mathbf{h} \cdot d\mathbf{S} = h \int_{A_h} U \, dA_h$$

where dA_h is the element of area on the plane having $\mathbf{h}/h$ as unit normal and the domain of integration, A_h, is just the projection of S on this plane, and $\nabla U \cdot \mathbf{h} = 0$

Set (1-4)

(1) Write out Div **V** in cylindrical coordinates.
(2) Write out $\nabla^2 F$ in spherical coordinates and in cylindrical coordinates.

Set (1-5)

(1) Prove that the divergence of a constant vector is zero.
(2) Derive the equation of heat conduction *with sources* where heat is *produced* in the medium at a rate, H, per unit volume per unit time.

Set (1-6)

(1) Write out curl **V** in cylindrical coordinates.
(2) Prove the *identity*

$$\nabla \times \nabla F = 0$$

for any function F.
(3) Prove the *identity*

$$\nabla \cdot (\nabla \times \mathbf{V}) = 0$$

for any vector **V**.

Set (1-7)

(1) Show that for any surface S bounding a region τ in which **V** is a continuous vector function,

$$\int_\tau \nabla \times \mathbf{V} \, d\tau = - \int_S \mathbf{V} \times d\mathbf{S}$$

(2) Show that the volume enclosed by any surface, S, is given by:

$$\tau = \frac{1}{3} \int_S \mathbf{r} \cdot d\mathbf{S}$$

where **r** is the position vector.
(3) Show that *any* vector field, $\mathbf{V(r)}$, can be represented as a sum $\mathbf{V}_1(\mathbf{r}) + \mathbf{V}_2(\mathbf{r})$ where

$$\nabla \cdot \mathbf{V}_1 = 0, \quad \text{and} \quad \nabla \times \mathbf{V}_2 = 0$$

TABLE 1-1. Vector Identities.

(1) $\mathbf{U} \times \mathbf{V} \times \mathbf{W} = \mathbf{V}(\mathbf{U} \cdot \mathbf{W}) - \mathbf{W}(\mathbf{U} \cdot \mathbf{V})$

(2) $\nabla(aF) = a\nabla F$

(3) $\nabla(aF + bG) = a\nabla F + b\nabla G$

(4) $\nabla(FG) = F\nabla G + G\nabla F$

(5) $\nabla \cdot (a\mathbf{U} + b\mathbf{V}) = a\nabla \cdot \mathbf{U} + b\nabla \cdot \mathbf{V}$

(6) $\nabla \cdot (F\mathbf{U}) = F\nabla \cdot \mathbf{U} + \mathbf{U} \cdot \nabla F$

(7) $\nabla \times (a\mathbf{U} + b\mathbf{V}) = a\nabla \times \mathbf{U} + b\nabla \times \mathbf{V}$

(8) $\nabla \cdot (\mathbf{U} \times \mathbf{V}) = \mathbf{V} \cdot \nabla \times \mathbf{U} - \mathbf{U} \cdot \nabla \times \mathbf{V}$

(9) $\nabla \cdot (\nabla \times \mathbf{V}) = 0$

(10) $\nabla \times \nabla F = 0$

(11) $\nabla \times (F\mathbf{U}) = F\nabla \times \mathbf{U} + \mathbf{U} \times \nabla F$

(12) $\nabla \times \nabla \times \mathbf{U} = \nabla(\nabla \cdot \mathbf{U}) - \nabla^2 \mathbf{U}$

(13) $\nabla \times (\mathbf{U} \times \mathbf{V}) = \mathbf{U}\nabla \cdot \mathbf{V} - \mathbf{V}\nabla \cdot \mathbf{U} + (\mathbf{V} \cdot \nabla)\mathbf{U} - (\mathbf{U} \cdot \nabla)\mathbf{V}$

(14) $\nabla(\mathbf{U} \cdot \mathbf{V}) = (\mathbf{U} \cdot \nabla)\mathbf{V} + (\mathbf{V} \cdot \nabla)\mathbf{U} + \mathbf{U} \times (\nabla \times \mathbf{V}) + \mathbf{V} \times (\nabla \times \mathbf{U})$

[a] The student may prove these as exercises and refer to them in future applications. Here $\mathbf{U}$, $\mathbf{V}$, $\mathbf{W}$ refer to any three vector fields and F and G are scalar functions. Constants are: a, b, c, etc.

[b] The scalar operators, ∇^2, and say $(\mathbf{U} \cdot \nabla)$ applied to a vector field $\mathbf{V}$ can easily be written in terms of the *rectangular* components of $\mathbf{V}$, and $\mathbf{U}$, i.e.,

$$\nabla^2 \mathbf{V} = \mathbf{1}_x \nabla^2 V_x + \mathbf{1}_y \nabla^2 V_y + \mathbf{1}_z \nabla^2 V_z$$

$$(\mathbf{U} \cdot \nabla)\mathbf{V} = \mathbf{1}_x\left(U_x \frac{\partial}{\partial x}V_x + U_y \frac{\partial}{\partial y}V_x + U_z \frac{\partial}{\partial z}V_x\right) + \mathbf{1}_y\left(U_x \frac{\partial}{\partial x}V_y + U_y \frac{\partial}{\partial y}V_y + U_z \frac{\partial}{\partial z}V_y\right)$$

$$+ \mathbf{1}_z\left(U_x \frac{\partial}{\partial x}V_z + U_y \frac{\partial}{\partial y}V_z + U_z \frac{\partial}{\partial z}V_z\right)$$

REFERENCES

L. Brand, *Vector Analysis*, 3rd ed. (John Wiley & Sons, Inc., New York, 1961).

E. A. Kraut, *Fundamentals of Mathematical Physics* (McGraw-Hill Book Company, Inc., New York, 1967).

E. Kreyszig, *Advanced Engineering Mathematics* (John Wiley & Sons, Inc., New York, 1962).

A. Kryala, *Theoretical Physics: Applications of Vectors, Matrices, Tensors and Quaternions*, W. B. Saunders Company, Philadelphia (1967).

H. Lass, *Vector and Tensor Analysis* (McGraw-Hill Book Company, Inc., New York, 1950).

P. M. Morse and H. Feshbach, *Methods of Theoretical Physics* (McGraw-Hill Book Company, Inc., New York, 1953).

L. P. Eisenhart, *An Introduction to Differential Geometry* (Princeton University Press, Princeton, New Jersey, 1947). Consult this book for the most complete discussion of the extension of coordinate transformations into the general subject of differential geometry and tensors.

CHAPTER TWO

Matrix algebra and transformations in linear vector spaces; dyadics

The matrix concept arises naturally in the consideration of linear relations among two sets of variables. We take this as our starting point for the study of matrices.

SYSTEMS OF LINEAR EQUATIONS AND MATRIX NOTATION; MULTIPLICATION OF MATRICES

Suppose we have a set of *linear relations* giving a set of quantities, y_i, $i = 1, 2, \ldots n$, in terms of another set of quantities x_j, $j = 1, 2, \ldots m$, such as:

(2.1)
$$\begin{cases} y_1 = a_{11}x_1 + a_{12}x_2 + \ldots + a_{1m}x_m \\ y_2 = a_{21}x_1 + a_{22}x_2 + \ldots + a_{2m}x_m \\ \quad\vdots \qquad \vdots \qquad \vdots \qquad\qquad \vdots \\ y_n = a_{n1}x_1 + a_{n2}x_2 + \ldots + a_{nm}x_m \end{cases}$$

where the a_{ij}, $i = 1, 2, \ldots n$; $j = 1, 2, \ldots m$ are a set of coefficients.

We construct the *simplified notation* for such a set of relations as,

(2.2)
$$Y = AX$$

by the following rules. We write

(2.3)
$$\begin{pmatrix} y_1 \\ y_2 \\ \vdots \\ y_n \end{pmatrix} = \begin{pmatrix} a_{11} & a_{12} \ldots a_{1m} \\ a_{21} & a_{22} \ldots a_{2m} \\ \vdots & \vdots & \vdots \\ a_{n1} & a_{n2} \ldots a_{nm} \end{pmatrix} \begin{pmatrix} x_1 \\ x_2 \\ \vdots \\ x_m \end{pmatrix}$$

and call each of the arrays in the parentheses a *matrix*, with the left member being Y, a column matrix, or "vector", and similarly for the last parenthesis, X, which, however, has m "components" whereas Y has n "components." The other array A, is an "n by m" matrix, having n rows and m columns, or a matrix of *order $n \times m$*.

In order that Eq. (2.3) be equivalent to Eq. (2.1) we require a *row-column multiplication rule*, i.e., to construct y_3 we multiply Row 3 of A into the single column of X, *term for term* and add the result, i.e.,

(2.4)
$$y_3 = \sum_{j=1}^{m} a_{3j} x_j$$

It could be that *another* set of y's, say y_i', $i = 1, 2, \ldots n$ is constructed from *another* set of x's, say x_j', $j = 1, 2, \ldots m$ by a set of linear relations having the *same coefficients*, a_{ij}. These would appear exactly as Eq. (2.1) with all x_i and y_j carrying primes. These too could be written in the form of Eq. (2.3). But, by a further generalization of our notation, we can represent *both* sets of linear relations in the *symbolic form* of Eq. (2.2) if we write,

(2.5)
$$\begin{pmatrix} y_1 & y_1' \\ y_2 & y_2' \\ \vdots & \vdots \\ y_n & y_n' \end{pmatrix} = \begin{pmatrix} a_{11} & a_{12} \ldots a_{1m} \\ a_{21} & a_{22} \ldots a_{2m} \\ \vdots & \vdots & \vdots \\ a_{n1} & a_{n2} \ldots a_{nm} \end{pmatrix} \begin{pmatrix} x_1 & x_1' \\ x_2 & x_2' \\ \vdots & \vdots \\ x_m & x_m' \end{pmatrix}$$

where the left member now corresponds to Y, A is unchanged and the last parenthesis now plays the role of X.

We now need the following rule: to obtain say the *third* element, y_3, of the *first column of* Y, we multiply the *third row of A* into the *first column of* X, i.e., just getting Eq. (2.4) again. To obtain say the *fourth element* of the *second column of* Y, y_4', we multiply the *fourth row of A* into the *second column of X*, thus,

(2.6)
$$y_4' = \sum_{j=1}^{m} a_{4j} x_j'$$

Thus, to generalize, consider two matrices,

(2.7)
$$A = \begin{pmatrix} a_{11} & a_{12} \ldots a_{1m} \\ a_{21} & a_{22} \ldots a_{2m} \\ \vdots \\ a_{n1} & a_{n2} \ldots a_{nm} \end{pmatrix}$$

and

$$(2.8) \qquad B = \begin{pmatrix} b_{11} & b_{12} \ldots b_{1n} \\ b_{21} & b_{22} \ldots b_{2n} \\ \vdots \\ b_{r1} & b_{r2} \ldots b_{rn} \end{pmatrix}$$

where *B has a number of columns, n, equal to the number of rows of A*, then *B* and *A* are con-*formable*, and,

$$(2.9) \qquad C = BA$$

can be defined, where the elements c_{ij} of *C* are given by:

$$(2.10) \qquad c_{ij} = \sum_{l=1}^{n} b_{il}a_{lj}, \qquad \begin{pmatrix} i = 1, 2, \ldots r \\ j = 1, 2, \ldots m \end{pmatrix}$$

following our *row-column rule,* and *C* is an $r \times m$ matrix.

For *square matrices A* and *B, n = m = r* in Eq. (2.7) and (2.8), we see that it is *possible* to apply the row-column rule to define the product *AB* as well as *BA*. However it is readily verified that these are not necessarily the same, i.e., let

$$(2.11) \qquad c'_{ij} = \sum_{l=1}^{m} a_{il}b_{lj}$$

be the elements of,

$$(2.12) \qquad C' = AB$$

and comparison to Eq. (2.10) above shows these to be different for any arbitrary square matrices, i.e.,

$$(2.13) \qquad AB \neq BA$$

and we say *matrix multiplication is not commutative.*

GENERAL RULES OF MATRIX ALGEBRA

Building on the general motivations provided by *linear relations* as outlined above we formu-late the following basic rules for the algebra of matrices; multiplications having already been defined.

Definition of n × m matrix (or order of a matrix)

An $n \times m$ array of elements, *n* rows and *m* columns, all of whose elements are the same kind of quantities,† i.e., real numbers, or complex numbers, etc.

---

† The elements of a matrix may themselves be matrices, in which case we call the matrix a *partitioned matrix.*

Addition

Let $A = (a_{ij})$, $B = (b_{ij})$ indicate the elements of two matrices A and B of the *same order*, then the elements, c_{ij}, of the sum $A + B = C$ are just $a_{ij} + b_{ij}$, i.e., c_{13} for example is $a_{13} + b_{13}$.

Subtraction

With A and B as above in the definition of addition we have for $C = A - B$, the elements $c_{ij} = a_{ij} - b_{ij}$.

Null matrix

Every element of the null matrix is zero.

Equality of matrices

Two matrices can be equal only if they are of the same order, then $A = B$ if $a_{ij} = b_{ij}$ for all i and j in A and B; $A - B =$ null matrix.

Multiplication of a matrix by a scalar

If a matrix A is multiplied by a scalar, (real or complex) *every* element of A is multiplied by the scalar, i.e., the elements of αA are αa_{ij}.

Transpose of a matrix

Given a matrix, of order $n \times m$,

$$(2.14) \qquad A = \begin{pmatrix} a_{11} & a_{12} & a_{13} \ldots a_{1m} \\ a_{21} & a_{22} \ldots\ldots\ldots a_{2m} \\ \vdots \\ a_{n1} & a_{n2} \ldots\ldots\ldots a_{nm} \end{pmatrix}$$

We define the *transpose of A*, or *A transpose*, denoted by $\tilde{A}$ as,

$$(2.15) \qquad \tilde{A} = \begin{pmatrix} a_{11} & a_{21} & a_{31} \ldots a_{n1} \\ a_{12} & a_{22} & a_{32} \ldots a_{n2} \\ \vdots \\ a_{1m} & a_{2m} \ldots\ldots\ldots a_{nm} \end{pmatrix}$$

simply by interchanging rows for columns, rth row $\rightarrow$ rth column, etc. Note that $\tilde{A}$ is an $(m \times n)$ matrix where as A was $(n \times m)$.

Identity matrix

The *square* matrix all of whose elements are zero *except* for the "main diagonal," where they are unity, is the identity matrix, i.e., for a (4×4) matrix,

(2.16)
$$I = \begin{pmatrix} 1 & 0 & 0 & 0 \\ 0 & 1 & 0 & 0 \\ 0 & 0 & 1 & 0 \\ 0 & 0 & 0 & 1 \end{pmatrix}$$

for example.

Inverse of a matrix

Given a *square* matrix A we define the matrix, denoted symbolically as A^{-1} and called A *inverse*, or the *inverse of* A, by:

(2.17)
$$AA^{-1} = A^{-1}A = I$$

Here we must note that A, A^{-1}, and I must *all* be square and of the *same order*, $(n \times n)$. Also we comment that not all square matrices possess an inverse.

Determinant of a matrix

The *determinant*, $|A|$, of a matrix, A, is a scalar, or number, defined as

(2.18)
$$|A| = \sum \pm (a_{1i}a_{2j}a_{3k}\ldots)$$

and has meaning *only for square matrices*. The sum here is constructed such that *every* term has as a factor an element, a_{1i}, from the first row, an element, a_{2j}, from the second row, etc. Note that these are indicated in normal order, $1, 2, \ldots n$. The column indices, i, j, k, etc. appear as some permutations of this normal order. Starting with i, j, k in normal order we assign a plus sign to the term; if an *even* number of permutations of adjacent indices of columns, $i, j, k, \ldots$ generates a term, then that term also gets a plus sign. If however an *odd* number of such permutations of adjacent indices generates the term then a *minus* sign is given to the term.

We will not elaborate on the properties of determinants, or methods for evaluating them. We presume the student to be fully familiar with these topics.

Adjoint matrix

For a matrix A whose elements are complex we can define the *complex conjugate*, A^*, as the matrix whose elements are just the complex conjugates, a_{ij}^*, of those of A, a_{ij}. If we also, at the same time, *transpose* the complex conjugate we form the adjoint, $A\dagger$,

(2.19)
$$A\dagger = (\tilde{A}^*) = (\tilde{A})^*$$

Note that the operations, $\sim$ and $*$, are interchangeable.

Trace of a matrix

For a *square* matrix we can define the trace as

(2.20)
$$\text{Tr } A = \sum_{i=1}^{n} a_{ii}$$

i.e., the sum of the elements on the *main* diagonal.

IMPORTANT OPERATIONS WITH MATRICES

Here we *list* some important rules for operations with matrices and leave to the student the proofs, or demonstrations, as exercises.

Transpose of a product of matrices

(2.21)
$$\widetilde{AB} = \tilde{B}\tilde{A}$$

Meaning that, if we multiply A times B (these being properly conformable so that the product exists), then transpose the result, we obtain the *same* final matrix by first transposing *each* matrix and then multiplying B into A by our *same* row column rule.

Determinant of a product of matrices

(2.22)
$$|AB| = |A|\,|B|$$

provided A and B are square matrices.

Determinant of the transpose of a matrix

(2.23)
$$|\tilde{A}| = |A|$$

Thus if we calculate the determinant of A or $\tilde{A}$ we get the same number.

Evaluation of the inverse of a matrix

Given a *square* matrix A, which has an inverse, (we say it is *nonsingular*) we form the *cofactor matrix*, $\bar{A}$, whose elements, $\bar{a}_{ij}$, are the transposed cofactors of the respective a_{ij}, i.e., the cofactor of a_{ij} is the *determinant*† of the matrix formed by deleting the row (i) and column (j) of A, containing a_{ij} and multiplying by $(-1)^{i+j}$. Then from the definition of the inverse and the definition of the determinant we can show that,

(2.24)
$$A^{-1} = \frac{\bar{A}}{|A|}$$

If $|A|$ is zero we say A is *singular*, and A^{-1} does not exist.

For details of one method of evaluation of the inverse see the discussion of Gaussian elimination.

---

† The determinant so formed *without* the sign factor, $(-1)^{i+j}$, is called the *minor* of a_{ij}.

Associative property of matrix multiplication

At the beginning of this chapter we defined multiplication of two matrices and pointed out that it is not commutative in general, $AB \neq BA$, even for *square* matrices A and B. However it is *associative*, for conformable matrices,

$$(2.25) \qquad\qquad (AB)C = A(BC)$$

which we leave to the student to prove. It is this property which proves most useful in the following way.

"SOLUTIONS" OF A MATRIX EQUATION

Given the matrix equation

$$(2.26) \qquad\qquad A = BC$$

we can solve for B or C, provided certain conditions are met. Obviously B and C must be conformable for this equation to have meaning. If all matrices are of the same order, and B is nonsingular, then multiplying *from the left* by B^{-1} gives

$$(2.27) \qquad\qquad B^{-1}A = (B^{-1}B)C = IC$$

where we have applied the associative property, and the definition of B^{-1}.

Now from the form of I, the identity matrix defined earlier, we see that,

$$(2.28) \qquad\qquad IC = CI = C$$

for *any* square matrix C. Thus we have

$$(2.29) \qquad\qquad C = B^{-1}A$$

In exactly the same way we can *formally* solve for the matrix B by multiplying *from the right* on both sides of Eq. (2.26) with C^{-1} to obtain

$$(2.30) \qquad\qquad B = AC^{-1}$$

It is these *formal* manipulations which often prove so valuable in problem analysis. ,

EXECUTE PROBLEM SET (2-1)

SOLUTION OF SIMULTANEOUS LINEAR EQUATIONS

We return to Eq. (2.1), which is a set of n linear equations in m unknowns, $x_j, j = 1, 2, \ldots m$, the y_i assumed to be known. This set of equations is equivalent to the matrix equation, Eq. (2.2),

$$(2.2) \qquad\qquad Y = AX$$

Formally the solution of this set of equations is given by

$$(2.31) \qquad\qquad X = A^{-1}Y$$

provided that A^{-1} exists.

It has already been pointed out that A^{-1} exists only for A being a *square matrix* and $|A| \neq 0$. Thus for A to be square the number of unknowns, m, must be the same as the number of equations, n. Also, as we know from elementary algebra, Det $A = |A|$ will in general vanish if *any* row of A can be expressed as a linear combination of the other rows of A. Furthermore, since $|\tilde{A}| = |A|$ we can make a similar statement about the *columns* of A.

In terms of our equations, y_j as functions of the x_i, the above statements about the rows of A, etc., are equivalent to saying that no linear relation exists among the y_j alone. The equations are *linearly independent*.

Here then we see that for n linearly independent equations in n unknowns the solution process boils down to finding the inverse, A^{-1}, of a given $n \times n$ matrix, A, of the coefficients. There are a variety of methods for constructing such an inverse, we present one method as follows, for *other methods* the student should consult the references at the end of the chapter.

GAUSSIAN ELIMINATION FOR MATRIX INVERSION

We can see the basis for this method of matrix inversion in the fundamental laws of algebra. Both sides of an equation may be multiplied by any quantity without altering the equality, and equals may be added to equals, i.e., corresponding sides of two equations may be added to produce a new equation.

With this in mind we write our matrix equation as,

$$(2.32) \qquad\qquad AX = IY$$

since IY is just the same as Y. Now note that if A^{-1} were known we could multiply by A^{-1} to have,

$$(2.33) \qquad\qquad IX = A^{-1}Y$$

so we now proceed as follows. Writing out Eq. (2.32) we have

$$(2.34)\quad
\begin{pmatrix}
a_{11} & a_{12} & a_{13}\ldots a_{1n} \\
a_{21} & a_{22} & a_{23}\ldots a_{2n} \\
\vdots & & \vdots \\
& & \\
a_{n1} & a_{n2} & \ldots\ldots a_{nn}
\end{pmatrix}
\begin{pmatrix}
x_1 \\ x_2 \\ \vdots \\ \\ x_n
\end{pmatrix}
=
\begin{pmatrix}
1 & 0 & 0 & 0\ldots \\
0 & 1 & 0 & 0\ldots \\
0 & 0 & 1 & \ldots\ldots \\
\vdots & & \\
0 & \ldots\ldots\ldots .1
\end{pmatrix}
\begin{pmatrix}
y_1 \\ y_2 \\ y_3 \\ \vdots \\ y_n
\end{pmatrix}
$$

Multiplying the *first equation* (i.e. for y_1) by a number, say a_{21}/a_{11} would produce a new *matrix* equation in which every element a_{1j} here would be replaced by

$$(2.35) \qquad\qquad a'_{1j} = a_{1j}\frac{a_{21}}{a_{11}}$$

and on the right we would have just a_{21}/a_{11} in place of 1 in the $(1, 1)$ position of the identity matrix.

We could then *subtract* this new first equation from the existing second equation and the resulting *new second* equation would have the first term in x_1 *missing*. The corresponding matrix equation could then be, keeping our *original* first equation,

$$(2.36)\quad
\begin{pmatrix}
a_{11} & a_{12} & a_{13}\ldots a_{1n} \\
0 & a_{22}^{(1)} & a_{23}^{(1)}\ldots a_{2n}^{(1)} \\
a_{31} & a_{32} & \ldots\ldots a_{3n} \\
\vdots & & \vdots \\
a_{n1} & \ldots\ldots\ldots a_{nn}
\end{pmatrix}
\begin{pmatrix}
x_1 \\ x_2 \\ x_3 \\ \vdots \\ x_n
\end{pmatrix}
=
\begin{pmatrix}
1 & 0 & 0 & 0\ldots 0 \\
b_{21}^{(1)} & 1 & 0 & 0\ldots 0 \\
0 & 0 & 1 & \ldots\ldots 0 \\
\vdots & & & \vdots \\
0 & \ldots\ldots\ldots .1
\end{pmatrix}
\begin{pmatrix}
y_1 \\ y_2 \\ y_3 \\ \vdots \\ y_n
\end{pmatrix}
$$

where,

(2.37)
$$a_{2j}^{(1)} = a_{2j} - a_{1j}\frac{a_{21}}{a_{11}}$$

and,

(2.38)
$$b_{21}^{(1)} = -a_{21}/a_{11}$$

So that we now have a *zero* in the (2, 1) position of A. We repeat this process on the third row to obtain in the *third row* of A, zero in the (3, 1) position and in general

(2.39)
$$a_{31}^{(1)} = 0, \qquad a_{3j}^{(1)} = a_{3j} - a_{1j}a_{31}/a_{11}, \qquad j = 1, 2, \ldots n$$

while the elements of the *third* row on the right become, $b_{31}^{(1)}, 0, 1, 0, 0 \ldots$ with

(2.40)
$$b_{31}^{(1)} = -a_{31}/a_{11}$$

We continue until we have

(2.41)
$$\begin{pmatrix} a_{11} & a_{12} \cdots \cdots a_{1n} \\ 0 & a_{22}^{(1)} \; a_{23}^{(1)} \ldots a_{2n}^{(1)} \\ 0 & a_{32}^{(1)} \; a_{33}^{(1)} \ldots a_{3n}^{(1)} \\ 0 & \\ 0 & a_{n2}^{(1)} \cdots \cdots a_{nn}^{(1)} \end{pmatrix} \begin{pmatrix} x_1 \\ x_2 \\ \vdots \\ \\ x_n \end{pmatrix} = \begin{pmatrix} 1 & 0 & 0 & 0 \ldots 0 \\ b_{21}^{(1)} & 1 & 0 & 0 \ldots 0 \\ b_{31}^{(1)} & 0 & 1 & 0 \ldots 0 \\ \\ b_{n1}^{(1)} \cdots \cdots \cdots 1 \end{pmatrix} \begin{pmatrix} y_1 \\ y_2 \\ y_3 \\ \vdots \\ y_n \end{pmatrix}$$

Now we start over and repeat the process using $a_{22}^{(1)}$ and operating on the rows below the second row, i.e., multiply the second row here by $a_{ij}^{(1)}/a_{22}^{(1)}$, *on both sides*, then subtract from the ith row and the result will be,

(2.42)
$$a_{ij}^{(2)} = a_{ij}^{(1)} - a_{ij}^{(1)}\frac{a_{i2}^{(1)}}{a_{22}^{(1)}}, \qquad j = 2, 3, \ldots n$$

for the new ith row of A and, on the right we will have the matrix

(2.43)
$$\begin{pmatrix} 1 & 0 & 0 & 0 & 0 \ldots \\ b_{21}^{(1)} & 1 & 0 & 0 & 0 \ldots \\ b_{31}^{(1)} & b_{32}^{(2)} & 1 & 0 & 0 \ldots \\ \vdots \\ b_{n1}^{(1)} & b_{n2}^{(2)} \cdots \cdots \cdots 1 \end{pmatrix}$$

This process is continued until we have obtained *all zeros below the main diagonal in A*. We then have,

(2.44)
$$\begin{pmatrix} a_{11} & a_{12} \cdots \cdots a_{1n} \\ 0 & a_{22}^{(1)} \; a_{23}^{(1)} \ldots a_{2n}^{(1)} \\ 0 & 0 \quad a_{33}^{(2)} \ldots a_{3n}^{(2)} \\ \vdots \\ \vdots \\ 0 \cdots \cdots \cdots a_{nn}^{(n-1)} \end{pmatrix} \begin{pmatrix} x_1 \\ x_2 \\ x_3 \\ \vdots \\ x_n \end{pmatrix} = \begin{pmatrix} 1 & 0 & 0 & 0 \ldots \\ b_{21}^{(1)} & 1 & 0 & 0 & 0 & 0 \\ b_{31}^{(2)} & b_{32}^{(2)} & 1 & 0 \ldots \\ b_{41}^{(3)} \cdots \cdots \cdots \\ b_{n1}^{(n-1)} \cdots \cdots \cdots 1 \end{pmatrix} \begin{pmatrix} y_1 \\ y_2 \\ y_3 \\ \vdots \\ y_n \end{pmatrix}$$

where

$$(2.45) \qquad a_{ij}^{(k)} = a_{ij}^{(k-1)}\left(1 - \frac{a_{kj}^{(k-1)}}{a_{kk}^{(k-1)}}\right), \qquad j = k, k+1, \ldots n$$

and

$$(2.46) \qquad b_{kj}^{(k-1)} = b_{kj}^{(k-2)}\left(1 - \frac{a_{kj}^{(k-1)}}{a_{kk}^{(k-1)}}\right) \qquad j = k, k+1, \ldots n$$

Then we begin at the bottom right corner and carry out a similar process going up through the rows. We multiply the last equation by $a_{n-1,n}^{(n-2)}/a_{nn}^{(n-1)}$ and subtract from the next row above, thus making the last element in this row zero. In fact it now has *all* zeros except in the main diagonal position. Continuing in this manner we finally have "A" in a *diagonal* form. Dividing each equation by the value of *this diagonal element* we then have the *form*,

$$(2.47) \qquad \begin{pmatrix} 1 & 0 & 0 & 0\ldots \\ 0 & 1 & 0 & 0 & 0 \\ 0 & 0 & 1\ldots\ldots \\ & \vdots & & \\ 0\ldots\ldots\ldots1 \end{pmatrix}\begin{pmatrix} x_1 \\ x_2 \\ x_3 \\ \vdots \\ x_n \end{pmatrix} = \begin{pmatrix} b_{11} & b_{12}\ldots b_{1n} \\ b_{21} & b_{22}\ldots\ldots \\ b_{31} & b_{32}\ldots\ldots \\ \vdots & \\ b_{n1}\ldots\ldots\ldots b_{nn} \end{pmatrix}\begin{pmatrix} y_1 \\ y_2 \\ y_3 \\ \vdots \\ y_n \end{pmatrix}$$

so that the matrix with elements b_{ij} here is just A^{-1}.

We call attention to the fact that if at any stage in the process a *zero* is formed on the diagonal in the "A" matrix it must be "removed" by interchanging rows, i.e., any two equations can always be interchanged in our set of equations.

Example (1)

Invert the matrix,

$$(2.48) \qquad A = \begin{pmatrix} 2 & 1 & 2 \\ 0 & 1 & 2 \\ 1 & 2 & 1 \end{pmatrix}$$

We begin by writing the two arrays

$$
\begin{array}{ccc}
(A) & & (B)
\end{array}
$$

$$
\begin{array}{ccc}
2 \quad 1 \quad 2 & \qquad & 1 \quad 0 \quad 0 \\
0 \quad 1 \quad 2 & & 0 \quad 1 \quad 0 \\
1 \quad 2 \quad 1 & & 0 \quad 0 \quad 1
\end{array}
$$

Now divide the first row of each by 2 and subtract the result from the last row, this gives,

$$
\begin{array}{ccc}
(A) & & (B)
\end{array}
$$

$$
\begin{array}{ccc}
2 \quad 1 \quad 2 & \qquad & 1 \quad 0 \quad 0 \\
0 \quad 1 \quad 2 & & 0 \quad 1 \quad 0 \\
0 \quad \tfrac{3}{2} \quad 0 & & -\tfrac{1}{2} \quad 0 \quad 1
\end{array}
$$

Since a zero has appeared on the main diagonal in the last equation we interchange the second and last equations to get

$$
\begin{array}{ccc} (A) \\ 2 & 1 & 2 \\ 0 & \frac{3}{2} & 0 \\ 0 & 1 & 2 \end{array}
\qquad
\begin{array}{ccc} (B) \\ 1 & 0 & 0 \\ -\frac{1}{2} & 0 & 1 \\ 0 & 1 & 0 \end{array}
$$

Now divide the second equation by 1.5 and subtract from the last to get,

$$
\begin{array}{ccc} (A) \\ 2 & 1 & 2 \\ 0 & \frac{3}{2} & 0 \\ 0 & 0 & 2 \end{array}
\qquad
\begin{array}{ccc} (B) \\ 1 & 0 & 0 \\ -\frac{1}{2} & 0 & 1 \\ \frac{1}{3} & 1 & -\frac{2}{3} \end{array}
$$

Then subtract the last equation from the first to arrive at

$$
\begin{array}{ccc} (A) \\ 2 & 1 & 0 \\ 0 & \frac{3}{2} & 0 \\ 0 & 0 & 2 \end{array}
\qquad
\begin{array}{ccc} (B) \\ \frac{2}{3} & -1 & \frac{2}{3} \\ -\frac{1}{2} & 0 & 1 \\ \frac{1}{3} & 1 & -\frac{2}{3} \end{array}
$$

Divide the second equation by 1.5 and subtract from the first to obtain the diagonal form of A,

$$
\begin{array}{ccc} (A) \\ 2 & 0 & 0 \\ 0 & \frac{3}{2} & 0 \\ 0 & 0 & 2 \end{array}
\qquad
\begin{array}{ccc} (B) \\ 1 & -1 & 0 \\ -\frac{1}{2} & 0 & 1 \\ \frac{1}{3} & 1 & -\frac{2}{3} \end{array}
$$

Now divide the first equation by 2, the second by 1.5 and the last by 2 to give

$$
\begin{array}{ccc} (A) \\ 1 & 0 & 0 \\ 0 & 1 & 0 \\ 0 & 0 & 1 \end{array}
\qquad
\begin{array}{ccc} (B) \\ \frac{1}{2} & -\frac{1}{2} & 0 \\ -\frac{1}{3} & 0 & \frac{2}{3} \\ \frac{1}{6} & \frac{1}{2} & -\frac{1}{3} \end{array}
$$

and therefore

(2.49)
$$
A^{-1} = \begin{pmatrix} \frac{1}{2} & -\frac{1}{2} & 0 \\ -\frac{1}{3} & 0 & \frac{2}{3} \\ \frac{1}{6} & \frac{1}{2} & -\frac{1}{3} \end{pmatrix}
$$

is A^{-1}, the inverse of the matrix, A. As a check we multiply A and A^{-1} here to yield, (the original A of course)

$$(2.50) \qquad AA^{-1} = \begin{pmatrix} 1 & 0 & 0 \\ 0 & 1 & 0 \\ 0 & 0 & 1 \end{pmatrix}$$

which shows that indeed we have formed the proper inverse for A.

We caution the student that in applying techniques such as this to *large* matrices, where *many* arithmetic operations must be performed, the "round-off" errors introduced in the calculations due to carrying only a finite number of digits in each number, can accumulate to produce *large* errors in the final inverse matrix. In our simple example here we obtained an *exact* inverse, but generally we would find that upon using the A^{-1} so formed to compute AA^{-1} there would exist small "residuals" in the off-diagonal positions, and the diagonal elements would differ slightly from unity, but such is the nature of numerical computation. Round-off errors are a serious problem in all such iterative techniques.

EXECUTE PROBLEM SET (2-2)

VECTORS IN N DIMENSIONS AND ORTHOGONAL TRANSFORMATIONS

In Chapter One we represented a general vector $\mathbf{V}$ in terms of its components on a set of *orthogonal* unit vectors† as,

$$(2.51) \qquad \mathbf{V} = \sum_{i=1}^{N} \mathbf{1}_i V_i$$

which we have now generalized to N dimensions by simply letting the index i run from 1 to N. Certainly we can not visualize such an N-dimensional *orthogonal space* but we require that‡

$$(2.52) \qquad \mathbf{1}_i \cdot \mathbf{1}_j = \delta_{ij} = \begin{cases} 1, i = j \\ 0, i \neq j, j = 1, 2, \ldots N \end{cases}$$

Now suppose we *rotate* this set of axes in some fashion, maintaining the *orthogonality* of the unit vectors. If we denote the unit vectors in the new position by $\mathbf{1}_i'$, where

$$(2.53) \qquad \mathbf{1}_i' \cdot \mathbf{1}_j' = \delta_{ij}$$

as above then we can express the components of $\mathbf{V}$ in the new system as we did in the old, thus,

$$(2.54) \qquad \mathbf{V} = \sum_{i=1}^{N} \mathbf{1}_i' V_i'$$

where V_i' is the orthogonal projection of $\mathbf{V}$ on the line directed as $\mathbf{1}_i'$.

We find the relation between these new components and the old components as follows. If we "dot" $\mathbf{1}_k'$ into Eq. (2.54) the result is V_k' as we have just said above, but we can also "dot" $\mathbf{1}_k'$ into Eq. (2.51). On the left the result must be,

$$(2.55) \qquad \mathbf{1}_k' \cdot \mathbf{V} = V_k'$$

† These can be constructed on a set of linearly independent vectors by the Schmidt orthonormalization process. See the appendix for details.

‡ δ_{ij} is called the "Kronecker Delta."

but on the right we have,

(2.56) $$V'_k = \sum_{i=1}^{N} (\mathbf{1}'_k \cdot \mathbf{1}_i) V_i$$

Here, from the definition of the dot or scalar product,

(2.57) $$a_{ki} = \mathbf{1}'_k \cdot \mathbf{1}_i = \cos \theta_{k'i}$$

is just the cosine of the angle, $\theta_{k'i}$, between these two unit vectors.

Since we can do this with *each* unit vector $\mathbf{1}_{k'}$ we see that there results the *system of linear equations*,

(2.58) $$V'_k = \sum_{i=1}^{N} a_{ki} V_i, \qquad k = 1, 2, \ldots N$$

which can be written as the *matrix equation*,

(2.59) $$V' = AV$$

where we call A the *transformation matrix*. This is one example of a *matrix operator*; A operates on V to produce V'. That is, we may consider ourselves as "fixed" to the coordinate frame and we view $\mathbf{V}$ as being rotated, not the axes. After all only the *relative* orientation of vectors is to be exhibited anyway.

Obviously, in the present case we could dot $\mathbf{1}_k$, of the original set, into $\mathbf{V}$ in the two representations of Eqs. (2.51) and (2.54) to yield

(2.60) $$V_k = \sum_{i=1}^{N} (\mathbf{1}_k \cdot \mathbf{1}_{i'}) V'_i, \qquad k = 1, 2, \ldots N$$

or, in matrix notation,

(2.61) $$V = A^{-1} V'$$

where A^{-1} is the inverse of A.

Now we call to attention the fact that the *length* of the vector $\mathbf{V}$, or its square, is the *same* whether we look at it from the original set of axes or the new set, that is,

(2.62) $$V^2 = \mathbf{V} \cdot \mathbf{V}$$

in vector notation, is the *same* in both representations. To write V^2 in matrix notation we have,

(2.63) $$V^2 = \tilde{V} V$$

with V as a single column matrix and $\tilde{V}$ as a single row matrix.

In *matrix notation* the requirement of the *same* V^2 reads as,

(2.64) $$\tilde{V}' V' = \tilde{V} V$$

Here let us now substitute the equation of transformation for V to V', Eq. (2.59). We have,

(2.65) $$(\widetilde{AV})(AV) = \tilde{V} V$$

but by the rule for the *transpose of a product* this is also,

(2.66) $$\tilde{V}(\tilde{A}A)V = \tilde{V} V$$

where we have used the associative property. Obviously this can only be an equality, or identity, if

(2.67)
$$\tilde{A}A = I$$

is the identity matrix. But this means that we must require

(2.68)
$$\tilde{A} = A^{-1}$$

since *only* A^{-1} multiplied into A gives the identity matrix.

Thus we come to an important conclusion: *any linear transformation which preserves the lengths of vectors must be represented by a transformation matrix, A, having an inverse equal to its transpose.*

The matrix A, as we built it from the dot products of unit vectors above automatically had this property, i.e., note Eqs. (2.56) and (2.60). We call such transformations *orthogonal transformations.*

EIGENVECTOR CONCEPT AND THE EIGENVALUE PROBLEM; SIMILARITY TRANSFORMATIONS

Here we want to generalize the idea of the orthogonal transformations; we wish to consider a general *linear transformation matrix A* and then ask the question, "do vectors, **X**, exist such that under the transformation A the *directions* of these vectors are unchanged?" That is, the effect of A on **X** is simply to alter its *length* by some factor, say λ.

We state this requirement mathematically as,

(2.69)
$$x_i' = \lambda x_i = \sum_{j=1}^{N} a_{ij}x_j, \qquad i = 1, 2, \ldots N$$

where the a_{ij} are the elements of A, x_i' are the transformed vector components, x_j the original components, and λ is the factor by which the length of **X** is changed. Note that **X** is unchanged in *direction* since

(2.70)
$$\frac{x_i'}{x_i} = \lambda = \text{constant}, \qquad i = 1, 2, \ldots N$$

In matrix form we write Eq. (2.69) as

(2.71)
$$AX = \lambda I X$$

where I is the identity matrix, or $[\lambda I]$ is the diagonal matrix with λ as each diagonal element, all off-diagonal elements being zero.

Rearranging Eq. (2.71) we have

(2.72)
$$(A - \lambda I)X = 0$$

where X is the column vector whose elements are the components of the sought for *eigenvector*, or proper vector, which is *not rotated* by A. Now this is a system of N linear *homogeneous* equations to be solved for the x_i, and we know a solution can only exist if the determinant of coefficients is zero, i.e.,

(2.73)
$$\begin{vmatrix} a_{11} - \lambda & a_{12} & a_{13} \ldots a_{1N} \\ a_{21} & a_{22} - \lambda & a_{23} \ldots a_{2N} \\ \vdots & & \\ a_{N1} & \ldots \ldots \ldots \ldots a_{NN} - \lambda \end{vmatrix} = 0$$

When this is expanded we find that we have a polynomial in λ of degree N, having N roots, $\lambda_1, \lambda_2, \ldots \lambda_N$, we call the *eigenvalues of A*. We call this polynomial,

$$(2.74) \qquad \lambda^N + \alpha_1 \lambda^{N-1} + \alpha_2 \lambda^{N-2} + \ldots + \alpha_N = 0$$

the characteristic equation of A.

Now we show how the matrix A can be given *another representation*.† Consider some transformation T applied to the vector X to produce X'',

$$(2.75) \qquad TX = X''$$

then multiply by the inverse, so,

$$(2.76) \qquad X = T^{-1}X''$$

Now we have, operating with A,

$$(2.77) \qquad AX = AT^{-1}X''$$

but if X is an eigenvector, Eq. (2.71) applies, and this is

$$(2.78) \qquad \lambda IX = AT^{-1}X''$$

We then operate on this with T, noting that

$$(2.79) \qquad T\lambda I = \lambda IT$$

since λ is a scalar, and I is the identity matrix. Thus,

$$(2.80) \qquad \lambda ITX = TAT^{-1}X''$$

or in view of Eq. (2.75),

$$(2.81) \qquad \lambda IX'' = (TAT^{-1})X''$$

Therefore we *must* interpret

$$(2.82) \qquad (TAT^{-1}) = A''$$

as the representation of the transformation matrix A in the double-prime coordinate frame.

Next we show that a particular choice for the matrix T will yield A'' as a *diagonal* matrix. Assume,

$$(2.83) \qquad T^{-1} = (X_1 X_2 X_3 \ldots X_N)$$

where each X_i here is a *column* vector, being the *eigenvector* of A corresponding to the eigenvalue λ_i, i.e.,

$$(2.84) \qquad AX_i = \lambda_i X_i$$

We must require linearly independent X_i and distinct λ_i so that T will have an inverse, i.e., $|T| \neq 0$.

Now operate on T^{-1} with A to obtain,

$$(2.85) \qquad AT^{-1} = A(X_1 X_2 \ldots X_N) = (\lambda_1 X_1 \lambda_2 X_2 \ldots \lambda_N X_N)$$

---

† This is sometimes called a representation in a *new basis set*, i.e., the unit vector $\mathbf{1}_i'$.

from the definition of A, Eq. (2.84). The right member here can be written as,

(2.86)
$$
\begin{pmatrix}
\lambda_1 x_{11} & \lambda_2 x_{21} \ldots \lambda_N x_{N1} \\
\lambda_1 x_{12} & \lambda_2 x_{22} \ldots \ldots \ldots \\
\lambda_1 x_{13} & \vdots \qquad \vdots \\
\vdots & \vdots \qquad \vdots \\
\lambda_1 x_{1N} & \lambda_2 x_{2N} \ldots \lambda_N x_{NN}
\end{pmatrix}
=
\begin{pmatrix}
x_{11} & x_{21} \ldots x_{N1} \\
x_{12} & x_{22} \ldots x_{N2} \\
\vdots \\
\vdots \\
x_{1N} \ldots \ldots x_{NN}
\end{pmatrix}
\begin{pmatrix}
\lambda_1 & 0 & 0 & 0 \\
0 & \lambda_2 & 0 & 0 \\
\vdots & 0 & \lambda_3 \ldots \\
\vdots \\
0 \ldots \ldots \ldots \lambda_N
\end{pmatrix}
$$

where x_{ji} is the ith component of vector j, or

(2.87)
$$AT^{-1} = T^{-1} \,(\text{Diagonal } \lambda_i)$$

Then multiplying from the left by T we obtain,

(2.88)
$$TAT^{-1} = (\text{Diagonal } \lambda_i) = A''$$

Hence in the double prime system A'' is a diagonal matrix with elements λ_i as the main diagonal.

The transformation of A into A'' above, or *any* transformation of a square matrix A, by some other square matrix T, according to the rule,

(2.89)
$$A'' = TAT^{-1}$$

is called a *similarity transformation.*

CAYLEY–HAMILTON THEOREM

This famous theorem states that the *matrix A itself satisfies its own characteristic equation,* Eq. (2.74). We exhibit the meaning of this and at the same time indicate the nature of the proof of the theorem as follows.

In the representation in the double-prime system, A'' is the diagonal matrix with elements $\lambda_1, \lambda_2, \ldots \lambda_N$ along the main diagonal as just shown. In this frame of reference, or representation

(2.90)
$$|D - \lambda I| = 0$$

where we now use the *notation $D = A''$* to emphasize that A'' is a diagonal matrix. Then the characteristic polynomial is just,

(2.91)
$$(\lambda_1 - \lambda)(\lambda_2 - \lambda) \ldots (\lambda_N - \lambda) = 0$$

If indeed a matrix does satisfy its own characteristic polynomial then D here should satisfy,

(2.92)
$$(D - \lambda_1 I)(D - \lambda_2 I) \ldots (D - \lambda_N I) = 0$$

meaning that this must be the *null matrix,* and indeed by inspection it is. Thus D does indeed satisfy *its* own characteristic polynomial, which is the *same* polynomial as that for A, i.e., we now see, expanding Eq. (2.92),

(2.93)
$$D^N + \alpha_1 D^{N-1} + \alpha_2 D^{N-2} + \ldots + \alpha_N = 0$$

as a *true* equation. [Note: here we have used the fundamental theorems of algebra that if two polynomials of the *same* degree and leading coefficient unity have all roots in common then all coefficients of the polynomials are identical.] The α_i here are the same as in Eq. (2.74).

Note that this also follows from our matrix formulation. We have,

(2.94) $$TAT^{-1} - \lambda TIT^{-1} = D - \lambda I$$

or

(2.95) $$T(A - \lambda I)T^{-1} = D - \lambda I$$

Then

(2.96) $$|T| \cdot |A - \lambda I| \cdot |T^{-1}| = |D - \lambda I|$$

But

(2.97) $$|T||T^{-1}| = |TT^{-1}| = |I| = 1$$

so

(2.98) $$|A - \lambda I| = |D - \lambda I|$$

and the two characteristic polynomials are identical.

Now note in Eq. (2.93) that for example

(2.99) $$D^k = (TAT^{-1})^k = (TAT^{-1})(TAT^{-1})\ldots(TAT^{-1})$$

with k factors. By the associative property this is also

(2.100) $$D^k = TA(T^{-1}T)A(T^{-1}T)\ldots AT^{-1} = TA^kT^{-1}$$

Thus Eq. (2.93) reads,

(2.101) $$TA^NT^{-1} + \alpha_1 TA^{N-1}T^{-1} + \ldots + \alpha_N = 0$$

Hence multiply from the left by T^{-1}, the right by T and obtain,

(2.102) $$A^N + \alpha_1 A^{N-1} + \alpha_2 A^{N-2} + \ldots + \alpha_N = 0$$

which is the *Cayley–Hamilton* equation for A.

CONSTRUCTION OF THE CHARACTERISTIC EQUATION

For a matrix A of large order the expansion of the determinant, $|A - \lambda I|$, to form the characteristic equation, say an expansion by minors applied iteratively, becomes quite a chore. Here is outlined a simpler method.

Choose any known vector Y_0, a column matrix, with elements $y_{01}, y_{02} \ldots y_{0N}$ then define

(2.103) $$Y_1 = AY_0, \qquad Y_2 = AY_1 = A^2 Y_0, \text{etc.}, \qquad Y_N = A^N Y_0$$

Thus if we multiply Eq. (2.102) by Y_0, from the *right*, we obtain:

(2.104) $$Y_N + \alpha_1 Y_{N-1} + \alpha_2 Y_{N-2} + \ldots + \alpha_N Y_0 = 0$$

This is a set of simultaneous equations,

(2.105) $$y_{Ni} + \alpha_1 y_{N-1,i} + \alpha_2 y_{N-2,i} \ldots + \alpha_N y_{0i} = 0, \qquad i = 1, 2, \ldots N$$

with all *known* coefficients, y_{ji}, to solve simultaneously for the α_k. Once we know the α_k we put these into the characteristic equation and solve for the eigenvalues as the roots, $\lambda_1, \lambda_2 \ldots \lambda_N$.

DETERMINATION OF THE EIGENVECTORS

Once the eigenvalues have been found, say by constructing the coefficients α_i of the characteristic equation as described above then solving the characteristic equation, Eq. (2.74), for the $\lambda_j, j = 1, 2, \ldots N$, then we can return to Eq. (2.69), written here again,

$$(2.69) \qquad \sum_{j=1}^{N} a_{ij}x_j = \lambda x_i, \qquad i = 1, 2, \ldots N$$

and solve for the components of the eigenvectors. That is, if here we put λ_1 for λ then we have N *homogeneous* equations to solve for the components x_j, of the *first* eigenvector. Obviously, since these are homogeneous equations these x_i, will be determined only to within a multiplying factor and therefore an *auxiliary condition* on the x_i can be imposed. For example we can take

$$(2.106) \qquad \sum_{j=1}^{N} x_j^2 = 1$$

and the eigenvectors are a set of *unit vectors*. Note that for each choice of $\lambda, \lambda = \lambda_1, \lambda = \lambda_2$, etc. in Eq. (2.69) we generate N equations in N unknowns for the successive eigenvectors corresponding to λ_1, λ_2 etc.†

EXECUTE PROBLEM SET (2-3)

PRINCIPAL AXIS TRANSFORMATIONS

As a special application of the concepts of orthogonal transformations and eigenvectors we consider the following example as typical of those encountered in physics.

Example (2)

Principal axes of the dielectric tensor. In the study of electromagnetic phenomena in crystalline materials we find the electric displacement vector **D** related to the electric field vector by an equation of the form,

$$(2.107) \qquad D = \varepsilon E$$

where D and E are column vectors and ε is *sometimes a symmetric* 3×3 square matrix, i.e.,

$$(2.108) \qquad \varepsilon_{ij} = \varepsilon_{ji},$$

This ε matrix is called the dielectric tensor of the medium. *If ε is symmetric* then our Eq. (2.107) can be put into a very simple form by an *orthogonal* transformation of the coordinate axes.

Let A be an orthogonal transformation matrix which rotates the coordinate axes, i.e., then

$$(2.109) \qquad A^{-1} = \tilde{A}$$

Applying this to E we get

$$(2.110) \qquad E' = AE$$

† These unit eigenvectors are said to define the *principal axes* of the matrix A.

in the new axes. Operating on D in Eq. (2.107) we get

(2.111)
$$D' = A(\varepsilon E)$$

in the new axes. Here we substitute Eq. (2.109) with the inverse operation,

(2.112)
$$E = A^{-1}E'$$

and have

(2.113)
$$D' = (A\varepsilon A^{-1})E'$$

so that we see the dielectric tensor in the new axes generated by a similarity transformation, i.e.,

(2.114)
$$\varepsilon' = A\varepsilon A^{-1}$$

Now note that if we operate from the left with A^{-1} and from the right with A we get

(2.115)
$$A^{-1}\varepsilon' A = \varepsilon$$

for the matrix ε. But since A is an *orthogonal* transformation we have Eq. (2.109) applying, thus

(2.116)
$$\tilde{A}\varepsilon' A = \varepsilon$$

Now *if ε is symmetric*, i.e., Eq. (2.108) apply, then

(2.117)
$$\tilde{\varepsilon} = \varepsilon$$

thus now take the transpose of Eq. (2.116); we get

(2.118)
$$\tilde{\varepsilon} = \widetilde{A\varepsilon' A} = \tilde{\varepsilon'}\tilde{A}\,\tilde{\tilde{A}} = \tilde{A}\tilde{\varepsilon'} A$$

which implies that,

(2.119)
$$\tilde{\varepsilon'} = \varepsilon'$$

by comparing the end result in Eq. (2.118) with Eq. (2.116). So if ε is symmetric in one coordinate frame it is symmetric in *every* coordinate frame. (Only orthogonal coordinate frames are considered.)

If we choose as the transformation matrix A that formed of the *eigenvectors of the ε matrix* as T above, then we can transform ε into a *diagonal ε' matrix*. This is the *principal axis transformation.*

The advantages of the principal axis representation are obvious, for then, in this particular set of coordinate axes, the dielectric properties of the medium are represented by *three* numbers instead of 9, or 6 for the *symmetric* case here. Here it is important to note that this is only possible for the symmetric tensor, for as seen in the last problem set only in this case are the eigenvectors orthogonal, i.e., we have an orthogonal coordinate system in the principal axes system.

INFINITESIMAL ORTHOGONAL TRANSFORMATIONS

We have seen in the problem set just executed above that if A, B, C represent orthogonal transformations then $AB \neq BA$, $ABC \neq CBA \neq CAB$, etc. but *all* of these do represent orthogonal transformations. Thus for example we might consider the three successive orthogonal transformations, in three dimensions,

$$(2.120) \qquad C = \begin{pmatrix} \cos\phi & \sin\phi & 0 \\ -\sin\phi & \cos\phi & 0 \\ 0 & 0 & 1 \end{pmatrix}$$

$$(2.121) \qquad B = \begin{pmatrix} 1 & 0 & 0 \\ 0 & \cos\theta & \sin\theta \\ 0 & -\sin\theta & \cos\theta \end{pmatrix}$$

$$(2.122) \qquad A = \begin{pmatrix} \cos\psi & \sin\psi & 0 \\ -\sin\psi & \cos\psi & 0 \\ 0 & 0 & 1 \end{pmatrix}$$

which correspond to the successive rotations depicted in *Fig. 2-1*, and define the famed *Euler Angles*, which are often used to describe rigid-body rotations in mechanics.

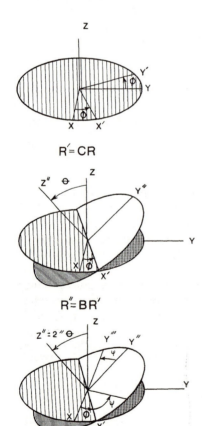

$$R' = CR$$

$$R'' = BR'$$

$$R''' = AR''$$

FIGURE 2.1.   Rotations Defining the Euler Angles.

Observe that first operating on a vector, say the position vector $\mathbf{r}$, we rotate the axes through the angle ϕ *about the z* axis with C. This gives a new set of axes as shown. Next we operate on this result with B which is a rotation about the *new x* axis, giving rise to yet another set of axes. Finally, we operate with A on the last result to rotate these new axes about their z axis through the angle ψ. These then, in the order ABC yield the axes indicated in the last configuration indicated, but *in any other order do not yield this configuration.*

But now let us consider what happens if we stipulate that all these rotations are to be *infinitesimal*, i.e., $\theta \to d\theta$, a differential, etc. We can look at this first in a very general way as follows.

Let D represent a general orthogonal transformation (in N dimensions) then we have, for a vector X,

(2.123)
$$X' = DX$$

and the requirement

(2.124)
$$D^{-1} = \tilde{D}$$

for orthogonality. If X' is to differ from X by only some infinitesimal amount

(2.125)
$$X' = IX + \varepsilon X$$

where I is the identity matrix and ε is a square matrix *all* of whose elements are *zero or infinitesimal*. Thus we require for an infinitesimal transformation

(2.126)
$$D = I + \varepsilon$$

and for orthogonality

(2.127)
$$(\tilde{I} + \tilde{\varepsilon})(I + \varepsilon) = I$$

Expanding this and *neglecting second-order terms*, $\tilde{\varepsilon}\varepsilon$, we see that we must have

(2.128)
$$\tilde{\varepsilon} = -\varepsilon$$

Thus ε has the form,

(2.129)
$$\varepsilon = \begin{pmatrix} 0 & \varepsilon_{12} & \varepsilon_{13}\ldots\varepsilon_{1N} \\ -\varepsilon_{12} & 0 & \varepsilon_{23}\ldots\varepsilon_{2N} \\ -\varepsilon_{13} & -\varepsilon_{23} & 0 \ldots \;\vdots \\ \vdots & & \vdots \\ \vdots & & \vdots \\ -\varepsilon_{1N}\cdots & & \cdots\cdots 0 \end{pmatrix}$$

or in general

(2.130)
$$\varepsilon_{ij} = -\varepsilon_{ji} \quad \text{all } i, j$$

We call such a matrix an *antisymmetric matrix*. (Note all $\varepsilon_{ii} = 0$).

For the *special case of three dimensions* this leads to a very useful result. Writing Eq. (2.125) now in the form

(2.131)
$$\begin{pmatrix} x_1 + \delta x_1 \\ x_2 + \delta x_2 \\ x_3 + \delta x_3 \end{pmatrix} = \left[\begin{pmatrix} 1 & 0 & 0 \\ 0 & 1 & 0 \\ 0 & 0 & 1 \end{pmatrix} + \begin{pmatrix} 0 & d\Omega_3 & -d\Omega_2 \\ -d\Omega_3 & 0 & d\Omega_1 \\ d\Omega_2 & -d\Omega_1 & 0 \end{pmatrix} \right] \begin{pmatrix} x_1 \\ x_2 \\ x_3 \end{pmatrix}$$

where we have chosen a special notation for the three distinct elements of the ε matrix (distinct except with regard to *sign*). From this we see,

(2.132)
$$\begin{cases} \delta x_1 = x_2 d\Omega_3 - x_3 d\Omega_2 \\ \delta x_2 = x_3 d\Omega_1 - x_1 d\Omega_3 \\ \delta x_3 = x_1 d\Omega_2 - x_2 d\Omega_1 \end{cases}$$

and these look like the components of the *cross product* of the vector $\mathbf{r}$ and a vector $d\Omega$ with the components, $d\Omega_1, d\Omega_2, d\Omega_3$, i.e.,

(2.133)
$$d\mathbf{r} = \mathbf{r} \times d\Omega$$

We can now examine the successive rotations C, B, A discussed previously if we let the three angles ϕ, θ, ψ all be infinitesimal. Thus we write

(2.134)
$$C \rightarrow I + \begin{pmatrix} 0 & d\phi & 0 \\ -d\phi & 0 & 0 \\ 0 & 0 & 0 \end{pmatrix} = I + (d\phi)$$

(2.135)
$$B \rightarrow I + \begin{pmatrix} 0 & 0 & 0 \\ 0 & 0 & d\theta \\ 0 & -d\theta & 0 \end{pmatrix} = I + (d\theta)$$

(2.136)
$$A \rightarrow I + \begin{pmatrix} 0 & d\psi & 0 \\ -d\psi & 0 & 0 \\ 0 & 0 & 0 \end{pmatrix} = I + (d\psi)$$

and we can readily verify that

(2.137)
$$ABC = CAB = BAC$$

to *first order in the differentials* $d\phi, d\theta, d\psi$.

The quantity $d\Omega$ above "appears" to have the character of a *vector* but actually it is not truly a vector. This is seen as follows. If we perform an *inversion* of the axes $x_1 \rightarrow -x_1$, $x_2 \rightarrow -x_2, x_3 \rightarrow -x_3$ then sure enough $\mathbf{r}$ and $d\mathbf{r}$ *both* change sign but $d\Omega$ does *not*. In fact the cross product of any two vectors *also* has this quirk of *not* changing sign in an inversion. Thus such quantities have properties different from those of ordinary vectors and are termed *pseudovectors*, or in the case of $d\Omega$ here, *axial vectors*.

The proper study and classification of scalars, pseudoscalars, vectors, etc. is a part of the task of tensor calculus which we will not undertake in this text, but the student may pursue these topics further in the references.

EXECUTE PROBLEM SET (2-4)

HERMITIAN OPERATORS

If instead of a *real* N-dimensional *vector space* in which we consider a vector X having *real* components $x_i, i = 1, 2, \ldots N$, we allow these vectors now to have *complex* components then it is only natural that we consider transformation matrices, A, which operate on such vectors to have complex elements. A *very special class* of such complex matrix operators are those we call *Hermitian Operators*, or *Hermitian Matrices*. These are those having the property,

(2.138)
$$A\dagger = A$$

i.e., a Hermitian matrix is *self-adjoint*. Recall that $A\dagger$ is just $\tilde{A}^*$.

Now suppose we re-examine the sequence of topics already discussed in terms of real matrix operators—now in terms of these complex operators. We can look at the *eigenvalue problems* in much the same way.

(2.139)
$$AX = \lambda X$$

but now A and X are complex. But, we can easily prove that if A is Hermitian then all of its eigenvalues are *real*, i.e., $\lambda_1, \lambda_2, \ldots \lambda_N$ are all real numbers. It is this characteristic that makes Hermitian operators valuable in physics because we associate *observable values* of physical variables with the eigenvalues of Hermitian operators, i.e., in quantum mechanics. [The *notation* most often used in the realm of quantum mechanics is that originally devised by Dirac. Two kinds of vectors are defined, *bra* and *ket* vectors. These correspond to row vectors and column vectors, respectively. Thus, we may write

$$V = |V\rangle$$

as the ket vector V, or also the bra vector,

$$V\dagger = \langle V^*|$$

Thus, the "dot" product of a vector and its *complex* conjugate is represented by

$$|V|^2 = \langle V^*|V\rangle$$

We only mention this notation here since its application is limited primarily to quantum theory.]

EXECUTE PROBLEM SET (2-5)

DYADICS AND MATRICES

We introduced the concept of a matrix and formed the basic rule for multiplication of matrices from consideration of a *linear* relation between two sets of quantities. However, the "need" for the matrix idea, and all the algebraic apparatus we have been considering also arises naturally in certain *physical problems* formulated in a *vector* notation. This leads to an entirely different *notation* and a slightly different point of view. We are led to define quantities we call *dyadics*. This is best introduced by an example.

MOMENTUM FLUX "TENSOR"

Let $\mathbf{V}(x, y, z, t)$ denote the volume flux density vector, or "velocity" of a fluid at any point x, y, z at any instant t in some region of space. Then consider an element of surface $\mathbf{ds}$ bounding some region. From Chapter One we know that

(2.140)
$$\delta Q/\delta t = -\mathbf{V} \cdot \mathbf{ds}$$

is the *volume* of fluid, per unit time, flowing *into* the region through ds. Here, we have $\mathbf{ds}$ as $\mathbf{n}\,ds$ with the unit normal vector, $\mathbf{n}$, directed *outward* from the region. Now, if each volume element of fluid carries with it some intrinsic property, such as, say a certain *mass per unit volume*, ρ, (the mass density), then the quantity of mass per unit time entering through ds is obviously just

(2.141) $$\rho(\delta Q/\delta t) = -(\rho \mathbf{V}) \cdot \mathbf{ds}$$

and we then call $\rho \mathbf{V}$ the *mass flux vector*. Note this is indeed a vector and is the mass per unit time, per unit area, flowing in the direction of $\mathbf{V}$.

We can do the same thing with any *scalar density* quantity, such as ρ, to form other *flux vectors*, i.e., energy flux tensor with ρ replaced by the intrinsic energy per unit volume, etc. However, we can also do the same thing with *vector densities*. Thus, note that at any point in the fluid $\rho \mathbf{V}$ is the *momentum per unit volume*, i.e., if we considered a volume element $d\tau$ it would have a mass $\rho\, d\tau$ and multiplied by $\mathbf{V}$ this would yield $(\rho\, d\tau)\mathbf{V}$ as the linear momentum of the volume element. Dividing by $d\tau$ we have just $\rho\mathbf{V}$ as momentum density. Thus, if volume δQ goes into the region it carries with it the linear momentum $(\rho\mathbf{V})\delta Q$, or the momentum per unit time entering the region across ds is given by:

(2.142) $$(\rho\mathbf{V})(\delta Q/\delta t) = -(\rho\mathbf{V}\mathbf{V}) \cdot \mathbf{ds}$$

Here note that the left member is a *vector* as is the right member but if we *lift out* and look at the quantity

(2.143) $$M = \rho\mathbf{V}\mathbf{V}$$

it is *not* a vector. It is the *algebraic product of two vectors* and is called a dyadic. This particular dyadic is called the momentum flux tensor and plays a major role in the study of hydrodynamics.

ALGEBRA OF DYADICS

We can consider dyadics in general by writing a general dyadic as the algebraic product of two vectors $\mathbf{A}$ and $\mathbf{B}$. Thus, with

(2.144) $$\begin{cases} \mathbf{A} = \sum_{i=1}^{N} \mathbf{1}_i A_i \\[2mm] \mathbf{B} = \sum_{i=1}^{N} \mathbf{1}_i B_i \end{cases}$$

we see the dyadic $(\mathbf{AB})$ as,

(2.145) $$(\mathbf{AB}) = \sum_{i=1}^{N}\sum_{j=1}^{N} \mathbf{1}_i \mathbf{1}_j (A_i B_j)$$

and we note immediately that in general this is *not* symmetric in the indices, i.e., $(\mathbf{BA}) \neq (\mathbf{AB})$.

Now let us dot this into some vector $\mathbf{X}$,

(2.146) $$\mathbf{X} = \sum_{k=1}^{N} \mathbf{1}_k x_k$$

thus,

(2.147) $$(\mathbf{AB}) \cdot \mathbf{X} = \sum_{i=1}^{N}\sum_{j=1}^{N}\sum_{k=1}^{N} \mathbf{1}_i (\mathbf{1}_j \cdot \mathbf{1}_k) A_i B_j x_k$$

But since,

(2.148)
$$1_j \cdot 1_k = \delta_{jk} = \begin{cases} 1, j = k \\ 0, j \neq k \end{cases}$$

because the 1_i are an orthogonal basis set of unit vectors, we see that,

(2.149)
$$(\mathbf{AB}) \cdot \mathbf{X} = \sum_{i=1}^{N} \sum_{k=1}^{N} 1_i A_i B_k x_k$$

so that the ith element of the dot product looks like,

(2.150)
$$\{(\mathbf{AB}) \cdot \mathbf{X}\}_i = \sum_{k=1}^{N} (A_i B_k) x_k$$

which looks exactly like the row-column rule of matrix multiplication of the $N \times N$ matrix with elements $(A_i B_j) = a_{ij}$ into the column vector X with elements x_k.

Thus, we can proceed to show that sure enough dyadics do indeed have the same algebraic properties as matrices, i.e., we just treat $\mathbf{AB}$ as a matrix, say M with elements $m_{ij} = A_i B_j$ and we have the correspondences

(2.151)
$$(\mathbf{AB}) \cdot \mathbf{X} \rightleftarrows MX$$

and we can define M^{-1}, $\tilde{M}$, etc. just in a complete one-to-one manner with the matrix symbolism. The reason for using dyadic notation in lieu of matrix notation is obvious from our introductory example. When we begin the formulation of a physical problem in a vector notation we often are led naturally to dyadics.

Example (3)

The stress "tensor" of the theory of elasticity, σ, can be represented as a matrix or as a dyadic. It is most easily *visualized* as a dyadic. From the definition of a *stress* as the force per unit area acting on a surface we see that *two* vectors are required to represent the total description of the stress. One vector $\mathbf{n}$ to specify the orientation of the surface, this being the outward unit normal vector, and one vector $\mathbf{f}$ specifying the *magnitude* and *direction* of the force per unit area on *this* surface, thus, we write

(2.152)
$$\sigma = \mathbf{fn}$$

as a dyadic for the *stress tensor*. If we dot this *into* an area element, $\mathbf{ds}$, we see

(2.153)
$$\sigma \cdot \mathbf{ds} = \mathbf{f}(\mathbf{n} \cdot \mathbf{ds})$$

and if $\mathbf{n}$ happens to be parallel to $\mathbf{ds}$ then this is just the vector force, in magnitude and direction acting on $\mathbf{ds}$.† We also see

(2.154)
$$1_i \cdot \sigma = (1_i \cdot \mathbf{f})\mathbf{n} = f_i \mathbf{n}$$

as the component of the force, parallel to the i axis, acting on unit area having normal vector, $\mathbf{n}$.

---

† If $\mathbf{n}$ is *not* parallel to $\mathbf{ds}$ then $\mathbf{f}(\mathbf{n} \cdot \mathbf{ds})$ is the force on an area of magnitude, $ds \cos \theta$, where θ is the angle between $\mathbf{n}$ and $\mathbf{ds}$.

MATRIX REPRESENTATION OF SYSTEMS OF ORDINARY LINEAR DIFFERENTIAL EQUATIONS

While we do not intend to branch out, here, into a general discussion of ordinary differential equations the student should be aware of the existence of the application of matrix notation in this area and its advantages. We describe a single example which should suffice to convey the general *idea*.

First of all, we must *define* differentiation of a matrix. Thus, consider $U(t)$ to be some matrix whose elements are *each* a function of the parameter t. Since the *difference* in two matrices is already defined we see

(2.155) $$\Delta U = U(t + \Delta t) - U(t)$$

as a matrix whose elements are $u_{ij}(t + \Delta t) - u_{ij}(t)$. Also, since multiplication of a matrix by a scalar is defined we see

(2.156) $$\frac{\Delta U}{\Delta t} = \frac{U(t + \Delta t) - U(t)}{\Delta t}$$

as a matrix whose elements are just $[u_{ij}(t + \Delta t) - u_{ij}(t)]/\Delta t$. Thus, in the limit as $\Delta t \to 0$ we see

(2.157) $$\frac{dU}{dt} = \left(\frac{du_{ij}}{dt}\right)$$

i.e., *each* element is differentiated with respect to the parameter t.

Example (4)

Represent the second-order differential equation

(2.158) $$\frac{d^2y}{dt^2} + a\frac{dy}{dt} + by = F$$

where a, b, and F are functions of t, in matrix form in terms of y and

(2.159) $$v = (dy/dt) + qy$$

where q is also a function of t.

Taking the derivative in Eq. (2.159) we can find,

(2.160) $$\frac{d^2y}{dt^2} = \frac{dv}{dt} - q\frac{dy}{dt} - y\dot{q}$$

where $\dot{q} = dq/dt$, or,

(2.161) $$\frac{d^2y}{dt^2} = \frac{dv}{dt} - qv + q^2y - \dot{q}y$$

Thus, we can write the *pair* of *first-order equations*,

(2.162) $$\begin{cases} \dfrac{dy}{dt} = v - qy \\[2mm] \dfrac{dv}{dt} = (q - a)v + (q^2 + \dot{q} + aq - b)y + F \end{cases}$$

where the first equation is just Eq. (2.159) rearranged while the second is obtained by using Eq. (2.160) to eliminate d^2y/dt^2 in Eq. (2.158). This *pair* of equations can then be written as the matrix equation,

(2.163)
$$\frac{d}{dt}\begin{pmatrix} y \\ v \end{pmatrix} = \begin{pmatrix} -q & 1 \\ (q^2 + \dot{q} + aq - b) & (q - a) \end{pmatrix}\begin{pmatrix} y \\ v \end{pmatrix} + \begin{pmatrix} 0 \\ F \end{pmatrix}$$

Quite obviously, the resulting form here depends on the *form* specified for v as well as the original differential equation.

It should be obvious to the student that this type of representation can readily be extended to higher-order equations and to *systems* of equations in more than one dependent variable. Such representations have great utility not only for "problem solving" but also, for general investigations of *classes* of equations.

EXECUTE PROBLEM SET (2-7)

PROBLEMS

Set (2-1)

(1) Show that if A is an $n \times m$ matrix then with the $m \times m$ identity matrix:

$$AI = A$$

and with the $n \times n$ identity matrix

$$IA = A$$

(2) Show that for any $n \times n$ *square* matrix, B

$$IB = BI = B$$

where I is $n \times n$.

(3) Prove Eq. (2.21)

$$\widetilde{AB} = \tilde{B}\tilde{A}$$

(4) Verify Eq. (2.25),

$$(AB)C = A(BC)$$

for conformable matrices.

(5) Verify Eqs. (2.22) and (2.23)

$$|AB| = |A| \cdot |B|$$
$$|\tilde{A}| = |A|$$

(6) Find the inverse of the matrix

$$A = \begin{pmatrix} 2 & 1 & 2 \\ 0 & 1 & 2 \\ 1 & 2 & 1 \end{pmatrix}$$

by applying the definition of Eq. (2.24)

(7) Find the inverse of the general 3×3 matrix,

$$A = \begin{pmatrix} a_{11} & a_{12} & a_{13} \\ a_{21} & a_{22} & a_{23} \\ a_{31} & a_{32} & a_{33} \end{pmatrix}$$

by applying the definition of Eq. (2.24) and show that

$$A^{-1}A = AA^{-1} = I$$

is then satisfied.

(8) The *rank* of an $n \times m$ matrix is the size, k, of the largest square submatrix, $k \times k$, having nonzero determinant obtained by deleting rows and columns. What is the rank of

$$M = \begin{pmatrix} 1 & 0 & 1 & 2 & 2 \\ 0 & 0 & 1 & 1 & 1 \\ 2 & 1 & 2 & 1 & 0 \\ 3 & 0 & 1 & 1 & 0 \end{pmatrix}$$

The *rank* determines the number of linearly independent equations in a set of linear relations.

Set (2-2)

(1) Find the inverse of the matrix

$$M = \begin{pmatrix} 1 & 1 & 0 & 3 \\ 0 & 0 & 3 & 1 \\ 1 & 2 & 1 & 0 \\ 2 & 1 & 0 & 1 \end{pmatrix}$$

by the Gaussian Elimination method.

Set (2-3)

(1) Show that if a real matrix A is *symmetric*, $a_{ij} = a_{ji}$, then it can always be put in *diagonal* form with a similarity transformation in which the transformation matrix, T, is an *orthogonal* transformation.

(2) Show that the eigenvectors of a real *symmetric* matrix are mutually orthogonal, if the eigenvalues are distinct.

(3) Find the eigenvalues and eigenvectors of the matrix

$$A = \begin{pmatrix} 2 & 1 & 2 \\ 0 & 1 & 2 \\ 2 & 2 & 1 \end{pmatrix}$$

(4) Find the eigenvalues and eigenvectors of the matrix

$$B = \begin{pmatrix} 1 & 2 & 1 \\ 2 & 3 & 2 \\ 1 & 2 & 0 \end{pmatrix}$$

(5) Prove that every *real* (3×3) orthogonal transformation has one real eigenvalue $+ 1$, and two complex eigenvalues whose product is $+ 1$, if det $= + 1$.

Set (2-4)

(1) Show that the equations

$$V_1 = a_{11}\frac{\partial U}{\partial x_1} + a_{12}\frac{\partial U}{\partial x_2} + \ldots + a_{1n}\frac{\partial U}{\partial x_n}$$

$$V_2 = a_{21}\frac{\partial U}{\partial x_1} + a_{22}\frac{\partial U}{\partial x_2} + \ldots + a_{2n}\frac{\partial U}{\partial x_n}$$

$$\vdots \quad \vdots \quad \vdots \quad \vdots \quad \vdots \quad \vdots$$

$$V_n = a_{n1}\frac{\partial U}{\partial x_1} + a_{n2}\frac{\partial U}{\partial x_2} + \ldots + a_{nn}\frac{\partial U}{\partial x_n}$$

where the x_i, $i = 1, 2, \ldots n$, are an orthogonal set of coordinates, can be put in the form,

$$V'_1 = a'_{11}(\partial U/\partial x'_1)$$

$$V'_2 = a'_{22}(\partial U/\partial x'_2)$$

$$\vdots \quad \vdots \quad \vdots$$

$$V'_n = a'_{nn}(\partial U/\partial x'_n)$$

by a principal axis transformation *if* the matrix, $(a_{ij}) = A$, is *symmetric*.

(2) Verify Eq. (2.137) for the matrices A, B, C defined in Eqs. (2.134), (2.135), and (2.136).

(3) For the matrices, C, B, A, defined in Eqs. (2.120), (2.121), and (2.122) take $\theta = \phi = \psi = \pi/2$, then show by an appropriate diagram various positions of a "rectangular block" with sides of unequal lengths, as a textbook for example, when C, B, A are applied in various orders.

(4) Show that dividing both sides of Eq. (2.132) by a "time increment," dt, we get

$$\frac{d\mathbf{r}}{dt} = \mathbf{r} \times \boldsymbol{\omega}$$

where we interpret $\boldsymbol{\omega} = d\Omega/dt$ as an "angular velocity." Can this be generalized to more than three dimensions?

Set (2-5)

(1) Show that the eigenvalues of a Hermitian matrix are real.

(2) The generalization of the eigenvalue problem to complex matrices, A, and complex eigenvectors, X, leads to the analog of the *similarity transformation* defined as

$$TAT^\dagger = A''$$

called a *unitary transformation* if

$$T^\dagger = T^{-1}$$

Show that every Hermitian matrix can be diagonalized by a unitary transformation.

(3) Show that $(AB)\dagger = B\dagger A\dagger$

(4) Show that if $C^{-1} = C\dagger$ (i.e., C is a unitary matrix) and A is Hermitian, then $C^{-1}AC$ is Hermitian.

Set (2-6)

(1) Consider a fluid of constant mass density, ρ, flowing in space with velocity field $\mathbf{v}(\mathbf{r}, t)$ where t = time parameter, and in which we assume a stress field, σ, exists. Consider a typical volume element, τ, and compute the net rate of increase of momentum in τ due to the fluid flow as a surface integral. Also compute the net force on the volume element τ due to the stress field, σ, as a surface integral. If $\mathbf{v}$ and σ are steady in time then by Newton's second law of motion these two quantities must be equal. Thus by transforming surface integrals to volume integrals derive the differential equation of equilibrium. (Use dyadic notation and apply the divergence theorem.)

Set (2-7)

(1) Defining $V = (dy/dt) - y$ write the second-order differential equation,

$$\frac{d^2y}{dt^2} + \frac{dy}{dt} + y = F(t)$$

in matrix form in terms of y and V.

REFERENCES

P. Dennery and A. Krzywicki, *Mathematics for Physicists* (Harper and Row, Publishers, Inc., New York, 1967).

J. Irving and N. Mullineux, *Mathematics in Physics and Engineering* (Academic Press Inc., New York, 1959).

H. Marganau and G. Murphy, *The Mathematics of Physics and Chemistry*, Vol. I (D. Van Nostrand Inc., Princeton, New Jersey, 1956).

J. Mathews and R. L. Walker, *Mathematical Methods of Physics* (W. A. Benjamin Inc., New York, 1965).

A. D. Michal, *Matrix and Tensor Calculus with Applications to Mechanics, Elasticity, and Aerodynamics* (John Wiley and Sons Inc., New York, 1947).

H. Goldstein, *Classical Mechanics* (Addison-Wesley Publishing Company, Inc., Reading, Massachusetts, 1959). Consult this book for a particularly lucid discussion of orthogonal transformations and the eigenvalue problem in terms of a concrete physical example.

A. Kyrala, *Theoretical Physics: Application of Vectors, Matrices, Tensors and Quaternions*, W. B. Saunders Company, Philadelphia (1967).

J. D. Jackson, *Mathematics for Quantum Mechanics* (W. A. Benjamin Inc., New York, 1962).

CHAPTER THREE

Introduction to boundary value problems and the special functions of mathematical physics

The mathematical formulation of a physical problem frequently takes the form of a partial differential equation, together with subsidiary conditions which the solution must satisfy. These "boundary conditions" are dictated by the physical nature of the problem.†

In this text we will not concern ourselves with proving that the solution of a boundary value problem, as described above, is unique. Instead we will assume that a solution of the partial differential equation, obtained by any means, which satisfies *all* the boundary conditions is the unique solution. Thus, throughout the text, we forego mathematical rigor in order to devote our attention to *methods* of solution.

PARTIAL DIFFERENTIAL EQUATIONS OF PHYSICS

The most commonly occurring partial differential equations in physical problems are linear. A *linear* partial differential equation is one having the properties that if $\phi_1(x, y, z, t)$ and $\phi_2(x, y, z, t)$ are two distinct solutions then $\phi_1 + \phi_2$ is a solution, and also $A\phi_1$, or $B\phi_2$, is a solution where A, or B, is constant.

---

† The requisite *number* and *type* of boundary conditions that *must* be given in order that a solution of the problem *exists* is a purely mathematical question. Some indication of this is indicated in Appendix I.

Typical partial differential equations of physics are:

(3.1) $$\nabla^2\phi = \rho$$ (Poisson's Equation)

(3.2) $$\nabla^2\phi = \kappa\frac{\partial\phi}{\partial t}$$ (Diffusion Equation)

(3.3) $$\nabla^2\phi = \frac{1}{c^2}\frac{\partial^2\phi}{\partial t^2}$$ (Wave Equation)

(3.4) $$\nabla^2\phi + \kappa^2\phi = 0$$ (Helmholtz Equation)

(3.5) $$-\frac{\hbar^2}{2m}\nabla^2\phi + V\phi = E\phi = i\hbar\frac{\partial\phi}{\partial t}$$ (Schrödinger Equation)

(3.6) $$\nabla\phi\cdot\nabla\phi + V = \frac{\partial\phi}{\partial t}$$ (Hamilton–Jacobi Equation)

(3.7) $$\nabla^4\phi = 0$$ (Stress Equation)

(3.8) $$\nabla^2\phi = 0$$ (Laplace Equation)

All of these except the Hamilton–Jacobi equation are linear. Furthermore all of these except the Poisson and Hamilton–Jacobi equations are also homogeneous. That is, the dependent function, ϕ, occurs to the same power in every term of the equation.

Initially we will develop methods for solving linear–homogeneous partial differential equations, then we will extend these methods to *nonhomogeneous* problems. Exact solutions of *nonlinear* partial differential equations cannot be obtained except under rather special conditions. We will describe a few such special cases and we also will present some approximate methods for such problems.

The methods we will develop initially for the linear–homogeneous partial differential equations will also be restricted to *homogeneous boundary conditions*. That is, the same condition applies to the function at all points of a particular boundary. This restriction will be explained more fully later. Also we will later extend our methods to nonhomogeneous boundary conditions.

In order to provide a definite pattern for the student to follow as we develop methods for solving boundary value problems we here set out a typical problem and its solution. Although this is an extremely simple problem it has essentially all the ingredients common to most boundary value problems.

Example (1)

The equation to be solved is,

(3.9) $$a\frac{\partial^2\phi}{\partial x^2} = \frac{\partial\phi}{\partial t} \qquad \begin{array}{l} 0 < x < L, \qquad t > 0 \\ a = \text{constant} \end{array}$$

with the conditions,

(3.10) $$\begin{cases} \phi(0, t) = \phi(L, t) = 0 \\ \phi(x, 0) = F(x) \end{cases}$$

We take as a trial solution of Eq. (3.9),

(3.11) $$\phi_\alpha = U(x)V(t)$$

where U is a function of x only while V is a function of t only. Substituting this into Eq. (3.9) and then dividing the equation through by aUV yields,

(3.12) $$\frac{1}{U}\frac{d^2U}{dx^2} = \frac{1}{aV}\frac{dV}{dt}$$

Here *total* derivatives replace partial derivatives because the functions are dependent each on a single variable.

Now in the above equation the left member is a function of x only while the right member is a function of t only; thus if we allow t to change while x does not we see that the left member does not change. Thus conclude that the right member is a constant. Therefore

(3.13) $$\frac{1}{U}\frac{d^2U}{dx^2} = \frac{1}{aV}\frac{dV}{dt} = -\alpha^2$$

where α^2 is some constant.

In this way we obtain the two *ordinary* differential equations,

(3.14) $$(d^2U/dx^2) + \alpha^2 U = 0$$

and

(3.15) $$(dV/dt) + a\alpha^2 V = 0$$

These are familiar equations. The general solution of the first is

(3.16) $$U = A \sin \alpha x + B \cos \alpha x, \qquad \alpha \neq 0$$

while that for the second is

(3.17) $$V = Ce^{-a\alpha^2 t} \qquad \alpha \neq 0$$

where A, B, and C are constants. Consequently

(3.18) $$\phi_\alpha = Ce^{-a\alpha^2 t}(A \sin \alpha x + B \cos \alpha x)$$

is a solution of our original partial differential equation. However this solution does *not* satisfy the boundary conditions of the problem.

We note that at $x = 0$ this solution would vanish as required if B were zero. Thus we take the unspecified constant B to be zero and have,

(3.19) $$\phi_\alpha = ACe^{-a\alpha^2 t} \sin \alpha x$$

Next we note that this would also vanish at $x = L$ as required if the unspecified parameter α is properly chosen, i.e.,

(3.20) $$\sin \alpha L = 0$$

which has an infinity of solutions

$$\alpha L = 0, \pm \pi, \pm 2\pi, \ldots, \pm n\pi, \ldots$$

Thus we have

(3.21)
$$\phi_n = B_n e^{-a\frac{n^2\pi^2}{L^2}t} \sin\frac{n\pi x}{L}, \qquad n = \pm 1, \pm 2, \ldots$$

as a *family* of solutions of our partial differential equation which satisfy all boundary conditions of the problem except that at $t = 0$. Here we have used the notation B_n for the constant AC.

Now at $t = 0$ we have,

(3.22)
$$\phi_n = B_n \sin\frac{n\pi x}{L}$$

and this certainly is *not* equal to $F(x)$ for an arbitrary function $F(x)$. However, since the differential equation is a linear homogeneous equation we can take as a solution the *sum* of any number of the distinct solutions ϕ_n, i.e.,

(3.23)
$$\phi(x, t) = \sum_{n=1}^{\infty} B_n e^{-a\frac{n^2\pi^2}{L^2}t} \sin\frac{n\pi x}{L}$$

This satisfies the partial differential equation and the first two of the boundary conditions. The last condition then requires that,

(3.24)
$$\phi(x, 0) = F(x) = \sum_{n=1}^{\infty} B_n \sin\frac{n\pi x}{L}$$

A set of constants, B_n, $n = 1, 2, \ldots$, can be found such that this is true if $F(x)$ satisfies the Dirichlet† conditions because this is just the Fourier sine series for $F(x)$. However to emphasize a certain point of view we proceed as follows.

Multiply each side of Eq. (3.24) by $\sin(m\pi x/L)$, where m is some particular integer, then integrate over x from zero to L. If the series is assumed to be *uniformly convergent* we can integrate term by term, thus:

(3.25)
$$\int_0^L F(x) \sin\frac{m\pi x}{L} dx = \sum_{n=1}^{\infty} B_n \int_0^L \sin\frac{m\pi x}{L} \sin\frac{n\pi x}{L} dx$$

Now we can readily show that

(3.26)
$$\int_0^L \sin\frac{m\pi x}{L} \sin\frac{n\pi x}{L} dx = \begin{cases} \dfrac{L}{2}, & n = m \\ 0, & n \neq m \end{cases}$$

Thus because of this *orthogonality* property of the trigonometric functions we find:

(3.27)
$$B_n = \frac{2}{L} \int_0^L F(x) \sin\frac{n\pi x}{L} dx$$

Finally then, if the constants B_n are specified by the above equation we see Eq. (3.23) as the complete solution of the boundary value problem.

EXECUTE PROBLEM SET (3-1)

† ... i.e., is sectionally continuous in x, having only a finite number of ordinary discontinuities in the interval given, $0 < x < L$.

SEPARATION OF VARIABLES AND SPECIAL FUNCTIONS

In Example (1) just discussed we saw that the partial differential equation in x and t was reduced to two *ordinary* differential equations by assuming the solution $\phi(x, t)$ to have the product form, $U(x) \cdot V(t)$. This is called *separation of variables* and the partial differential equation is said to be *separable*. Here we consider this method in some detail.

A linear, homogeneous partial differential equation for a function $\phi(x, y, z, t)$ can be looked upon as an operator equation. Thus

$$(3.28) \qquad O\{\phi(x, y, z, t)\} = 0$$

where O is a linear differential operator such as ∇^2, or $\nabla^2 - \partial/\partial t$, etc.

For a very *restricted* class of operators the substitution of

$$(3.29) \qquad \phi = U(x)V(y)W(z)T(t)$$

will reduce the single operator equation to a set of four operator equations. In our present considerations these will be four *ordinary* differential equations, say,

$$(3.30) \qquad \begin{cases} O_x\{U(x)\} = 0 \\ O_y\{V(y)\} = 0 \\ O_z\{W(z)\} = 0 \\ O_t\{T(t)\} = 0 \end{cases}$$

In our example problem we were able to recognize solutions of these *ordinary* differential equations immediately but in general the equations obtained by this separation process are not so simple and we resort to a power series solution procedure.

Thus we will construct a particular solution of the equation in x, say of the form,

$$(3.31) \qquad U(x) = \sum_{n=0}^{\infty} a_n x^{m+n}$$

In some cases particular ordinary differential equations occur so frequently and the series solution is used so much, that we give names to the series. We then tabulate values and properties of the functions represented by these series.

For example the ordinary differential equation,

$$(3.32) \qquad (dT/dt) + \alpha T = 0$$

occurs very often in physics. We might assume a series solution,

$$(3.33) \qquad T(t) = \sum_{n=0}^{\infty} a_n t^{m+n}$$

and then substitute this into the differential equation to obtain,

$$(3.34) \qquad a_0 m t^{m-1} + \sum_{k=1}^{\infty} \{(m+k)a_k + \alpha a_{k-1}\} t^{m+k-1} = 0$$

In order for this to hold for *all* values of t it is necessary for the coefficient of each power of t to vanish separately. Thus we see

$$(3.35) \qquad \begin{cases} \text{if } a_0 \neq 0 \\ \text{then } m = 0 \end{cases}$$

from the first term and thus

(3.36)
$$ka_k + \alpha a_{k-1} = 0$$

for each term of the series. These then yield

(3.37)
$$a_k = a_0 \left[\frac{(-1)^k \alpha^k}{k!} \right]$$

so that the solution is,

(3.38)
$$T(t) = a_0 \left[1 - \alpha t + \frac{\alpha^2 t^2}{2!} - \frac{\alpha^3 t^3}{3!} + \cdots \right]$$

The series which appears here occurs so often that we have given it the name, *exponential function*, with the argument, $-\alpha t$, i.e.,

(3.39)
$$T(t) = a_0 e^{-\alpha t}$$

Furthermore, using this series, and the differential equation we have developed extensive tables of values of this function, for various values of its argument, and also rather complete tabulations of properties of the function.

The series solution method just described is called by the name of its inventor, Frobenius. Using this method for the equation

(3.40)
$$\frac{d^2 U}{dx^2} + \alpha^2 U = 0$$

one can readily show that the solution contains two unspecified, but nonzero, constants as coefficients of *two distinct series*. These series represent the sine and cosine functions. Thus *we see the trigonometric and exponential functions simply as power series solutions of ordinary differential equations* which arise from some partial differential equation in the separation of variables process.

Of course other differential equations obtained in the separation of variables process can also be solved by this series method but most of these do not correspond to functions which are as familiar to us as the exponential and trigonometric functions. However, such is the case only because the trigonometric functions are also defined geometrically and are most often used in that context.

All of those functions represented by series in the manner just described we lump together under the general heading of *special functions*. We also include under this same heading some functions generated in other ways, but the majority of those to be considered in this text arise as just described.

The problems in set (3-2) serve as a model to illustrate how one can use the differential equation and the series solution to establish some properties of special functions.

EXECUTE PROBLEM SET (3-2)

FURTHER DISCUSSION OF THE SEPARATION OF VARIABLES

So far we have seen how the separation of variables technique reduces the solution of a *partial* differential equation to solving a set of *ordinary* differential equations. It is these which yield the *special functions*. Now we must point out certain very important points about this method. First of all we must in general include *all* possible values of the separation constants.

In our earlier example in Eqs. (3.14) and (3.15) we only took the solutions, Eqs. (3.16) and (3.17), corresponding to *nonzero* values of the separation constant α^2. Note that if in Eqs. (3.14) and (3.15) we put $\alpha^2 = 0$ there results:

(3.41) $$(d^2 U/dx^2) = 0$$

and

(3.42) $$\frac{dV}{dt} = 0$$

The solutions of these are then obviously,

(3.43) $$U = A_0 x + B_0$$

and

(3.44) $$V = C_0$$

where A_0, B_0, and C_0 are arbitrary constants. Thus in addition to the set of solutions ϕ_α as given in Eq. (3.18) for *nonzero* separation constant we also have

(3.45) $$\phi_0 = C_0(A_0 x + B_0) = A_0' x + B_0'$$

as another solution. Here A_0' and B_0' are used to replace $C_0 A_0$ and $C_0 B_0$.

Such special solutions prove useful and necessary in certain types of boundary value problems, in particular those with "mixed" boundary conditions such as in Problem Set (3-3), i.e., note that *different* kinds of conditions are specified at $x = 0$ and $x = L$.

EXECUTE PROBLEM SET (3-3)

SEPARATION OF VARIABLES IN MULTI-DIMENSIONAL CASES AND CHOICE OF COORDINATE SYSTEMS

Consider as an example the wave equation, Eq. (3.3), which we might write in rectangular coordinates, x, y, as

(3.46) $$\frac{\partial^2 \phi}{\partial x^2} + \frac{\partial^2 \phi}{\partial y^2} = \frac{1}{c^2} \frac{\partial^2 \phi}{\partial t^2}$$

or in plane-polar coordinates r, θ as,

(3.47) $$\frac{\partial^2 \phi}{\partial r^2} + \frac{1}{r} \frac{\partial \phi}{\partial r} + \frac{1}{r^2} \frac{\partial^2 \phi}{\partial \theta^2} = \frac{1}{c^2} \frac{\partial^2 \phi}{\partial t^2}$$

These equations are *completely* equivalent and our choice of one in preference to the other will be predicated entirely on which offers the greatest simplicity in the solution of a given problem. *Both* are separable.

In general if the *domain of definition* of the solution is bounded by straight line segments we will prefer rectangular coordinates, but if the domain is some segment, or section of a circle, or circular annulus, then the plane polar coordinates will be chosen; thus in general, the *shape of the domain*, or the *geometry* of the problem dictates the coordinate system to be used.

Now observe the separation of variables technique in these two cases. In Eq. (3.46) suppose ϕ to have the form,

$$(3.48) \qquad \phi = W(x, y)T(t)$$

When this is substituted into Eq. (3.46) there results,

$$(3.49) \qquad \frac{1}{W}\left(\frac{\partial^2 W}{\partial x^2} + \frac{\partial^2 W}{\partial y^2}\right) = \frac{1}{c^2}\frac{1}{T}\frac{\partial^2 T}{\partial t^2} = -\alpha^2$$

Here we have indicated that since the left member is a function of x and y, but not t, while the right member is a function of t only, each member must be equal to a constant which we have denoted by $-\alpha^2$. In this way we now have the two equations:

$$(3.50) \qquad \frac{d^2 T}{dt^2} + \alpha^2 c^2 T = 0$$

and

$$(3.51) \qquad \frac{\partial^2 W}{\partial x^2} + \frac{\partial^2 W}{\partial y^2} + \alpha^2 W = 0$$

The first equation is an *ordinary* differential equation but the second is still a *partial* differential equation. Thus, now assume

$$(3.52) \qquad W(x, y) = U(x)V(y)$$

and substitute into Eq. (3.51). This yields:

$$(3.53) \qquad \frac{1}{U}\frac{d^2 U}{dx^2} = -\left(\frac{1}{V}\frac{d^2 V}{dy^2} + \alpha^2\right) = -\beta^2$$

where we have indicated each side of this equation equal to a separation constant, $-\beta^2$, since the left member is a function of x only, while the right member is a function of y only. Thus we have now,

$$(3.54) \qquad \frac{d^2 U}{dx^2} + \beta^2 U = 0$$

and

$$(3.55) \qquad \frac{d^2 V}{dy^2} + (\alpha^2 - \beta^2)V = 0$$

as the other two ordinary differential equations following from our original *partial* differential equation.

The three ordinary differential equations above are just those we have already discussed so extensively. Now consider the *same* partial differential equation in plane polar coordinates. Assume a solution of Eq. (3.47) to be of the form,

$$(3.56) \qquad \phi = M(r, \theta)T(t)$$

and substitute into Eq. (3.47) to obtain

$$(3.57) \qquad \frac{1}{M}\left(\frac{\partial^2 M}{\partial r^2} + \frac{1}{r}\frac{\partial M}{\partial r} + \frac{1}{r^2}\frac{\partial^2 M}{\partial \theta^2}\right) = \frac{1}{c^2}\frac{1}{T}\frac{d^2 T}{dt^2} = -\alpha^2$$

Here we use our same arguments as before to obtain,

(3.58)
$$\frac{d^2 T}{dt^2} + \alpha^2 c^2 T = 0$$

and also,

(3.59)
$$\frac{1}{r} \frac{\partial}{\partial r}\left(r \frac{\partial M}{\partial r}\right) + \frac{1}{r^2} \frac{\partial^2 M}{\partial \theta^2} + \alpha^2 M = 0$$

where $-\alpha^2$ is the separation constant. Note that we have also elected to use the *identity*,

(3.60)
$$\frac{1}{r} \frac{\partial}{\partial r}\left(r \frac{\partial}{\partial r}\right) = \frac{\partial^2}{\partial r^2} + \frac{1}{r} \frac{\partial}{\partial r}$$

Just as in the rectangular case we have a *partial* differential equation remaining. Assume now that,

(3.61)
$$M(r, \theta) = R(r)\Omega(\theta)$$

and substitute into Eq. (3.59) to obtain,

(3.62)
$$\frac{r}{R} \frac{d}{dr}\left(r \frac{dR}{dr}\right) + \alpha^2 r^2 = -\frac{1}{\Omega} \frac{d^2 \Omega}{d\theta^2} = \beta^2$$

where by our usual argument β^2 is a *constant*. Thus we have now,

(3.63)
$$\frac{d^2 \Omega}{d\theta^2} + \beta^2 \Omega = 0$$

and

(3.64)
$$r \frac{d}{dr}\left(r \frac{dR}{dr}\right) + (\alpha^2 r^2 - \beta^2)R = 0$$

While the equation for Ω is that now familiar to us as yielding the trigonometric functions, for real β, this last equation is *not* so familiar. This equation is *Bessel's equation*. Later we will familiarize ourselves with the special functions which can be generated from this by the Frobenius series technique. For the moment the point we wish to emphasize is that the *choice of coordinate system determines the kinds of special functions* one will encounter in a given problem.

EXECUTE PROBLEM SET (3-4)

GENERAL DIFFERENTIAL EQUATION OF SECOND-ORDER; POWER SERIES SOLUTIONS

In a previous section we listed several of the more important partial differential equations occurring in physical problems. These yield, with few exceptions, ordinary differential equations of *second order* by the separation of variables procedure. Thus we now consider the general equation:

(3.65)
$$\frac{d^2 y}{dx^2} + P(x)\frac{dy}{dx} + Q(x)y = R(x)$$

where P, Q, and R are known functions of x. This can also be written in the form,

$$(3.66) \qquad (x - x_0)^2 \frac{d^2 y}{dx^2} + (x - x_0)p(x)\frac{dy}{dx} + q(x)y = r(x)$$

If $R(x)$ is identically zero the equation is *homogeneous*.

It has already been noted that the Frobenius method can be employed to construct power series solutions of some second-order ordinary differential equations. The general conditions under which such solutions are possible are contained in the theorem due to Fuchs, here stated without proof:

Theorem

"The Eq. (3.66), with $r = 0$, has a general solution in the form of a *linear combination* of convergent series of the type,

$$(3.67) \qquad y = \sum_{n=0}^{\infty} a_n(x - x_0)^{m+n}$$

provided that $p(x)$ and $q(x)$ can each be expanded in a Taylor series in x about the point $x = x_0$."

When $p(x)$ and $q(x)$ are expandable in Taylor series about x_0 the point x_0 is called a *regular point* of the equation. Note that $P(x)$ and $Q(x)$ need *not* have such expansions at x_0 even though $p(x)$ and $q(x)$ do have such expansions.

If $P(x_0)$ and $Q(x_0)$ are finite then $x = x_0$ is an *ordinary point*. If $P(x_0)$ and/or $Q(x_0)$ are infinite but $p(x_0)$ and/or $q(x_0)$ are finite then $x = x_0$ is a *regular singular point* and by the above theorem a general solution exists as power series about $x = x_0$. If however either $p(x)$ or $q(x)$ does not have a Taylor series expansion at $x = x_0$ then a general solution in such series does not exist. However in some cases a *particular* solution may exist as a *single* such power series even when the general solution as *two* such series does not exist.

In some cases an equation may not admit power series solutions of the type stated above but a simple change of the independent variable yields another differential equation which does admit such solutions. One of the simplest cases is that in which $p(x)$ and/or $q(x)$ are infinite at $x = 0$, in this case a solution can sometimes be found as a power series in $1/x$.

Sometimes it is helpful in determining the behavior of the solution of an ordinary differential equation for large x to construct *asymptotic solutions* for the equation. These are power series solutions, usually multiplied by exponential functions, such as,

$$(3.68) \qquad y \approx e^{\alpha x} \sum_{n=0}^{\infty} a_n \left(\frac{1}{x}\right)^{n+m}$$

which are actually *divergent series*.

Even though a solution such as that above may be divergent it can still be valuable in estimating the behavior of y at large x. The constants α, m, and the a_n are determined so that the differential equation is satisfied. Then if we let y_N represent the truncated form of Eq. (3.68) corresponding to the first N terms of the sum, we will usually find that

$$y - y_N \to 0, \qquad \text{as } x \to \infty$$

and

$$x^N(y - y_N) \to 0, \qquad \text{as } x \to \infty$$

Furthermore it can usually be shown that $y - y_N$ is of the order of the $(N + 1)$st term provided that term is smaller than the Nth term. Hence if the $(N + 1)$st term is sufficiently small y_N is a good approximation to y. However the addition of more terms will ultimately destroy the approximation, beyond some N, because the terms *must increase* since the series is *divergent*.

STURM–LIOUVILLE EQUATION AND ORTHOGONAL FUNCTIONS

The general linear, homogeneous, second-order, ordinary differential equation, as Eq. (3.65) with $R = 0$, can be put in the form,

$$(3.69) \qquad \frac{d}{dx}\left(h(x)\frac{dy}{dx}\right) - g(x)y = \lambda\omega(x)y$$

where the substitutions,

$$(3.70) \qquad h = \exp\int P(x)\,dx, \qquad Q(x) = [-g(x) - \lambda\omega(x)]/h(x)$$

are employed. Equation (3.69) is the *Sturm–Liouville form* of the second-order equation.

In general this equation has a solution $y_a(x)$ for each value of the *constant* parameter λ, say λ_a. The parameter λ_a is called the *eigenvalue* corresponding to the *eigenfunction*, y_a. If more than one solution, y_a, exists for a given eigenvalue the eigenvalue is said to be *degenerate*. Here we only consider nondegenerate cases.

We introduce this special form of the second-order equation here because in this form we can readily see the origin of *orthogonal functions*. Thus let $y_i(x)$ and $y_j(x)$ be the solutions of Eq. (3.69) corresponding to the eigenvalues λ_i and λ_j, respectively; that is:

$$(3.71) \qquad \frac{d}{dx}\left(h\frac{dy_i}{dx}\right) - gy_i = \lambda_i\omega y_i$$

and

$$(3.72) \qquad \frac{d}{dx}\left(h\frac{dy_j}{dx}\right) - gy_j = \lambda_j\omega y_j$$

Now multiply Eq. (3.71) by y_j and Eq. (3.72) by y_i then subtract the second from the first to obtain,

$$(3.73) \qquad \frac{d}{dx}(hy_j'y_i - hy_jy_i') = (\lambda_j - \lambda_i)\omega y_iy_j$$

where primes denote differentiation with respect to x. This is then integrated over the domain $x_1 \le x \le x_2$, in which the solutions are to be valid, to give,

$$(3.74) \qquad h(y_j'y_i - y_jy_i')\Big|_{x_1}^{x_2} = (\lambda_j - \lambda_i)\int_{x_1}^{x_2} y_iy_j\omega\,dx$$

It is here that we consider the, as yet unspecified, boundary conditions of the solutions. Let these be of a form such that,

$$(3.75) \qquad h(y_j'y_i - y_jy_i')\Big|_{x_1}^{x_2} = 0$$

Then, since by hypothesis λ_i and λ_j are distinct, we see that we must have,

(3.76)
$$\int_{x_1}^{x_2} y_i(x)y_j(x)\omega(x)\,dx = 0, \qquad i \neq j$$

We say that the functions y_i and y_j are *orthogonal with weight function* $\omega(x)$. This is just the same property already noted for the trigonometric functions. We also see here that for $i = j$ the integral,

(3.77)
$$\int_{x_1}^{x_2} y_j^2(x)\omega(x)\,dx = N_j$$

which we call the *norm*, is not required to vanish. In fact if $\omega > 0$ over the whole domain we see that the integral *cannot* vanish. Furthermore it can be shown that $\omega > 0$ in the interval $x_1 < x < x_2$ is *necessary* in order that all the eigenvalues, λ_j, be *real*.

It is important to note that the orthogonality property of these functions arises not only because of the form of the differential equation but also because of the *boundary conditions*, i.e., Eq. (3.75). The general form of the boundary conditions most often encountered in physical problems which satisfies Eq. (3.75) is,

(3.78)
$$ay + b\frac{dy}{dx} = 0, \qquad \begin{cases} x = x_1 \\ x = x_2 \end{cases}$$

with the particular cases of either $a = 0$ or $b = 0$, being of frequent occurrence.

The *norm*, defined in Eq. (3.77), will have a different form depending upon the specific boundary conditions imposed on the functions.

Most of the special functions with which we are concerned in physical problems can be represented as solutions of a Sturm–Liouville equation. A few of the more important such equations appear in Problem Set 3-5. Here it should be noted that Bessel's equation can be represented as a Sturm–Liouville equation in two different ways.

EXPANSIONS IN ORTHOGONAL FUNCTIONS; COMPLETE SETS OF ORTHOGONAL FUNCTIONS

In the solution of the example problem above we found that linear superposition of solutions led to representing the initial function, $F(x)$, by an infinite series of sine functions, i.e., Eq. (3.24). The undetermined coefficients of the series were then determined by the orthogonality property of the trigonometric functions. In much the same way a function can be represented in other kinds of orthogonal functions.

Thus if $F(x)$ has a finite number of ordinary discontinuities in the interval $x_1 < x < x_2$ then the series

(3.79)
$$F(x) = \sum_{j=1}^{\infty} A_j y_j(x)$$

is uniformly convergent to $F(x)$ at all points at which $F(x)$ is continuous, provided that the $y_j(x)$ form a *complete set of orthogonal functions*.

To see the meaning of this multiply both sides of Eq. (3.79) by $y_i(x)\omega(x)\,dx$ and integrate over $x_1 \leq x \leq x_2$. Since by hypothesis the series is uniformly convergent the order of summation and integration can be interchanged to yield,

$$(3.80) \qquad \int_{x_1}^{x_2} F(x) y_i(x) \omega(x) \, dx = \sum_{j=1}^{\infty} A_j \int_{x_1}^{x_2} y_i(x) y_j(x) \omega(x) \, dx$$

and then because of the orthogonal property all integrals on the right vanish except that for $j = i$ and this integral is just the *norm*; thus,

$$(3.81) \qquad A_i = \frac{1}{N_i} \int_{x_1}^{x_2} F(x') y_i(x') \omega(x') \, dx'$$

Here then the coefficients in the series are determined. However we note that if we substitute this for the A_j back in Eq. (3.79) and again reverse the order of summation and integration we obtain

$$(3.82) \qquad F(x) = \int_{x_1}^{x_2} F(x') \left\{ \sum_{j=1}^{\infty} \frac{y_j(x) y_j(x')}{N_j} \right\} \omega(x') \, dx'$$

which shows that

$$(3.83) \qquad \sum_{j=1}^{\infty} \frac{y_j(x) y_j(x')}{N_j} = \begin{cases} 0, & x \neq x' \\ \infty, & x = x' \end{cases} = \delta_\omega(x - x')$$

must be the *delta function* for weight function $\omega(x')$ on the interval $x_1 < x < x_2$. Such delta functions are discussed extensively in subsequent chapters.

Equation (3.83), with Eq. (3.82), is the criterion for the $y_j(x)$ to form a *complete set* of orthogonal functions.

The nature of the representation of a function $F(x)$ by a series of orthogonal functions is most clearly seen as follows. Select K equally spaced points in the interval $x_a \leq x \leq x_b$ at $x_a, x_a + \Delta x, x_a + 2\Delta x, \ldots$, etc. to x_b, with $\Delta x = (x_b - x_a)/(K - 1)$. Then *approximate* the function $F(x)$ at any one of these K points by the *finite* series,

$$(3.84) \qquad F(x_i) \simeq \sum_{j=1}^{N} A_j y_j(x_i)$$

Here we make the approximation "as good as possible" in the least-squares sense† by choosing the A_j so that the mean-square error is as small as possible, that is with weight function, $\omega(x_i)\Delta x$, so,

$$(3.85) \qquad \sum_{i=1}^{K} \left(F(x_i) - \sum_{j=1}^{N} A_j y_j(x_i) \right)^2 \omega(x_i)\Delta x = \text{minimum}$$

Thus, since this is a function of the A_j to be extremized,‡ we differentiate with respect to each A_j and set the result equal to zero. This yields,

$$(3.86) \qquad \sum_{i=1}^{K} F(x_i) y_k(x_i) \omega(x_i)\Delta x - \sum_{j=1}^{N} A_j \sum_{i=1}^{K} y_k(x_i) y_j(x_i) \omega(x_i)\Delta x_i = 0$$

for $k = 1, 2, \ldots N$.

† See Chapter XIII.
‡ See Chapter XI.

Here, if we now take the limit as $K \to \infty$ and $\Delta x \to 0$, we see this approaching the limit,

(3.87)
$$\int_{x_a}^{x_b} F(x)y_k(x)\omega(x)\,dx - \sum_{j=1}^{N} A_j \int_{x_a}^{x_b} y_k(x)y_j(x)\omega(x)\,dx = 0$$

so by orthogonality the coefficients A_i are as given in Eq. (3.81) above.

Thus the representation of $F(x)$ by the series of orthogonal functions is seen to imply a "convergence in the mean" in the least-squares sense.

EXECUTE PROBLEM SET (3-5)

PROBLEMS

Set (3-1)

(1) Show by direct integration that:

$$\int_{0}^{L} \sin\frac{n\pi x}{L} \sin\frac{m\pi x}{L}\,dx = \begin{cases} \dfrac{L}{2}, & n = m \\ 0, & n \neq m \end{cases}$$

$$\int_{0}^{L} \cos\frac{n\pi x}{L} \cos\frac{m\pi x}{L}\,dx = \begin{cases} \dfrac{L}{2}, & n = m \\ 0, & n \neq m \end{cases}$$

$$\int_{-L}^{L} \sin\frac{n\pi x}{L} \cos\frac{m\pi x}{L}\,dx = 0, \qquad \text{all } n, m$$

$$\int_{0}^{L} \sin\frac{n\pi x}{L} \cos\frac{m\pi x}{L}\,dx = \begin{cases} 0, & n, m \text{ both even or both odd} \\ \dfrac{L}{\pi}\dfrac{2n}{n^2 - m^2}, & n \text{ even, } m \text{ odd; or } n \text{ odd, } m \text{ even} \end{cases}$$

(2) Solve the boundary value problem:

$$\frac{\partial^2 U}{\partial x^2} = \frac{1}{c^2}\frac{\partial^2 U}{\partial t^2}, \qquad \begin{array}{l} 0 < x < L, \qquad t > 0 \\ \\ c = \text{constant} \end{array}$$

with

$$\frac{\partial U}{\partial x}(0, t) = \frac{\partial U}{\partial x}(L, t) = 0$$

$$U(x, 0) = Ax, \qquad A = \text{constant}$$

$$\frac{\partial U}{\partial t}(x, 0) = 0$$

Set (3-2)

(1) Solve the differential equation

$$\frac{d^2U}{dx^2} + \alpha^2 U = 0$$

by the method of Frobenius series to obtain the series representations of the trigonometric and hyperbolic functions.

(2) Show that Problem (1) above can be solved by factoring the operator,

$$(D^2 + \alpha^2) = (D + i\alpha)(D - i\alpha), \qquad i = \sqrt{-1}$$

and solving the two first-order equations

$$\frac{dU}{dx} + i\alpha U = 0$$

$$\frac{dU}{dx} - i\alpha U = 0$$

by Frobenius series, thereby establishing the relationships between exponential functions and the trigonometric and hyperbolic functions.

(3) Show that solely from the series representations one can establish the differentiation formulas,

$$\frac{d}{dx} \sin \alpha x = \alpha \cos \alpha x$$

$$\frac{d}{dx} \cos \alpha x = -\alpha \sin \alpha x$$

(4) Show simply by using the differential equation in Problem (1) and the fact that $\sin \alpha x$ and $\cos \alpha x$ *each* satisfy this linear equation, it follows that

$$\sin^2 \alpha x + \cos^2 \alpha x = \text{constant}$$

(Can you then show from results of previous problems that the constant is unity?)

(5) Show that if U_α is a solution of

$$\frac{d^2U_\alpha}{dx^2} + \alpha^2 U_\alpha = 0, \qquad 0 < x < L$$

and U_β is a solution of

$$\frac{d^2U_\beta}{dx^2} + \beta^2 U_\beta = 0, \qquad 0 < x < L$$

then for suitable conditions imposed on U_α and U_β at $x = 0$ and $x = L$ we must have,

$$\int_0^L U_\alpha U_\beta \, dx = \begin{cases} 0, & \alpha \neq \beta \\ \text{constant}, & \alpha = \beta \end{cases}$$

Set (3-3)

(1) Solve the boundary value problem:

$$a\frac{\partial^2 U}{\partial x^2} = \frac{\partial U}{\partial t}, \qquad \begin{matrix} 0 < x < L, & t > 0 \\ a = \text{constant} \end{matrix}$$

with

$$\frac{\partial U}{\partial x}(L, t) = K = \text{constant}$$

$$U(0, t) = 0$$

$$U(x, 0) = 0$$

(2) Show that the solution of Problem (1) above is the *same* as the solution of the following problem in the range of x, $0 < x < L$

$$a\frac{\partial^2 U}{\partial x^2} = \frac{\partial U}{\partial t}, \qquad \begin{matrix} -L < x < L, & t > 0 \\ a = \text{constant} \end{matrix}$$

$$\frac{\partial U}{\partial x}(L, t) = K$$

$$\frac{\partial U}{\partial x}(-L, t) = K, \qquad U(x, 0) = 0$$

This illustrates the important role of *symmetry* in such problems.

(3) Show that the solution ϕ_α given in Eq. (3.18) has the limiting *form* of ϕ_0, in Eq. (3.45), as $\alpha \to 0$.

Set (3-4)

(1) Solve by separation of variables

$$\frac{\partial^2 \phi}{\partial r^2} + \frac{1}{r}\frac{\partial \phi}{\partial r} + \frac{1}{r^2}\frac{\partial^2 \phi}{\partial \theta^2} = 0$$

(Hint: use the change of variable $r = e^s$)

(2) Solve the boundary value problem,

$$\nabla^2 \phi = 0$$

for the domain $0 < r < a$, $0 < \theta < \pi$
(two-dimensional) subject to the conditions,

$$\phi = \phi_0 \quad \text{on } r = a, \qquad\qquad 0 \le \theta \le \pi$$

$$\phi = 0 \quad \text{on } \theta = 0, \theta = \pi, \qquad 0 < r \le a$$

(3) Solve the boundary value problem,

$$\nabla^2 \phi = 0$$

for the two-dimensional case in the domain $0 \le x \le a, 0 \le y \le \infty$ with

$$\phi(0, y) = \phi(a, y) = 0, \qquad 0 < y < \infty$$

$$\lim_{y \to \infty} \phi(x, y) = 0$$

and

$$\phi(x, 0) = Kx, \qquad K = \text{constant}, \qquad 0 < x < a$$

(4) Solve the *inhomogeneous* boundary value problem,

$$\frac{\partial^2 \psi}{\partial x^2} + \frac{\partial^2 \psi}{\partial y^2} + p = 0, \qquad \begin{array}{l} -a < x < a \\[2mm] -b < y < b, \qquad p = \text{constant} \end{array}$$

with

$$\psi = 0 \quad \text{on } x = \pm a, \qquad -b < y < +b$$

and

$$\psi = 0 \quad \text{on } y = \pm b, \qquad -a < x < a$$

Hint: Use the substitution $\psi = U(x, y) + (p/2)(a^2 - x^2)$ to formulate a *homogeneous* boundary value problem for the function $U(x, y)$. ($p = $ constant.)

Set (3-5)

(1) Carry out the separation of variables procedure for the Helmholtz equation in: (three dimensions)
(a) cylindrical coordinates.
(b) spherical coordinates.
(2) Show that Bessel's equation,

$$x^2 \frac{d^2 y}{dx^2} + x \frac{dy}{dx} + (\beta^2 x^2 - n^2)y = 0$$

can be put into the Sturm–Liouville form in *two* ways, one with eigenvalue $+n^2$, and one with eigenvalue $-\beta^2$.
(3) Put the associated Legendre equation,

$$(1 - x^2)\frac{d^2 y}{dx^2} - 2x\frac{dy}{dx} + \left\{ l(l + 1) - \frac{m^2}{1 - x^2} \right\} y = 0$$

into the standard Sturm–Liouville form.
(4) Show that if $F(x)$ is sectionally continuous on $-L < x < L$ and bounded, then if the series

$$F(x) = a_0 + \sum_{n=1}^{\infty} a_n \cos \frac{n\pi x}{L} + \sum_{n=1}^{\infty} b_n \sin \frac{n\pi x}{L}$$

is uniformly convergent on $-L < x < L$ we have:

$$a_0 = \frac{1}{2L} \int_{-L}^{L} F(x') \, dx'$$

$$a_n = \frac{1}{L} \int_{-L}^{L} F(x') \cos \frac{n\pi x'}{L} dx', \qquad n > 0$$

$$b_n = \frac{1}{L} \int_{-L}^{L} F(x') \sin \frac{n\pi x'}{L} dx', \qquad n > 0$$

(5) Show that if $F(x)$ is an *even* function, $F(x) = F(-x)$, sectionally continuous, and bounded on $-L < x < L$, then

$$F(x) = a_0 + \sum_{n=1}^{\infty} a_n \cos \frac{n\pi x}{L}$$

with (for uniform convergence on $-L < x < L$)

$$a_0 = \frac{1}{L} \int_{0}^{L} F(x') dx'$$

$$a_n = \frac{2}{L} \int_{0}^{L} F(x') \cos \frac{n\pi x'}{L} dx'$$

(6) Show that if $F(x)$ is an *odd* function, $F(x) = -F(-x)$, sectionally continuous, and bounded on $-L < x < L$, then

$$F(x) = \sum_{n=1}^{\infty} b_n \sin \frac{n\pi x}{L}$$

with (for uniform convergence on $-L < x < L$)

$$b_n = \frac{2}{L} \int_{0}^{L} F(x') \sin \frac{n\pi x'}{L} dx'$$

(7) Expand e^{-ax} in a Fourier series on the interval $-L$ to L as in Problem (4).
(8) Expand $\sinh x$ in a Fourier series on the interval $0 < x < L$.

REFERENCES

R. V. Churchill, *Fourier Series and Boundary Value Problems* (McGraw-Hill Book Company, Inc., New York, 1941).

J. Irving and N. Mullineux, *Mathematics in Physics and Engineering* (Academic Press, Inc., New York, 1959).

E. A. Kraut, *Fundamentals of Mathematical Physics* (McGraw-Hill Book Company, Inc., New York, 1967).

P. M. Morse and H. Feshbach, *Methods of Theoretical Physics* (McGraw-Hill Book Company, Inc., New York, 1953).

CHAPTER FOUR

Useful properties of some special functions of mathematical physics

In the previous chapter we have indicated the origins of the special functions and our present motivation for their study. In this chapter we will list, essentially without any derivation or demonstration, the more useful properties of the most commonly used special functions. We omit the trigonometric and hyperbolic functions here when actually these should head the list since they arise from the simplest second-order equation we could consider. However the student is certainly familiar with these and they are therefore omitted. At the end of this chapter are a few problems intended to broaden the students acquaintance with the material of this chapter.

BESSEL FUNCTIONS

The differential equation is:

(4.1) $$x^2 y'' + xy' + (\beta^2 x^2 - v^2)y = 0$$

Case 1, β real: Regular Bessel functions

The linearly independent solutions in the domain $0 \leq x < \infty$ are:

$$J_v(\beta x), \qquad J_{-v}(\beta x)$$

or,

$$J_v(\beta x), \qquad Y_v(\beta x)$$

for v *not* an integer or zero; but only,

$$J_n(\beta x), \; Y_n(\beta x), \quad \text{for } n = 0, \pm 1, \pm 2, \ldots$$

Here†

(4.2)
$$J_v(\beta x) = \sum_{k=0}^{\infty} \frac{(-1)^k \left(\frac{\beta x}{2}\right)^{v+2k}}{k! \Gamma(v + k + 1)}$$

is the Bessel function of the first kind of order v, and

(4.3)
$$Y_v(\beta x) = \frac{\cos v\pi J_v(\beta x) - J_{-v}(\beta x)}{\sin v\pi}$$

is the Bessel function of the second kind of order v. In this last equation we use L'Hospitals rule for v an integer, or zero, to yield,

(4.4)
$$Y_n(\beta x) = \frac{1}{\pi} \left[\frac{\partial J_v(\beta x)}{\partial v} - (-1)^n \frac{\partial J_{-v}(\beta x)}{\partial v} \right]_{v=n}$$

which is,

(4.4a)
$$Y_n(\beta x) = \frac{2}{\pi} \left\{ [\ln \beta x - \ln 2 + \gamma] J_n(\beta x) - \frac{1}{2} \sum_{s=0}^{n-1} \frac{(n - s - 1)!}{s!} \left(\frac{\beta x}{2}\right)^{-n+2s} \right.$$
$$\left. - \frac{1}{2} \sum_{s=0}^{\infty} \frac{(-1)^s \left(\frac{\beta x}{2}\right)^{n+2s}}{s!(s + n)!} [\phi(s) + \phi(s + n)] \right\}$$

where

$$\gamma = \lim_{n \to \infty} \left(1 + \frac{1}{2} + \frac{1}{3} \ldots + \frac{1}{n} - \ln n \right) \approx 0.577216$$

is Euler's constant and $\phi(p) = 1 + \frac{1}{2} + \ldots + 1/p$ with $\phi(0) \triangleq 0$.

Note that in the above we have, since $\Gamma(-n + k + 1)$ is infinite for every integer k less than the integer n,

(4.5)
$$(-1)^n J_{-n}(\beta x) = J_n(\beta x)$$

for all integers n.

Graphs

Graphs of these functions are indicated in the following sketches.

† In this equation $\Gamma(m)$ denotes the Gamma function defined elsewhere in this chapter.

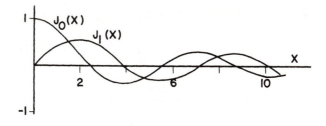

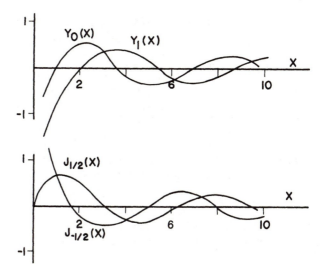

FIGURE 4.1.   Bessel Functions of First and Second Kinds.

Asymptotic forms

For very *large* βx:

$J_n(\beta x)$

$$\approx \left(\frac{2}{\pi\beta x}\right)^{\frac{1}{2}}\left\{\left[1 - \frac{(4n^2 - 1^2)(4n^2 - 3^2)}{2!(8\beta x)^2} + \frac{(4n^2 - 1^2)(4n^2 - 3^2)(4n^2 - 5^2)(4n^2 - 7^2)}{4!(8\beta x)^4} - \ldots\right]\cos\phi\right.$$

(4.6)

$$\left. - \left[\frac{4n^2 - 1^2}{8\beta x} - \frac{(4n^2 - 1^2)(4n^2 - 3^2)(4n^2 - 5^2)}{3!(8\beta x)^3} + \ldots\right]\sin\phi\right\}$$

where,

$$\phi = \beta x - (2n + 1)\frac{\pi}{4} \quad \text{and} \quad |\beta x| \gg 1, |\beta x| \gg n$$

The asymptotic form for $Y_n(\beta x)$ for large x is exactly the same as that for $J_n(\beta x)$ *except* the factors $\sin\phi$ and $\cos\phi$ are *interchanged*. For very *small* βx:

(4.7)

$$\begin{cases} J_0(\beta x) \approx 1 - \left(\frac{\beta x}{2}\right)^2 \\ \\ J_n(\beta x) \approx \frac{1}{\Gamma(n + 1)}\left(\frac{\beta x}{2}\right)^n, \quad n > 0 \end{cases}$$

and

(4.8)

$$\begin{cases} Y_0(\beta x) \approx \dfrac{2}{\pi}(\ln \beta x - \ln 2 + \gamma) \\[2em] Y_n(\beta x) \approx -\dfrac{(n-1)!}{\pi}\left(\dfrac{2}{\beta x}\right)^n, \qquad n = 1, 2, \ldots \end{cases}$$

Roots

Note that all $J_n(\beta x)$ and $Y_n(\beta x)$ have an infinity of *real* roots, i.e., the equation

(4.9) $$J_n(z) = 0$$

has solutions at $z_1^{(n)}, z_2^{(n)}, \ldots$ which form in infinite set. We can show that

(4.10) $$z_j^{(n+1)} > z_j^{(n)} > z_j^{(n-1)}$$

except for the origin; and also,

(4.11) $$z_{j+1}^{(n)} > z_j^{(n)}$$

Similarly for roots of $Y_n(z)$. However we note $Y_n(z) \to -\infty$ at $z = 0$ whereas $J_n(z) = 0$ at $z = 0$ if $n = 1, 2, \ldots$. Also see from asymptotic forms that

(4.12) $$\lim_{j \to \infty} (z_{j+1}^{(n)} - z_j^{(n)}) = \pi$$

for roots of J_n or Y_n. The first few roots of some of the Bessel functions are listed in Table 4.1. See the references at the end of the chapter for more extensive tables.

TABLE 4-1.   Zeros of $J_n(x)$.

Root Number	$J_0(x)$	$J_1(x)$	$J_2(x)$	$J_3(x)$
(1)	2.4048	3.8317	5.1356	6.3802
(2)	5.5201	7.0156	8.4172	9.7610
(3)	8.6537	10.1735	11.6198	13.0152
(4)	11.7915	13.3237	14.7960	16.2235
(5)	14.9309	16.4706	17.9598	19.4094
(6)	18.0711	19.6159	21.1170	22.5827

Differentiation and recursion formulas

From the series forms see that for any real number, n, including non-integers: [with $B_n(\beta x)$ denoting $J_n(\beta x)$, $Y_n(\beta x)$ or a linear combination of these]

(4.13) $$nB_n(\beta x) + x\frac{d}{dx}B_n(\beta x) = \beta x B_{n-1}(\beta x)$$

or

(4.14) $$\frac{1}{\beta}\frac{d}{dx}((\beta x)^n B_n(\beta x)) = (\beta x)^n B_{n-1}(\beta x)$$

Also

(4.15)
$$nB_n(\beta x) - x \frac{d}{dx} B_n(\beta x) = \beta x B_{n+1}(\beta x)$$

or

(4.16)
$$\frac{1}{\beta} \frac{d}{dx} \left(\frac{1}{(\beta x)^n} B_n(\beta x) \right) = -\frac{B_{n+1}(\beta x)}{(\beta x)^n}$$

These combine to yield,

(4.17)
$$B_{n-1}(\beta x) + B_{n+1}(\beta x) = \frac{2n}{\beta x} B_n(\beta x)$$

or

(4.18)
$$B_{n-1}(\beta x) - B_{n+1}(\beta x) = \frac{2}{\beta} \frac{d}{dx} B_n(\beta x)$$

Special forms

Note that for half-integer order,

(4.19)
$$J_{\frac{1}{2}}(\beta x) = \left(\frac{2}{\pi \beta x} \right)^{\frac{1}{2}} \sin \beta x, \qquad J_{-\frac{1}{2}}(\beta x) = \left(\frac{2}{\pi \beta x} \right)^{\frac{1}{2}} \cos \beta x$$

Orthogonality

The solutions $y = J_n(\beta x)$ of Eq. (4.1) are orthogonal as:

(4.20)
$$\int_0^a J_n(\beta_{nj} x) J_n(\beta_{ni} x) x \, dx = \begin{cases} 0, & i \neq j \\ N_{ni}, & j = i \end{cases}$$

with norm,

(4.21)
$$N_{ni} = \frac{1}{2\beta_{ni}^2} \left\{ x^2 \left[\frac{dJ_n(\beta_{ni} x)}{dx} \right]^2 + (\beta_{ni}^2 x^2 - n^2)(J_n(\beta_{ni} x))^2 \right\} \Big|_0^a$$

where the β_{ni} are determined from the roots of

(4.22)
$$J_n(\beta_{ni} a) + h \frac{d}{dx} J_n(\beta_{ni} a) = 0$$

with $h = 0$ or $1/h = 0$ being special cases. For *completeness* the expansion of a function $f(x)$ in a series of the $J_n(\beta_{ni} x)$ on $0 \leq x < a$ must be summed over all the β_{ni}, $i = 1, 2, \ldots \infty$, for fixed n. (Another form of orthogonality also exists but is rarely used in physical problems. See Problem 2, Problem Set 3-5.)

Generating function

If one expands the form,

$$\exp\left[\frac{x}{2}\left(t - \frac{1}{t}\right)\right]$$

in series as,

(4.23) $e^{\frac{xt}{2}} \cdot e^{-\frac{x}{2t}} = \left\{1 + \frac{xt}{2} + \frac{1}{2!}\left(\frac{xt}{2}\right)^2 + \ldots\right\} \cdot \left\{1 - \left(\frac{x}{2t}\right) + \frac{1}{2!}\left(\frac{x}{2t}\right)^2 - \ldots\right\}$

then multiplying the series reveals the coefficient of t^n as $J_n(x)$.

Integral representations

Using the above results and the fact that

(4.24) $J_{-n}(x) = (-1)^n J_n(x)$

find,

(4.25) $\exp\left[\frac{x}{2}\left(t - \frac{1}{t}\right)\right] = J_0(x) + \left(t - \frac{1}{t}\right)J_1(x) + \ldots + \left\{t^n + (-1)^n\frac{1}{t^n}\right\}J_n(x) + \ldots$

Then put $t = e^{i\theta}$, $1/t = e^{-i\theta}$ and use

(4.26) $e^{i\theta} = \cos\theta + i\sin\theta, \qquad i = \sqrt{-1}$

to show that

(4.27) $\exp i(x\sin\theta) = J_0(x) + 2iJ_1(x)\sin\theta + 2J_2(x)\cos 2\theta + 2iJ_3(x)\sin 3\theta + \ldots$

We can then separate real and imaginary parts, multiply both sides by some $\sin m\theta$ or $\cos m\theta$ and integrate from 0 to π. The orthogonality of the trigonometric functions then yields:

(4.28) $J_{2n}(x) = \frac{1}{\pi}\int_0^\pi \cos(x\sin\theta)\cos 2n\theta\, d\theta$

and

(4.29) $J_{2n+1}(x) = \frac{1}{\pi}\int_0^\pi \sin(x\sin\theta)\sin(2n+1)\theta\, d\theta$

as well as,

(4.30) $\int_0^\pi \cos(x\cos\theta)\cos n\theta\, d\theta = 0, \qquad n\text{ odd}$

and

(4.31) $\int_0^\pi \sin(x\sin\theta)\sin n\theta\, d\theta = 0, \qquad n\text{ even}$

The above integrals can be combined into the single integral,

(4.32) $J_n(x) = \frac{1}{\pi}\int_0^\pi \cos(n\theta - x\sin\theta)\, d\theta$

Other forms are derivable from these and are included in the problems.

Hankel functions

Two functions particularly useful in some problems are

(4.33)
$$\begin{cases} H_n^{(1)}(\beta x) = J_n(\beta x) + i Y_n(\beta x) \\ H_n^{(2)}(\beta x) = J_n(\beta x) - i Y_n(\beta x) \end{cases}$$

where $i = \sqrt{-1}$. These satisfy Eq. (4.1) also. The *asymptotic forms* for *large x* are:

(4.34) $\quad H_n^{(1)}(\beta x) \approx \left(\dfrac{2}{\pi \beta x} \right)^{\frac{1}{2}} e^{i\left(\beta x - n\frac{\pi}{2} - \frac{\pi}{4}\right)} \left[1 - \dfrac{4n^2 - 1^2}{1!(8i\beta x)} + \dfrac{(4n^2 - 1^2)(4n^2 - 3^2)}{2!(8i\beta x)^2} - \cdots \right]$

and the same for $H_n^{(2)}(\beta x)$ *except* in the exponential i is replaced by $-i$.

Equations reducible to Bessel's equation:

The equation,

(4.35)
$$z'' + \frac{1 - 2c}{\xi} z' + \left[(ab\xi^{b-1})^2 + \frac{c^2 - n^2 b^2}{\xi^2} \right] z = 0$$

with $z' = dz/d\xi$ etc., reduces to Eq. (4.1) with the substitution,

(4.36)
$$z = y\xi^c, \qquad x = a\,\xi^b$$

Case 2: β pure imaginary, $\beta^2 < 0$; modified Bessel functions

For this case the differential equation becomes,

(4.37)
$$x^2 y'' + xy' - (\beta^2 x^2 + v^2)y = 0$$

Solutions are obviously $J_v(i\beta x)$ and $Y_v(i\beta x)$ but we instead define the functions,

(4.38)
$$I_v(\beta x) = (i)^{-v} J_v(i\beta x)$$

and

(4.39)
$$K_v(\beta x) = \frac{\pi}{2} \frac{I_{-v}(\beta x) - I_v(\beta x)}{\sin v\pi}$$

which are *linearly independent*. These are the *modified Bessel functions* of first and second kind, respectively.

For v an integer we have:

(4.40)
$$I_n(\beta x) = I_{-n}(\beta x)$$

and by L'Hospital's rule,

(4.41)
$$K_n(\beta x) = \frac{(-1)^n}{2} \left\{ \frac{\partial I_{-n}(\beta x)}{\partial n} - \frac{\partial I_n(\beta x)}{\partial n} \right\}$$

This yields:

(4.42)
$$K_0(\beta x) = -I_0(\beta x) \ln \frac{\beta x}{2} + \sum_{k=0}^{\infty} \frac{\left(\dfrac{\beta x}{2} \right)^{2k}}{k!} \phi(k)$$

where

$$\phi(k) = 1 + \frac{1}{2} + \ldots + \frac{1}{k} \ldots, \qquad \phi(0) = 0$$

Also,

$$K_n(\beta x) = (-1)^{n+1} \sum_{k=0}^{\infty} \frac{\left(\dfrac{\beta x}{2}\right)^{n+2k}}{k!(n+k)!} \left\{ \ln \frac{\beta x}{2} - \frac{1}{2}\phi(k) - \frac{1}{2}\phi(n+k) \right\}$$

(4.43)

$$+ \frac{1}{2} \sum_{k=0}^{n-1} \frac{(-1)^k (n-k-1)!}{k!} \left(\frac{\beta x}{2}\right)^{-n+2k}$$

for $n = 1, 2, \ldots$.

Asymptotic forms:

For *large* βx,

(4.44) $$I_n(\beta x) \approx \frac{e^{\beta x}}{\sqrt{2\pi \beta x}} \left[1 - \frac{4n^2 - 1^2}{1!(8\beta x)} + \frac{(4n^2 - 1^2)(4n^2 - 3^2)}{2!(8\beta x)^2} - \cdots \right]$$

and

(4.45) $$K_n(\beta x) \approx \frac{\pi e^{-\beta x}}{\sqrt{2\pi \beta x}} \left[1 + \frac{4n^2 - 1^2}{1!(8\beta x)} + \frac{(4n^2 - 1^2)(4n^2 - 3^2)}{2!(8\beta x)^2} + \cdots \right]$$

Graphs

The general character of these functions is indicated in the sketches in Fig. 4-2.

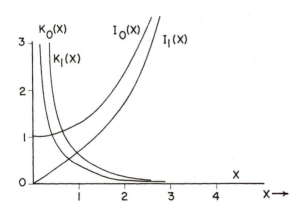

FIGURE 4.2. Modified Bessel Functions of First and Second Kinds.

Roots

Note that neither I_n nor K_n have any real roots except the origin.

Differentiation and recurrence relations

From the series relationships we find:

(4.46)
$$x\frac{d}{dx}I_n(\beta x) + nI_n(\beta x) = \beta x I_{n-1}(\beta x)$$

or

(4.47)
$$\frac{d}{dx}(x^n I_n(\beta x)) = \beta x^n I_{n-1}(\beta x)$$

Also,

(4.48)
$$x\frac{d}{dx}I_n(\beta x) - nI_n(\beta x) = \beta x I_{n+1}(\beta x)$$

or,

(4.49)
$$\frac{d}{dx}\left(\frac{1}{x^n}I_n(\beta x)\right) = \beta\frac{I_{n+1}(\beta x)}{x^n}$$

These combine to yield:

(4.50)
$$I_{n-1}(\beta x) - I_{n+1}(\beta x) = \frac{2n}{\beta x}I_n(\beta x)$$

and

$$I_{n-1}(\beta x) + I_{n+1}(\beta x) = \frac{2}{\beta}\frac{d}{dx}I_n(\beta x)$$

Similarly we have:

(4.51)
$$x\frac{d}{dx}K_n(\beta x) + nK_n(\beta x) = -\beta x K_{n-1}(\beta x)$$

or

(4.52)
$$\frac{d}{dx}(x^n K_n(\beta x)) = -\beta x^n K_{n-1}(\beta x)$$

Also,

(4.53)
$$x\frac{d}{dx}K_n(\beta x) - nK_n(\beta x) = -\beta x K_{n+1}(\beta x)$$

or,

(4.54)
$$\frac{d}{dx}\left[\frac{1}{x^n}K_n(\beta x)\right] = -\beta\frac{K_{n+1}(\beta x)}{x^n}$$

These yield

(4.55)
$$K_{n-1}(\beta x) - K_{n+1}(\beta x) = -\frac{2n}{\beta x}K_n(\beta x)$$

and

(4.56)
$$K_{n-1}(\beta x) + K_{n+1}(\beta x) = -\frac{2}{\beta}\frac{d}{dx}K_n(\beta x)$$

The series also give:

(4.57)
$$K_{-n}(\beta x) = K_n(\beta x)$$

for *all* values of n.

Special cases

Note that from the series:

(4.58)
$$I_{\frac{1}{2}}(\beta x) = \left(\frac{2}{\pi\beta x}\right)^{\frac{1}{2}} \sinh \beta x$$

and

(4.59)
$$I_{-\frac{1}{2}}(\beta x) = \left(\frac{2}{\pi\beta x}\right)^{\frac{1}{2}} \cosh \beta x$$

Also: ˙

(4.60)
$$K_n(\beta x) = \frac{\pi i}{2} e^{i\frac{n\pi}{2}} H_n^{(1)}(\beta x)$$

Orthogonality

Since the modified functions have no *real* roots (except the origin) there is no general orthogonality property.

Ber and bei, and ker and kei functions

In certain problems we find the functions $I_0(x\sqrt{i})$ and $K_0(x\sqrt{i})$ occurring. Here $i = \sqrt{-1}$. If in the general series we put $\beta = \sqrt{i} = e^{i\pi/4}$ we find, collecting real and imaginary parts,

(4.61)
$$I_0(x\sqrt{i}) = I_0\left(\frac{1+i}{\sqrt{2}}x\right) = \text{ber } x + i \text{ bei } x$$

where,

(4.62)
$$\text{ber } x = 1 - \frac{x^4}{2^2 4^2} + \frac{x^8}{2^2 4^2 6^2 8^2} - \cdots$$

and

(4.63)
$$\text{bei } x = \frac{x^2}{2^2} - \frac{x^6}{2^2 4^2 6^2} + \frac{x^{10}}{2^2 4^2 6^2 8^2 10^2} - \cdots$$

and similarly,

(4.64)
$$K_0(x\sqrt{i}) = \text{ker } x + i \text{ kei } x$$

TABLE 4-2.   Some Integrals Involving Bessel Functions.

(1) $\displaystyle\int_0^a x^{n+1} J_n(\xi_i x)\, dx = \frac{a^{n+1}}{\xi_i} J_{n+1}(\xi_i a), \qquad n > -1$

(2) $\displaystyle\int_0^\infty e^{-ax} J_n(bx)\frac{dx}{x} = \{(a^2 + b^2)^{\frac{1}{2}} - a\}^n / nb^n$

(3) $\displaystyle\int_0^\infty e^{-ax} J_n(bx) x^n\, dx = \frac{(2b)^n \Gamma(n + \frac{1}{2})}{\sqrt{\pi}(a^2 + b^2)^{n+\frac{1}{2}}}, \qquad n > -\frac{1}{2}$

(4) $\displaystyle\int_0^\infty J_n(x)\sin ax\, dx = \frac{\{\sin(n \sin^{-1} a)\}}{\sqrt{1 - a^2}}, \qquad |a| < 1, \qquad n > -2$

(5) $\displaystyle\int_0^\infty e^{-ax} J_n(bx)\, dx = \left\{\frac{(a^2 + b^2)^{\frac{1}{2}} - a}{b}\right\}^n \Big/ (a^2 + b^2)^{\frac{1}{2}}$

(6) $\displaystyle\int_0^\infty K_0(ax)\cos bx\, dx = \frac{\pi}{2}(a^2 + b^2)^{-\frac{1}{2}}$

(7) $\displaystyle\int_0^\infty J_\nu(ax) e^{-p^2 x^2} x^{\nu+1}\, dx = \frac{a^\nu}{(2p^2)^{\nu+1}} e^{-\frac{a^2}{4p^2}}, \qquad (2\nu + 2) > 0$

(8) $\displaystyle\int_0^\infty J_{\mu+1}(ax) J_\nu(bx) x^{\nu-\mu}\, dx = \begin{cases} 0, & a < b \\ \dfrac{(a^2 - b^2)^{\mu-\nu} b^\nu}{2^{\mu-\nu} a^{\mu+1} \Gamma(\mu - \nu + 1)}, & a \geq b \end{cases}$

(9) $K_0(ax) = \displaystyle\int_0^\infty \frac{\cos xt\, dt}{\sqrt{a^2 + t^2}}, \qquad x,\, a,\, \text{real, positive}$

(10) $J_\mu(x) J_\nu(x) = \dfrac{2}{\pi}\displaystyle\int_0^{\pi/2} J_{\mu+\nu}(2x \cos \theta)\cos(\mu - \nu)\theta\, d\theta, \qquad \mu + \nu > -1$

(11) $H_0^{(1)}(kx) = \dfrac{-i}{\pi}\displaystyle\int_{-\infty}^{+\infty} \frac{e^{ik\sqrt{x^2 + t^2}}\, dt}{\sqrt{x^2 + t^2}}, \qquad k,\, x\ \text{real, positive}$

(12) $H_0^{(2)}(kx) = \dfrac{i}{\pi}\displaystyle\int_{-\infty}^{+\infty} \frac{e^{-ik\sqrt{x^2 + t^2}}\, dt}{\sqrt{x^2 + t^2}}, \qquad k,\, x\ \text{real, positive}$

LEGENDRE FUNCTIONS

The differential equation is:

(4.65) $\qquad\qquad\qquad (1 - x^2)y'' - 2xy' + n(n + 1)y = 0$

This has linearly independent solutions in the domain $-1 < x < +1$ *with finite derivatives only for* $n = 0, 1, 2, 3, \ldots$. These are:

(4.66) $\qquad\qquad\qquad P_n(x) = \displaystyle\sum_{l=0}^{N} (-1)^l \frac{(2n - 2l)! x^{n-2l}}{2^n l!(n - 2l)!(n - l)!}$

where $N = n/2$ for n even, and $N = (n - 1)/2$ for n odd, and;

$$(4.67) \qquad Q_n(x) = \tfrac{1}{2}P_n(x) \ln \frac{1 + x}{1 - x} - \sum_{l=0}^{M} (-1)^{n-2l} \frac{(2n - 4l - 1)}{(2l + 1)(n - l)} P_{n-2l-1}^{(x)}$$

where $M = (n - 1)/2$ for n odd and $M = (n - 2)/2$ for n even, $n \geq 1$. For $n = 0$ the special forms,

$$(4.68) \qquad P_0(x) = 1$$

and

$$(4.69) \qquad Q_0(x) = \tfrac{1}{2} \ln \frac{1 + x}{1 - x}$$

apply. Here $P_n(x)$ is the Legendre polynomial of order n and $Q_n(x)$ is the Legendre function of second kind of order n.

The first few Legendre polynomials are:

$$(4.70) \qquad P_0(x) = 1, \qquad P_1(x) = x, \qquad P_2(x) = (3x^2 - 1)/2$$
$$P_3(x) = (5x^3 - 3x)/2, \qquad P_4(x) = (35x^4 - 30x^2 + 3)/8, \ldots$$

Note that for n even, $P_n(x)$ is an *even function*,

$$(4.71) \qquad P_{2k}(x) = P_{2k}(-x), \qquad k = 0, 1, 2, \ldots$$

while for n odd, $P_n(x)$ is an *odd function*,

$$(4.72) \qquad P_{2k+1}(x) = -P_{2k+1}(-x), \qquad k = 0, 1, 2, \ldots$$

Since the $Q_n(x)$ functions *rarely* arise in boundary value problems we largely neglect them in our discussion here.

Also note that with regard to even or odd character,

$$(4.73) \qquad \begin{cases} P_n(-1) = (-1)^n \\ P_n(+1) = 1, \qquad \text{all } n \end{cases}$$

Generating function

Expanding $(1 - 2xt + t^2)^{-\frac{1}{2}}$ in a power series in t shows that:

$$(4.74) \qquad (1 - 2xt + t^2)^{-\frac{1}{2}} = \sum_{n=0}^{\infty} P_n(x)t^n$$

Rodrigues formula

$$(4.75) \qquad P_n(x) = \frac{1}{2^n n!} \frac{d^n}{dx^n}(x^2 - 1)^n$$

Differentiation and recurrence formulas

From the generating function one can show,

$$(4.76) \qquad (n + 1)P_{n+1}(x) - (2n + 1)xP_n(x) + nP_{n-1}(x) = 0$$

also,

(4.77)
$$x\frac{d}{dx}P_n(x) - \frac{d}{dx}P_{n-1}(x) = nP_n(x)$$

or

(4.78)
$$\frac{d}{dx}[P_{n+1}(x) - P_{n-1}(x)] = (2n + 1)P_n(x), \qquad n \geq 1$$

Orthogonality

One can show that from the standard Sturm–Liouville form for the Legendre equation,

(4.79)
$$\int_{-1}^{1} P_n(x)P_m(x)\, dx = 0, \qquad n \neq m$$

and the *norm* is

(4.80)
$$\int_{-1}^{+1} P_n^2(x)\, dx = \frac{2}{2n + 1}$$

For *completeness* the expansion of a function $f(x)$ in a series of the $P_n(x)$ on $-1 < x < 1$ must be summed over all $n = 0, 1, 2, \ldots \infty$.

Expansion of practical utility

The exponential form $\exp ikz$, $i = \sqrt{-1}$, occurs in the representation of a plane wave traveling in the z direction with wavenumber k. It can be shown that: $(z = r \cos \theta)$

(4.81)
$$e^{ikz} = e^{ikr \cos \theta} = \sum_{n=0}^{\infty} (2n + 1)i^n P_n(\cos \theta)\mathcal{J}_n(kr)$$

where the *spherical Bessel function* is defined as,

(4.82)
$$\mathcal{J}_n(kr) = \left(\frac{\pi}{2kr}\right)^{\frac{1}{2}} J_{n+\frac{1}{2}}(kr)$$

Some useful integrals

(4.83)
$$\int_{-1}^{1} x^k P_n(x)\, dx = \begin{cases} 0, \ k < n \text{ or } (k - n) = \text{odd integer} \\[2mm] 2\dfrac{k(k - 1)(k - 2)\ldots(k - n + 2)}{(k + n + 1)(k + n - 1)\ldots(k - n + 3)}, \text{ for } k \geq n, \\[2mm] \text{with } k - n = \text{even integer} \end{cases}$$

(4.84)
$$\int_{0}^{1} x^\lambda P_m(x)\, dx = \begin{cases} \dfrac{(-1)^n \Gamma(n - \frac{\lambda}{2})\Gamma(\frac{1}{2} + \frac{\lambda}{2})}{2\Gamma(-\frac{\lambda}{2})\Gamma(n + \frac{3}{2} + \frac{\lambda}{2})}, \ m = 2n \\[4mm] \dfrac{(-1)^n \Gamma(n + \frac{1}{2} - \frac{\lambda}{2})\Gamma(1 + \frac{\lambda}{2})}{2\Gamma(n + 2 + \frac{\lambda}{2})\Gamma(\frac{1}{2} - \frac{\lambda}{2})}, \ m = 2n + 1 \end{cases}$$

Integral representation

With $x = \cos \theta$ we have,

$$(4.85) \qquad P_n(\cos \theta) = \frac{\sqrt{2}}{\pi} \int_0^\pi \frac{\sin(n + \frac{1}{2})\phi \, d\phi}{\sqrt{\cos \theta - \cos \phi}}$$

also

$$(4.86) \qquad P_n(\cos \theta) = \frac{1}{\pi} \int_0^\pi (\cos \theta + i \sin \theta \cos \phi)^n \, d\phi, \qquad i = \sqrt{-1}$$

ASSOCIATED LEGENDRE FUNCTIONS

The differential equation is:

$$(4.87) \qquad (1 - x^2)y'' - 2xy' + \left[n(n + 1) - \frac{m^2}{1 - x^2} \right] y = 0$$

however the change of *dependent* variable,

$$(4.88) \qquad y = (1 - x^2)^{\frac{m}{2}} u$$

yields

$$(4.89) \qquad (1 - x^2)u'' - 2(m + 1)xu' + (n - m)(n + m + 1)u = 0$$

Then if we differentiate Legendre's Eq. (4.66) m times we obtain this *same* form with

$$(4.90) \qquad u = \frac{d^m Y}{dx^m}$$

where Y is a solution of Legendre's equation,

$$(4.91) \qquad Y = AP_n(x) + BQ_n(x)$$

Thus we define the associated Legendre functions as:

$$(4.92) \qquad P_n^m(x) = (1 - x^2)^{\frac{m}{2}} \frac{d^m P_n(x)}{dx^m}$$

and

$$(4.93) \qquad Q_n^m(x) = (1 - x^2)^{\frac{m}{2}} \frac{d^m Q_n(x)}{dx^m}$$

Using *Rodrigues formula* we have,

$$(4.94) \qquad P_n^m(x) = (1 - x^2)^{\frac{m}{2}} \frac{d^{m+n}}{dx^{m+n}} \left[\frac{(x^2 - 1)^n}{2^n n!} \right]$$

The first few *associated Legendre polynomials* are:

TABLE 4-3.   The First Few Associated Legendre Polynomials are: [note $x = \cos\theta$]

$$P_1^1(x) = (1 - x^2)^{\frac{1}{2}} = \sin\theta$$

$$P_2^1(x) = 3x(1 - x^2)^{\frac{1}{2}} = 3\sin\theta\cos\theta = \tfrac{3}{2}\sin 2\theta$$

$$P_2^2(x) = 3(1 - x^2) = 3\sin^2\theta = \tfrac{3}{2}(1 - \cos 2\theta)$$

$$P_3^1(x) = \tfrac{3}{2}(5x^2 - 1)(1 - x^2)^{\frac{1}{2}} = \tfrac{3}{8}(\sin\theta + 5\sin 3\theta)$$

$$P_3^2(x) = 15x(1 - x^2) = \tfrac{15}{4}(\cos\theta - \cos 3\theta)$$

Differentiation and recursion formulas

These are readily derived from our prior formulas for the Legendre polynomials

(4.95)
$$(2n + 1)(1 - x^2)^{\frac{1}{2}}P_n^{m-1}(x) = P_{n+1}^m(x) - P_{n-1}^m(x)$$

also,

(4.96)
$$(1 - x^2)\frac{d}{dx}P_n^m(x) = (n + m)P_{n-1}^m(x) - nxP_n^m(x)$$

Orthogonality

Again from the standard Sturm–Liouville form we find:

(4.97)
$$\int_{-1}^{1} P_n^m(x)P_k^m(x)\,dx = 0 \qquad n \neq k$$

and the *norm*,

(4.98)
$$\int_{-1}^{1} \{P_n^m(x)\}^2\,dx = \frac{2(n + m)!}{(2n + 1)(n - m)!}$$

For *completeness* the expansion of a function $f(x)$ in a series of the $P_n^m(x)$ on $-1 < x < +1$ must be summed over the values $n = m, m + 1, m + 2, \ldots \infty$ for a fixed m.

SPHERICAL HARMONICS

Strictly speaking the functions usually referred to as "spherical harmonics" are *not* "special functions" in the same sense as are say the Legendre or Bessel functions because they do not result as solutions of an *ordinary* second-order equation. We list them here however because of their intimate connection to the associated Legendre Polynomials.

In the separation of variables in equations involving the Laplacian operator in spherical coordinates there frequently results the equation,

(4.99)
$$\frac{1}{\sin\theta}\frac{\partial}{\partial\theta}\left(\sin\theta\frac{\partial Y}{\partial\theta}\right) + \frac{1}{\sin^2\theta}\frac{\partial^2 Y}{\partial\phi^2} + n(n + 1)Y = 0$$

If separation of variables is then applied to this one obtains the equation for the trigonometric functions of ϕ, $\sin m\phi$, $\cos m\phi$, where m is the separation constant, and the equation for the associated Legendre functions in $\cos \theta$ as argument, i.e., $P_n^m(\cos \theta)$. For this reason the functions

$$\cos m\phi P_n^m(\cos \theta), \qquad \sin m\phi P_n^m(\cos \theta)$$

or the complex form,

$$e^{im\phi} P_n^m(\cos \theta)$$

with some multiplying *constant* attached are often defined and tabulated as the "spherical harmonics."

In many problems formulated in spherical coordinates it is convenient to consider a function, say $f(\theta, \phi)$, expanded as a double series in the functions $e^{im\phi} P_n^m(\cos \theta)$. This will take the form,

$$(4.100) \qquad f(\theta, \phi) = \sum_{n=0}^{\infty} \sum_{m=-n}^{n} A_{mn} e^{im\phi} P_n^{|m|}(\cos \theta)$$

where the expansion coefficient, A_{mn}, is to be treated as complex. It is quite evident that if one has all the properties of $\sin m\phi$, $\cos m\phi$, and $P_n^m(\cos \theta)$ readily at hand then any detailed tabulation of functions defined as say $Y_n^m(\phi, \theta)$ as suggested here is quite superfluous.

OTHER SPECIAL FUNCTIONS

Almost without exception the most frequently used special functions are the trigonometric, hyperbolic, Bessel, and Legendre functions. However there are a few *classic* problems in physics and engineering which require other functions. We list these here in more abbreviated fashion. For more extensive information on these and other functions the references at the end of this chapter should be consulted.

HERMITE POLYNOMIALS

The functions $He_n(x)$ satisfy

$$(4.101) \qquad y'' - xy' + ny = 0$$

and are represented by:

$$(4.102) \qquad He_n(x) = x^n - \frac{n!}{2!(n-2)!} x^{n-2} + 1 \cdot 3 \frac{n!}{4!(n-4)!} x^{n-4} - 1 \cdot 3 \cdot 5 \frac{n!}{6!(n-6)!} x^{n-6} \cdots$$

which terminates. The first few polynomials are:

$$He_0 = 1 \qquad\qquad He_3 = x^3 - 3x$$
$$He_1 = x \qquad\qquad He_4 = x^4 - 6x^2 + 3$$
$$He_2 = x^2 - 1 \qquad He_5 = x^5 - 10x^3 + 15x$$

These can also be represented as:

$$(4.103) \qquad He_n(x) = (-1)^n e^{\frac{x^2}{2}} \frac{d^n}{dx^n}\left(e^{-\frac{x^2}{2}}\right)$$

Differentiation and recursion formulas

(4.104)
$$He_{n+1}(x) = xHe_n(x) - \frac{d}{dx}He_n(x)$$

(4.105)
$$\frac{d}{dx}He_n(x) = nHe_{n-1}(x)$$

Generating function

(4.106)
$$e^{tx - \frac{t^2}{2}} = \sum_{n=0}^{\infty} He_n(x)\frac{t^n}{n!}$$

Integral representation

(4.107)
$$He_n(x)\frac{1}{\sqrt{2\pi}}\int_{-\infty}^{+\infty} (x + it)^n e^{-\frac{t^2}{2}}\,dt, \qquad i = \sqrt{-1}$$

Orthogonality relations

$$\int_{-\infty}^{+\infty} He_m(x)He_n(x)e^{-\frac{x^2}{2}}\,dx = \begin{cases} 0 & n \neq m \\ n!\sqrt{2\pi}, & m = n \end{cases}$$

LAGUERRE POLYNOMIALS (ASSOCIATED)

The functions denoted by $Y = L_n^{(\alpha)}(x)$ satisfy the equation:

(4.109)
$$xy'' + (\alpha + 1 - x)y' + ny = 0$$

and are represented by:

(4.110)
$$L_n^{(\alpha)}(x) = \sum_{k=0}^{n} \frac{\Gamma(n + \alpha + 1)}{(n - k)!(\alpha + k)!}\frac{(-x)^k}{k!}$$

or,

(4.111)
$$L_n^{(\alpha)}(x) = \frac{e^x x^{-\alpha}}{n!}\frac{d^n}{dx^n}(e^{-x}x^{n+\alpha})$$

Orthogonality relations

(4.112)
$$\int_0^{\infty} L_m^{(\alpha)}(x)L_n^{(\alpha)}(x)e^{-x}x^{\alpha}\,dx = \begin{cases} 0 & m \neq n \\ \dfrac{(n + \alpha)!}{n!}, & m = n \end{cases}$$

SPECIAL FUNCTIONS NOT ORIGINATING FROM FROBENIUS SERIES

We list here a few of the more common special functions which arise in physical and mathematical problems in contexts other than the Frobenius series solution of some second-order ordinary differential equations.

GAMMA FUNCTION (factorial function)

The gamma function can be viewed as a function of the complex variable (as can many of the functions already discussed here);however we describe it only as a function of a *real* argument x.

Define

$$(4.113) \qquad \Gamma(x) = \int_0^\infty e^{-t}t^{x-1}\, dt$$

as the gamma function. Most often $\Gamma(x)$ occurs for x an integer in physical problems, i.e., for n any *positive integer*,

$$(4.114) \qquad \Gamma(n + 1) = n! = n(n - 1)(n - 2)\ldots 1$$

$\Gamma(x)$ exists for x negative as well as positive but becomes *infinite* for x any *negative* integer. Note the *special values*

$$(4.115) \qquad \Gamma(1) = 0! = 1$$

and

$$(4.116) \qquad \Gamma(\tfrac{1}{2}) = \sqrt{\pi}$$

ERROR FUNCTIONS

The integral,

$$(4.117) \qquad \mathrm{erf}(x) = \frac{2}{\sqrt{\pi}} \int_0^x e^{-t^2}\, dt$$

is defined as the *error function of x*, and can be represented by the infinite series,

$$(4.118) \qquad \mathrm{erf}(x) = \frac{2}{\sqrt{\pi}}\left(x - \frac{x^3}{3\cdot 1!} + \frac{x^5}{5\cdot 2!} - \cdots\right)$$

Also defined is the *complementary error function*,

$$(4.119) \qquad \mathrm{erfc}(x) = 1 - \mathrm{erf}(x) = \frac{2}{\sqrt{\pi}} \int_x^\infty e^{-t^2}\, dt$$

where we note that

$$(4.120) \qquad \mathrm{erf}(\infty) = 1, \quad \text{and} \quad \mathrm{erfc}(0) = 0$$

In certain diffraction problems in physics the function,

$$(4.121) \qquad \frac{1}{1 + i}\,\mathrm{erf}\!\left(\frac{1 + i}{2}x\sqrt{\pi}\right) = C(x) + iS(x)$$

occurs, where $i = \sqrt{-1}$. Here,

$$(4.122) \qquad C(x) = \int_0^x \cos\frac{\pi t^2}{2}\, dt$$

$$(4.123) \qquad S(x) = \int_0^x \sin\frac{\pi t^2}{2}\, dt$$

are the *Fresnel integrals*.

EXPONENTIAL INTEGRAL

The integral defined as,

$$(4.124) \qquad -Ei(-x) = \int_x^\infty \frac{e^{-t}}{t}\, dt$$

is called the *exponential integral* and arises in many physical problems. With the argument $x = iy$, $i = \sqrt{-1}$, we define,

$$(4.125) \qquad Ei(iy) = Ci(y) + iSi(y) + i\frac{\pi}{2}$$

where

$$(4.126) \qquad Ci(y) = -\int_y^\infty \frac{\cos t}{t}\, dt = \gamma + \ln y - \int_0^y \frac{1 - \cos t}{t}\, dt$$

(γ = Euler's constant) is the *cosine integral* and,

$$(4.127) \qquad Si(y) = \int_0^y \frac{\sin t}{t}\, dt = \frac{\pi}{2} - \int_y^\infty \frac{\sin t}{t}\, dt$$

is the *sine integral.*

An important approximation for the exponential integral is:

$$(4.128) \qquad -Ei(-x) \approx -\gamma - \ln x, \qquad x \text{ small}$$

where again γ is *Euler's constant,*

$$(4.129) \qquad \gamma \approx 0.577215$$

which can be represented in general by:

$$(4.130) \qquad \gamma = \lim_{m \to \infty} \left(\sum_{l=1}^m \frac{1}{l} - \ln m \right)$$

ELLIPTIC INTEGRALS AND ELLIPTIC FUNCTIONS

There is a great profusion of elliptic integrals, here we list two.

Elliptic integral of the first kind

$$(4.131) \qquad K(\kappa, t) = \int_0^t \frac{dx}{\sqrt{(1 - x^2)(1 - \kappa^2 x^2)}}$$

This is called the incomplete integral of modulus κ and argument t. For $t = 1$, $K(\kappa, 1) = K(\kappa)$ is the *complete* elliptic integral of modulus κ.

Elliptic integral of the second kind

$$4.132) \qquad E(\kappa, t) = \int_0^t \sqrt{\frac{1 - \kappa^2 x^2}{1 - x^2}}\, dx$$

is the *incomplete* elliptic integral of the second kind with modulus κ and argument t. For the upper limit $t = 1$ this also becomes a *complete elliptic integral* with modulus κ.

For the incomplete elliptic integrals the inverse functions, say t as a function of K in Eq. (4.131) above defines an "elliptic function." These "Jacobian elliptic functions" however are *not* included in the scope of this text.

EXECUTE PROBLEM SETS

PROBLEMS

(1) Since v appears in Bessel's equation only as the square we see that $J_v(\beta x)$ and $J_{-v}(\beta x)$ must both be solutions. Form the Wronskian (see Appendix IX) and show that these are linearly independent solutions for v not an integer. Specifically show that the Wronskian is,

$$-2(\sin v\pi)/\pi\beta x$$

(2) Consider the linear combination,

$$AJ_v(\beta x) + BJ_{-v}(\beta x) = U_v(\beta x)$$

as a solution of Bessel's equation and show by forming the Wronskian with $J_v(\beta x)$ that the *only* choices for A and B which make $U_v(\beta x)$ linearly independent of $J_v(\beta x)$, and nontrivial, for *all* v including integers are

$$A = \frac{D \cos v\pi}{\sin v\pi}, \qquad B = -\frac{D}{\sin v\pi}$$

where D is an arbitrary constant. Thus see that for the choice $D = 1$ the function $U_v(\beta x) = Y_v(\beta x)$ as given in the text.

(3) Show that the functions,

$$B_n(\beta x) = J_n(\beta a)Y_n(\beta x) - J_n(\beta x)Y_n(\beta a)$$

are solutions of Bessel's equation of order n, and that if

$$B_n(\beta b) = 0$$

they form an orthogonal set on the interval $a < x < b$.

(4) Derive the expression for the *norm*

$$N_{nj} = \int_a^b B_n^2(\beta_{nj}x)x\,dx$$

for the B_n functions above, where the β_{nj} are appropriately defined.

(5) Show that

$$K_n(x)I_n'(x) - K_n'(x)I_n(x) = \frac{1}{x}$$

(6) Show that,

$$J_0(kr) = \frac{1}{2\pi}\int_0^{2\pi} e^{ikr\cos\theta}\,d\theta$$

(7) Show that,

$$\int_0^\infty \cos ax J_0(bx)\, dx = \begin{cases} (b^2 - a^2)^{-\frac{1}{2}}, & b > a \\ 0, & b < a \end{cases}$$

and

$$\int_0^\infty \sin ax J_0(bx)\, dx = \begin{cases} 0, & b > a \\ (a^2 - b^2)^{-\frac{1}{2}}, & b < a \end{cases}$$

(8) Show that the equation,

$$\left\{\frac{1}{r}\frac{d}{dr}\left(r\frac{d}{dr}\right)\right\}\left\{\frac{1}{r}\frac{d}{dr}\left(r\frac{d}{dr}\right) + a^2\right\}y = 0$$

has solutions in terms of Bessel functions.

(9) Deduce the specific expression for the *norm* in Eq. (4.21) of the Bessel functions for *each* of the special cases $h = 0$, $h = \infty$ in Eq. (4.22).

(10) Show that

$$J_{\frac{3}{2}}(x) = \sqrt{\frac{2}{\pi x}}\left(\frac{\sin x}{x} - \cos x\right)$$

(11) Show that

$$\int_{-1}^1 x^n P_n(x)\, dx = \frac{2^{n+1}(n!)^2}{(2n + 1)!}$$

(12) Show that for $m > n$ and $m - n$ even or zero

$$\int_{-1}^1 x^m P_n(x)\, dx = 2\frac{m(m - 1)(m - 2)\ldots(m - n + 2)}{(m + n + 1)(m + n - 1)\ldots(m - n + 3)}$$

(13) Expand the function,

$$f(x) = \begin{cases} x(1 - x), & 0 < x < 1 \\ 0, & -1 < x < 0 \end{cases}$$

in a series of Legendre polynomials.

(14) Show that

$$\int_{-1}^1 x P_n(x) P_{n-1}(x)\, dx = \frac{2n}{4n^2 - 1}$$

(15) Show that by choosing *special* values of t in the generating function for $He_n(x)$ one obtains,

$$\frac{1}{e}\cosh x\sqrt{2} = \sum_{n=0}^\infty \frac{2^n}{(2n)!} He_{2n}(x)$$

and

$$\frac{1}{e}\sinh x\sqrt{2} = \sum_{n=0}^\infty \frac{\sqrt{2^{2n}}}{(2n + 1)!} He_{2n+1}(x)$$

(16) Show that:

$$\Gamma(n + \tfrac{3}{2}) = \sqrt{\pi}\frac{1 \cdot 3 \cdot 5 \ldots (2n + 1)}{2^{n+1}}$$

(17) Show that:

$$\Gamma(x + 1) = x\Gamma(x)$$

(18) Show that:

$$\int_0^t e^{-\alpha x^2} x^2 \, dx$$

can be expressed in terms of an error function.

(19) Show that:

$$\int_0^x Ei(-mt) \, dt = xEi(-mx) - \frac{1 - e^{-mx}}{m}$$

(20) Show that:

$$I = \int_0^\phi \frac{\sin^2 \theta \, d\theta}{\sqrt{1 - k^2 \sin^2 \theta}}$$

can be expressed in terms of K and E (elliptic integrals).

REFERENCES

A. Erdelyi, W. Magnus, F. Oberhettinger, and F. G. Tricomi, *Higher Transcendental Functions* Vols. I, II, and III (McGraw-Hill Book Company, Inc., New York, 1953).

E. Jahnke and F. Emde, *Tables of Functions with Formulas and Curves* (Dover Publications, Inc., New York, 1945).

W. Magnus and F. Oberhettinger, *Formulas and Theorems for the Special Functions of Mathematical Physics* (Chelsea Publishing Company, New York, 1949).

P. M. Morse and H. Feshbach, *Methods of Theoretical Physics* (McGraw-Hill Book Company, Inc., New York, 1953).

I. N. Sneddon, *Special Functions of Mathematical Physics and Chemistry* (Interscience Publishers, Inc., New York, 1956).

E. T. Whittaker and G. N. Watson, *A Course in Modern Analysis* (Cambridge University Press, London, England, 1958).

CHAPTER FIVE

Solution of linear homogeneous boundary value problems; separation of variables method and eigenfunction concepts

In this chapter we develop, *by examples*, the basic techniques for solving linear homogeneous partial differential equations subject to *homogeneous* boundary conditions. We also develop by example the intimate connections between the separation of variables procedure and the concepts of *linear operators* and *eigenfunctions*. Finally we summarize with an outline of the basic format for solutions of this class of boundary value problems and close with suggestions for general techniques. Problems for the student are interspersed with the examples and should be executed in sequence.

Example (1)

Conduction of heat in a finite rectangular rod of length L immersed in a cooling bath and heated at one end.

We have the differential equation,

$$(5.1) \qquad k\left(\frac{\partial^2 T}{\partial x^2} + \frac{\partial^2 T}{\partial y^2} + \frac{\partial^2 T}{\partial z^2}\right) = C\rho\frac{\partial T}{\partial t}$$

where k, C, ρ are the thermal conductivity, specific heat, and mass density, respectively, and

T is the temperature at any point x, y, z in the rod at any time t. The rod is assumed to have some initial temperature, T_i, uniform throughout, and is immersed in a cooling bath at temperature, T_0. We also suppose the end at $z = 0$ has a continual, uniform supply of heat at the rate per unit area. Let the dimensions of the cross section be

$$-\frac{\overline{w}}{2} \le x \le +\frac{\overline{w}}{2}, \qquad -\frac{w}{2} \le y \le +\frac{w}{2}$$

The condition of constant rate of heat supply appears as:

(5.2)
$$\left(-k\frac{\partial T}{\partial z}\right)_{z=0} = q; \qquad \left\{ \begin{array}{c} t > 0 \\ -\dfrac{\overline{w}}{2} \le x \le +\dfrac{\overline{w}}{2} \\ -\dfrac{w}{2} \le y \le +\dfrac{w}{2} \end{array} \right\}$$

Note that we have chosen *rectangular coordinates* in our Eq. (5.1), because of the given shape of the rod.

Now we point out an important initial step in the solution of *any* physical problem. We express the *entire* problem in a *dimensionless form*. This achieves two important objectives. First of all it simplifies the mathematical manipulations because we have fewer symbols to keep track of and secondly our solution will be immediately applicable in *any* consistent set of units, and appears in its most compact and general form.

In our present example we do this as follows: (this basic procedure is applicable in essentially *every* type of boundary value problems).

Define the dimensionless coordinates,

(5.3)
$$\chi = \frac{x}{L}, \qquad \eta = \frac{y}{L}, \qquad \xi = \frac{z}{L}$$

and also define the dimensionless "temperature",

(5.4)
$$U = \frac{T - T_0}{T_i - T_0}$$

When these are substituted into the differential equation and it is rearranged it appears as:

(5.5)
$$\frac{\partial^2 U}{\partial \chi^2} + \frac{\partial^2 U}{\partial \eta^2} + \frac{\partial^2 U}{\partial \xi^2} = \frac{L^2 C \rho}{k} \frac{\partial U}{\partial t}$$

which suggests defining the dimensionless time variable as,

(5.6)
$$\tau = \frac{kt}{C\rho L^2}$$

so that now we have

(5.7)
$$\frac{\partial^2 U}{\partial \chi^2} + \frac{\partial^2 U}{\partial \eta^2} + \frac{\partial^2 U}{\partial \xi^2} = \frac{\partial U}{\partial \tau}$$

in the domain:

$$\begin{cases} -\beta \le \chi \le +\beta, & \beta = \dfrac{\bar{w}}{2L} \\[2ex] -\beta \le \eta \le +\beta, & \beta = \dfrac{w}{2L} \\[2ex] 0 \le \xi \le 1 \\[1ex] \tau > 0 \end{cases}$$

(5.8)

and subject to the boundary conditions:

$$\begin{cases} U = 0, \chi = \pm\beta, & \text{all } \eta, \xi, \tau > 0 \\ U = 0, \eta = \pm\beta, & \text{all } \chi, \xi, \tau > 0 \\ U = 0, \xi = 1, & \text{all } \chi, \eta, \tau > 0 \\ U = 1, \tau = 0, & \text{all } \chi, \eta, \xi \end{cases}$$

(5.9)

We also have the heating condition at $z = 0$ now in the form,

(5.10)
$$\left(\frac{\partial U}{\partial \xi}\right)_{\xi=0} = -\alpha, \qquad \alpha = \frac{qL}{k(T_i - T_0)}$$

Observe that now our boundary value problem, consisting of the partial differential Eq. (5.7); the domain, Eq. (5.8); and the boundary and initial conditions, Eqs. (5.9) and (5.10), contains only *three constants*, $\bar{\beta}$, β, and α. This is most certainly a great simplification of the problem at the outset.

The particular choice of dimensionless variables here is by no means unique. However we would like to emphasize that one should *always* select a set which eliminates as many constants as possible from the differential equation itself.

To construct the solution to the problem now for $U(\chi, \eta, \xi, \tau)$ we assume a form, just as in Chapter Three

(5.11)
$$U = X(\chi)N(\eta)Z(\xi)M(\tau)$$

as a trial solution. However we do this in a *sequence* of steps here.

Observe that if we were to assume a trial solution of the form

(5.12)
$$U = X(\chi)Q(\eta, \xi, \tau)$$

this would yield in the differential equation,

(5.13)
$$Q\frac{\partial^2 X}{\partial \chi^2} + X\left(\frac{\partial^2 Q}{\partial \eta^2} + \frac{\partial^2 Q}{\partial \xi^2} - \frac{\partial Q}{\partial \tau}\right) = 0$$

and we could then proceed with the separation of variables as we did in Chapter Three. Instead suppose we *assume directly* that

(5.14)
$$\frac{\partial^2 X}{\partial \chi^2} = -a^2 X$$

Here we say X is an *eigenfunction* of the *operator* $\partial^2/\partial \chi^2$ with *eigenvalue* $-a^2$. Obviously the solutions for $X(x)$ are trigonometric functions for real a.

Now we have,

(5.15)
$$-a^2 Q + \frac{\partial^2 Q}{\partial \eta^2} + \frac{\partial^2 Q}{\partial \xi^2} - \frac{\partial Q}{\partial \tau} = 0$$

remaining for our consideration upon dividing out X. We could then continue by assuming a form,

(5.16)
$$Q = N(\eta)Q'(\xi, \tau)$$

as a solution here and repeat the above process, this time assuming

(5.17)
$$\frac{\partial^2 N}{\partial \eta^2} = -b^2 N$$

so that N is an eigenfunction of the operator $\partial^2/\partial \eta^2$ with eigenvalue $-b^2$.

In the same way we construct Z as a solution of the *eigenvalue problem*,

(5.18)
$$\frac{\partial^2 Z}{\partial \xi^2} = -c^2 Z$$

and have finally, upon inserting *all* of the assumed forms for X, N, and Z into our *original* differential Eq. (5.7), the form:

(5.19)
$$-(a^2 + b^2 + c^2)M = \frac{\partial M}{\partial \tau}$$

Hence M is also the solution of an eigenvalue problem, with the operator being $\partial/\partial \tau$ and the eigenvalue being†

(5.20)
$$-\gamma^2 = -(a^2 + b^2 + c^2)$$

Now we can write down the general form for our trial solution as we originally set out to construct in Eq. (5.11). The solutions of our eigenvalue problems in X, N, and Z are:

(5.21)
$$X = \begin{cases} A \sin a\chi + B \cos a\chi, & a \neq 0 \\ A' \chi + B', & a = 0 \end{cases}$$

(5.22)
$$N = \begin{cases} C \sin b\eta + D \cos b\eta, & b \neq 0 \\ C'\eta + D', & b = 0 \end{cases}$$

(5.23)
$$Z = \begin{cases} E \sin c\xi + F \cos c\xi, & c \neq 0 \\ E'\xi + F', & c = 0 \end{cases}$$

and that for M is then:

(5.24)
$$M = \begin{cases} G\, e^{-(a^2 + b^2 + c^2)\tau}, & a^2 + b^2 + c^2 > 0 \\ G', & a^2 + b^2 + c^2 = 0 \end{cases}$$

In these equations A, B, A', B', C, etc. are constants to be determined. A *single* solution of our Eq. (5.7) is formed as the product of *one each* of the forms of X, N, Z, and M but these must be *consistent*, i.e., if we pick the second forms in *every* case for X, N, and Z then we *must* use the last form for M.

Of course the general solution, or more specifically *the solution*, of our boundary value problem will be formed as a *sum* of solutions of the type $XNZM$, since the differential equation is *linear*. But first we must examine the *boundary conditions*.

We begin with the fact, in Eq. (5.9), that $U = 0$ at $\chi = \pm \beta$. Since all χ dependence is in $X(\chi)$ we require that

(5.25)
$$X(\beta) = X(-\beta) = 0$$

----

† In view of our original Eq. (5.7) and we also see $M X N Z$ as an eigenfunction of the operator, $\partial^2/\partial \chi^2 + \partial^2/\partial \eta^2 + \partial^2/\partial \xi^2$ with eigenvalue $-(a^2 + b^2 + c^2)$.

Here we make an immediate simplification by noting that our problem is *symmetric* in χ and hence our solution should be an *even* function of χ. We use this fact to immediately discard the $\sin a\chi$ functions since these are *odd* functions. Thus we require,

$$(5.26) \qquad \cos(a\beta) = \cos(-a\beta) = 0$$

and thus determine the *spectrum* of the eigenvalue a as:

$$(5.27) \qquad a_n = \frac{2n+1}{2}\frac{\pi}{\beta}, \qquad n = 0, 1, 2, \ldots$$

In the same way we note that we have *symmetry* in η so that we discard the $\sin b\eta$ term of N, and using the boundary conditions on η in Eq. (5.9),

$$(5.28) \qquad \cos(b\beta) = \cos(-b\beta) = 0$$

determines the spectrum of the eigenvalue b as:

$$(5.29) \qquad b_m = \frac{2m+1}{2}\frac{\pi}{\beta}, \qquad m = 0, 1, 2, \ldots$$

When we turn to consider the boundary conditions on ξ and the form for Z we see that we do *not* have symmetry in the ξ domain, and in fact have "mixed" boundary conditions at $\xi = 0$ and $\xi = 1$. Therefore we can expect a solution corresponding to a zero separation constant to be required. Since the condition at $\xi = 1$ is $U = 0$ and the condition at $\xi = 0$ is independent of τ we form a solution dependent on τ and then later add to this another solution independent of τ. Thus we choose $E = 0$ in Eq. (5.23) and at $\xi = 1$ we set

$$(5.30) \qquad \cos c = 0$$

so that the spectrum of c is

$$(5.31) \qquad c_s = \frac{2s+1}{2}\pi, \qquad s = 0, 1, 2, \ldots$$

Then we have a particular solution for U as

$$(5.32) \qquad U_1 = \sum_{n=0}^{\infty}\sum_{m=0}^{\infty}\sum_{s=0}^{\infty} K_{nms}\, e^{-\gamma_{nms}^2 \tau}\cos a_n\chi \cos b_m\eta \cos c_s\xi$$

where $K_{nms} = BDFG$ is a combined constant and,

$$(5.33) \qquad \gamma_{n,m,s}^2 = \pi^2\left[\frac{(2n+1)^2}{4\beta^2} + \frac{(2m+1)^2}{4\beta^2} + \frac{(2s+1)^2}{4}\right]$$

This solution yields $\partial U/\partial\xi = 0$ at $\xi = 0$, so we now form another solution, U_2, independent of τ, which yields $\partial U/\partial\xi = -\alpha$ as required at $\xi = 0$, but which also satisfies all the other boundary conditions. Note in Eq. (5.24) that for a solution independent of τ we must have

$$(5.34) \qquad c = \pm\sqrt{-a^2-b^2} = \pm ic'$$

which is imaginary if a and b are the real numbers a_n and b_m above. Thus, in Eq. (5.23) the solution for Z becomes a linear combination of $\sinh c'\xi$ and $\cosh c'\xi$. A combination of these which vanishes at $\xi = 1$ is $\sinh c'(1 - \xi)$, so we have as the solution

$$(5.35) \qquad U_2 = \sum_{n=0}^{\infty}\sum_{m=0}^{\infty} L_{nm}\sinh\left[\sqrt{a_n^2+b_m^2}\,(1 - \xi)\right]\cos a_n\chi\cos b_m\eta$$

which vanishes at $\chi = \pm \bar{\beta}$ and $\eta = \pm \beta$, as well as at $\xi = 1$ but yields at $\xi = 0$

(5.36) $$\frac{\partial U_2}{\partial \xi}\bigg|_{\xi=0} = -\sum_{n=0}^{\infty}\sum_{m=0}^{\infty} L_{nm}\sqrt{a_n^2+b_m^2}\cosh\sqrt{a_n^2+b_m^2}\cos a_n\chi\cos b_m\eta$$

which is not zero. Here we set this equal to $-\alpha$ as required and employ the orthogonality property of the cosine functions to determine the L_{nm}. This gives:

(5.37) $$L_{nm} = \frac{(-1)^{n+m}\,4\alpha}{a_n b_m\,\bar{\beta}\beta\sqrt{a_n^2+b_m^2}\,\cosh\sqrt{a_n^2+b_m^2}}$$

Having the L_{nm} now known, we form the sum $U_1 + U_2$ as U and proceed to satisfy the one remaining condition at $\tau = 0$, namely $U = 1$. Thus, at $\tau = 0$,

(5.38) $$1 = \sum_{n=0}^{\infty}\sum_{m=0}^{\infty} L_{nm}\sinh\left[\sqrt{a_n^2+b_m^2}(1-\xi)\right]\cos a_n\chi\cos b_m\eta$$

$$+ \sum_{n=0}^{\infty}\sum_{m=0}^{\infty}\sum_{s=0}^{\infty} K_{nms}\cos a_n\chi\cos b_m\eta\cos c_s\xi$$

We then use the orthogonality property of the cosines to determine the K_{nms}. Thus, multiplying through by $\cos a_n\chi\cos b_m\eta\cos c_s\xi'\,d\chi d\eta d\xi$, integrating $-\bar{\beta}$ to $\bar{\beta}$ on χ, $-\beta$ to β on η and 0 to 1 on ξ we obtain

(5.39) $$K_{nms} = \frac{8(-1)^{n+m+s}}{a_n b_m c_s\bar{\beta}\,\beta} - 2L_{nm}\int_0^1 \sin h\left[\sqrt{a_n^2+b_m^2}(1-\xi)\right]\cos c_s\xi d\xi$$

or

(5.40) $$K_{nms} = \frac{8(-1)^{n+m+s}}{a_n b_m c_s\bar{\beta}\,\beta}\left[1 + \frac{(-1)^s c_s\,(\cos c_s + \cosh\sqrt{a_n^2+b_m^2})\alpha}{(c_s^2 + a_n^2 + b_m^2)\cosh\sqrt{a_n^2+b_m^2}}\right]$$

and with all constants thus determined the solution is:

(5.41) $$U = \sum_{n=0}^{\infty}\sum_{m=0}^{\infty} L_{nm}\sin h\left[\sqrt{a_n^2+b_m^2}(1-\xi)\right]\cos a_n\chi\cos b_m\eta$$

$$+ \sum_{n=0}^{\infty}\sum_{m=0}^{\infty}\sum_{s=0}^{\infty} K_{nms}\exp^{(-\partial_{m}\tau)}\cos a_n\chi\cos b_m\eta\cos c_s\xi$$

Of course this can be expressed back in terms of the original variables if desired, but this form is more compact and more general.

EXECUTE PROBLEM SET (5-1)

Example (2)

The potential distribution in a semi-infinite, grounded, right-circular cylinder with the potential specified in the end plane.

The differential equation is Laplace's equation and in view of the given geometry we select cylindrical coordinates, thus:

(5.42)
$$\frac{1}{r}\frac{\partial}{\partial r}\left(r\frac{\partial V}{\partial r}\right) + \frac{1}{r^2}\frac{\partial^2 V}{\partial \theta^2} + \frac{\partial^2 V}{\partial z^2} = 0$$

where $V(r, \theta, z)$ is the potential function. The boundary conditions are

(5.43)
$$\begin{cases} V(R, \theta, z) = 0 \\ V(r, \theta, 0) = F(r, \theta) \\ \lim_{z \to \infty} V(r, \theta, z) = 0 \end{cases}$$

Here R is the radius of the cylinder and $F(r, \theta)$ is the specified function in the end $z = 0$.

Here we call attention to an *important point about symmetry*. If the boundary conditions are completely symmetric in θ then there can be no θ dependence in the solution, and hence the term $\partial^2 V/\partial \theta^2$ could be deleted from the equation at the outset. However we here let $F(r, \theta)$ be a general function and later it will become even more evident that the *symmetry* of V depends upon the symmetry of F.

We again reduce our problem to *dimensionless form* by defining:

(5.44)
$$\rho = \frac{r}{R}, \qquad \xi = \frac{z}{R}, \qquad U = \frac{V}{F_m}$$

where F_m is the maximum value of F on the end $z = 0$. Now we have as our boundary value problem,

(5.45)
$$\frac{1}{\rho}\frac{\partial}{\partial \rho}\left(\rho\frac{\partial U}{\partial \rho}\right) + \frac{1}{\rho^2}\frac{\partial^2 U}{\partial \theta^2} + \frac{\partial^2 U}{\partial \xi^2} = 0$$

with boundary conditions,

(5.46)
$$\begin{cases} U(1, \theta, \xi) = 0 \\ U(\rho, \theta, 0) = f(\rho, \theta) \\ \lim_{\xi \to \infty} U(\rho, \theta, \xi) = 0 \end{cases}$$

where $f = F/F_m$ is a dimensionless function of ρ and θ.

We now dispense with any allusions to separation of variables procedure and use exclusively the eigenfunction format as developed in the last example. Of course we develop in *either* fashion a product solution, $\mathcal{R}(\rho), \Theta(\theta)Z(\xi)$ for $U(\rho, \theta, \xi)$.

Thus let

(5.47)
$$\frac{\partial^2 \Theta}{\partial \theta^2} = -a^2 \Theta$$

and

(5.48)
$$\frac{\partial^2 Z}{\partial \xi^2} = -b^2 Z$$

define the functions Θ and Z with eigenvalues $-a^2$ and $-b^2$. The Θ solutions will be, for real a,

(5.49)
$$\Theta = \begin{cases} A \sin a\theta + B \cos a\theta, & a \neq 0 \\ A'\theta + B', & a = 0 \end{cases}$$

Here we note that since the final solution $U(\rho, \theta, \xi)$ *must be single valued* we require

(5.50)
$$U(\rho, \theta + 2\pi, \xi) = U(\rho, \theta, \xi)$$

This *periodicity condition* is actually another, though usually not explicitly expressed, boundary condition which often occurs in physical problems. In our case it now shows us that the parameter a in Eq. (5.47) and Eq. (5.49), *must* be either zero or an *integer*. Also we see that in Eq. (5.49) we must have $A' = 0$.

Next we note that in view of the last boundary condition in Eq. (5.46) we must require $\bar{b}$ in Eq. (5.48) to be *pure imaginary*; then the Z functions are: $(\bar{b} = ib)$,

(5.51)
$$Z = \begin{cases} C e^{-b\xi} + D e^{+b\xi}, & b \neq 0 \\ C' + D'\xi, & b = 0 \end{cases}$$

Then the only acceptable term is the negative exponential so $D = 0$ and the solution for $b = 0$ is not required.

With these forms for Θ and Z substituted into our Eq. (5.45) there results

(5.52)
$$\frac{1}{\rho} \frac{\partial}{\partial \rho} \left(\rho \frac{\partial \mathcal{R}}{\partial \rho} \right) - \frac{n^2}{\rho^2} \mathcal{R} + b^2 \mathcal{R} = 0$$

for the function $\mathcal{R}(\rho)$. This is readily rearranged to read

(5.53)
$$\rho^2 \frac{d^2 \mathcal{R}}{d\rho^2} + \rho \frac{d\mathcal{R}}{d\rho} + (b^2 \rho^2 - n^2)\mathcal{R} = 0$$

which is readily recognized as Bessel's equation with the general solution

(5.54)
$$\mathcal{R} = L J_n(b\rho) + N Y_n(b\rho), \qquad b \neq 0$$

Immediately we see that we must exclude the functions $Y_n(b\rho)$ by taking the constant $N = 0$ because $Y_n(b\rho)$ has a singularity on the axis $\rho = 0$, and our solution must be finite everywhere.

Thus we now see as a *possible* solution of the problem,

(5.55)
$$U = (A \sin n\theta + B \cos n\theta) e^{-b\xi} J_n(b\rho)$$

however we have yet to fit two of our boundary conditions. If in this function we put $\rho = 1$ we see that it does indeed vanish for *all* θ, ξ, *if*:

(5.56)
$$J_n(b) = 0$$

Thus we define a *spectrum* of b as b_{nj} where these are the roots of the above equation, i.e., b_{n1} is the first root of J_n etc. Then we form the *double sum*,

(5.57)
$$U = \sum_{j=1}^{\infty} \sum_{n=0}^{\infty} (A_{nj} \sin n\theta + B_{nj} \cos n\theta) e^{-b_{nj}\xi} J_n(b_{nj}\rho)$$

as our solution. This now satisfies the differential equation, the periodicity condition on θ, the vanishing as $\xi \to \infty$ and the vanishing on $\rho = 1$; in fact every requirement *except* that at $\xi = 0$. Therefore now put $\xi = 0$ and require:

(5.58)
$$f(\rho, \theta) = \sum_{j=1}^{\infty} \sum_{n=0}^{\infty} (A_{nj} \sin n\theta + B_{nj} \cos n\theta) J_n(b_{nj}\rho)$$

Here we apply the orthogonal properties of the trigonometric and Bessel functions as follows. Multiply both sides by $\sin m\theta$ and integrate from zero to 2π; there results

$$(5.59) \qquad \int_0^{2\pi} f(\rho, \theta) \sin m\theta \, d\theta = \sum_{j=1}^{\infty} \pi A_{mj} J_m(b_{mj}\rho)$$

because of the orthogonality. Then we multiply this by $J_m(b_{me}\rho)\rho$ and integrate from zero to one on ρ. This yields,

$$(5.60) \qquad \int_0^1 \int_0^{2\pi} f(\rho, \theta) \sin m\theta J_m(b_{me}\rho) \, d\theta \, \rho \, d\rho = \pi A_{me} N_{me}$$

because of the orthogonality property of the Bessel functions. Here N_{me} is the *norm*,

$$(5.61) \qquad N_{me} = \int_0^1 J_m^2(b_{me}\rho)\rho \, d\rho = \tfrac{1}{2} J_{m+1}^2(b_{me})$$

Thus A_{me} is explicitly evaluated in terms of the function $f(\rho, \theta)$.

In the same way we begin by multiplying Eq. (5.58) by $\cos m\theta$, performing the integration 0 to 2π, and repeating the above procedure for the Bessel function. There results,

$$(5.62) \qquad \int_0^1 \int_0^{2\pi} f(\rho, \theta) \cos m\theta J_m(b_{me}\rho)\rho \, d\theta \, d\rho = \pi B_{me} N_{me}$$

for $m \neq 0$. The particular case of $m = 0$ appears as:

$$(5.63) \qquad \int_0^1 \int_0^{2\pi} f(\rho, \theta) J_0(b_{0e}\rho)\rho \, d\theta \, d\rho = 2\pi B_{0e} N_{0e}$$

Thus *all* the constants A_{nj}, B_{nj} in Eq. (5.57) above are determined and this is the solution of the problem.

Here we again call attention to the question of *symmetry*. Note that if the function $f(\rho, \theta)$ is *independent* of θ then the integrals on θ in Eqs. (5.60) and (5.62) are *all* zero, *but* the integral on θ in Eq. (5.63) does *not* vanish. Hence all A_{nj}, B_{nj} are zero except B_{0j}, and the solution for *axial symmetry*, $f = f(\rho)$, a function of ρ only, is:

$$(5.64) \qquad U(\rho, \xi) = \sum_{j=1}^{\infty} B_j e^{-b_j \xi} J_0(b_j \rho)$$

where the B_j are just the B_{0j} above and the $b_{0j} \to b_j$.

EXECUTE PROBLEM SET (5-2)

Example (3)

Propagation of acoustic waves in the infinite domain *bounded internally* by the circular cylinder, $r = R$, having a sinusoidal temporal perturbation of R specified. This gives rise to *steady waves* in the outside medium.

Here the differential equation is the wave equation† but because of the symmetry of the problem there is no Z dependence (variation along the axis). Thus we take,

$$(5.65) \qquad \frac{1}{r}\frac{\partial}{\partial r}\left(r\frac{\partial V}{\partial r}\right) + \frac{1}{r^2}\frac{\partial^2 V}{\partial \theta^2} = \frac{1}{c^2}\frac{\partial^2 V}{\partial t^2}$$

as the differential equation of the problem.

In *all* problems of steady wave propagation it is *usually* most convenient to develop the solution in a *complex form*; then having completed the solution we can take either the real or imaginary portion as applying to the physical situation. It is important that in carrying out this program *all* parts of the problem be treated *consistently*. Thus if now we assume a *trial* solution of the form, with V' complex,

$$(5.66) \qquad V = V'(r, \theta)\, e^{i(\omega t + \gamma)}$$

then this form must be imposed in all boundary conditions as well as in the differential equations.

Thus we have:

$$(5.67) \qquad \frac{1}{r}\frac{\partial}{\partial r}\left(r\frac{\partial V'}{\partial r}\right) + \frac{1}{r^2}\frac{\partial^2 V'}{\partial \theta^2} + \frac{\omega^2}{c^2}V' = 0$$

as our differential equation and the boundary condition,‡

$$(5.68) \qquad \frac{dR}{dt} = \frac{\partial V(R, \theta, t)}{\partial r} = f(\theta)\, e^{i(\omega t + \gamma)}$$

for V, becomes,

$$(5.69) \qquad \frac{\partial V'(R, \theta)}{\partial r} = f(\theta)$$

for the conditions on V' at $r = R$.

For the conditions on V, or now V', as we let $r \to \infty$ we require that the solution correspond to a *wave* traveling *outward* from the axis. This requires a bit of explanation.

In wave solutions represented in *complex form* as:

$$(5.70) \qquad V = A(\mathbf{r}, t)\, e^{i\phi(\mathbf{r}, t)}, \quad (A \text{ and } \phi \text{ real})$$

we call A the *amplitude* and ϕ the *phase* of the wave. Thus a surface of *constant phase* is given by,

$$(5.71) \qquad \phi(\mathbf{r}, t) = \phi_0 = \text{constant}$$

Then

$$(5.72) \qquad \frac{d\phi}{dt} = 0 = \left(\frac{\partial \phi}{\partial t}\right)_{\phi_0} + \mathbf{v}\cdot(\nabla\phi)_{\phi_0}$$

and the velocity $\mathbf{v}$, defined to be *normal* to the surface of constant phase is the phase velocity $\mathbf{v}$, whose magnitude is given by,

$$(5.73) \qquad v = -\left\{\frac{\frac{\partial \phi}{\partial t}}{|\nabla\phi|}\right\}_{\phi_0}$$

† V is the *velocity potential* of the fluid medium.
‡ *Normal* velocity of fluid and solid surface are equal.

In most wave phenomena v is *not* a constant. In the solution of our present problem we will see just how this is employed in imposing the "outward wave condition" above.

We deviate from our usual format in the present problem and leave our equations in *dimensional* form. This will make the outward wave condition more clear and meaningful.

Now the eigenfunctions for the operator $\partial^2/\partial\theta^2$ in Eq. (5.67) are again the trigonometric functions and because of the single valued requirement in θ must be of the form,

(5.74)
$$\Theta = A \sin n\theta + B \cos n\theta, \qquad n = 0, 1, 2, \ldots$$

Thus the r dependence of V' is now determined by,

(5.75)
$$\frac{1}{r}\frac{\partial}{\partial r}\left(r\frac{\partial \mathscr{R}}{\partial r}\right) - \frac{n^2}{r^2}\mathscr{R} + \frac{\omega^2}{c^2}\mathscr{R} = 0$$

where $V' = \mathscr{R}\Theta$. Here we emphasize again that $\mathscr{R}$ is now *complex*.

This is Bessel's equation of order n and we have readily at hand *two* complex solutions in the form of the Hankel functions, $H_n^{(1)}[(\omega/c)r]$, and $H_n^{(2)}[(\omega/c)r]$, or,

(5.76)
$$\mathscr{R} = CH_n^{(1)}\left(\frac{\omega}{c}r\right) + DH_n^{(2)}\left(\frac{\omega}{c}r\right)$$

It is here that we examine the outward wave property.

From the previous chapter we see that for *very large* r these functions have the *asymptotic forms*,

(5.77)
$$H_n^{(1)}\left(\frac{\omega}{c}r\right) \approx \left(\frac{2c}{\pi\omega r}\right)^{\frac{1}{2}} e^{i\left(\frac{\omega}{c}r - \frac{n\pi}{2} - \frac{\pi}{4}\right)}$$

and

(5.78)
$$H_n^{(2)}\left(\frac{\omega}{c}r\right) \approx \left(\frac{2c}{\pi\omega r}\right)^{\frac{1}{2}} e^{-i\left(\frac{\omega}{c}r - \frac{n\pi}{2} - \frac{\pi}{4}\right)}$$

Thus when multiplied by the complex time factor as in Eq. (5.66) we see as the coefficient of i in the exponent *either*

(5.79)
$$\phi^{(1)} = \frac{\omega}{c}r - \frac{n\pi}{2} - \frac{\pi}{4} + \omega t + \gamma$$

if we use $H_n^{(1)}$, or

(5.80)
$$\phi^{(2)} = -\frac{\omega}{c}r + \frac{n\pi}{2} + \frac{\pi}{4} + \omega t + \gamma$$

if we use $H_n^{(2)}$. Using the first of these for ϕ in Eq. (5.73) (and neglecting θ dependence) we get,

(5.81)
$$v = \frac{-(\partial\phi^{(1)}/\partial t)}{\partial\phi^{(1)}/\partial r} = -c$$

which is *negative*, while for the second we get in the same way $+c$. Thus if we use $H_n^{(2)}$ in forming our solution we will satisfy the *outgoing* wave condition, whereas with $H_n^{(1)}$ we would have an *incoming* wave at large r.

Thus we have as our trial solution now,

(5.82)
$$V'(r, \theta) = \sum_{n=0}^{\infty} (A_n \sin n\theta + B_n \cos n\theta)H_n^{(2)}\left(\frac{\omega}{c}r\right)$$

This satisfies the differential equation and the outgoing wave condition. We still have to satisfy the condition in Eq. (5.69) on the surface of the cylinder, thus we require:

$$(5.83) \qquad f(\theta) = \sum_{n=0}^{\infty} \frac{\partial}{\partial r} H_n^{(2)}\left(\frac{\omega}{c}R\right)(A_n \sin n\theta + B_n \cos n\theta)$$

Hence we can now use the orthogonality properties of $\sin n\theta$ and $\cos n\theta$ just as in our previous problems to obtain explicit expressions for A_n and B_n

$$(5.84) \qquad A_n = \frac{1}{\pi\left[\partial H_n^{(2)}\left(\frac{\omega}{c}R\right)\Big/\partial r\right]} \int_0^{2\pi} f(\theta) \sin n\theta \, d\theta, \qquad n > 0$$

$$(5.85) \qquad B_n = \frac{1}{\pi\left[\partial H_n^{(2)}\left(\frac{\omega}{c}R\right)\Big/\partial r\right]} \int_0^{2\pi} f(\theta) \cos n\theta \, d\theta, \qquad n > 0$$

and

$$(5.86) \qquad B_0 = \frac{1}{2\pi\left[\partial H_0^{(2)}\left(\frac{\omega}{c}R\right)\Big/\partial r\right]} \int_0^{2\pi} f(\theta) \, d\theta$$

all of which are *complex* constants.

With these constants the solution for $V'(r, \theta)$, Eq. (5.82), is now complete and can be used back in Eq. (5.66) to provide the complete *complex* description of the wave in the infinite medium.† Then either the real or the imaginary part of this represents one physical solution of the problem.

Example (4)

Sphere in a uniform stream. This is the classic problem of flow of an ideal incompressible fluid around a solid sphere (*steady flow.*)

The velocity potential U satisfies

$$(5.87) \qquad \frac{1}{r^2}\frac{\partial}{\partial r}\left(r^2\frac{\partial U}{\partial r}\right) + \frac{1}{r^2 \sin\theta}\frac{\partial}{\partial \theta}\left(\sin\theta\frac{\partial U}{\partial \theta}\right) = 0$$

which is Laplace's equation in spherical coordinates with the ϕ dependent parts deleted since we have *axial symmetry* in the problem. We require the velocity, $\mathbf{v} = \nabla U$, finite everywhere in the fluid, a uniform flow at infinite distance from the sphere,

$$(5.88) \qquad U \to vz = vr \cos\theta, \quad \text{as} \quad r \to \infty$$

and zero velocity *normal* to the surface of the sphere,

$$(5.89) \qquad \left(\frac{\partial U}{\partial r}\right)_R = 0$$

where R is the radius of the sphere, which we assume centered at the origin.

† This is of course a *steady state solution,* $-\infty < t < +\infty$.

Since this problem is so simple dimensionally we omit the process of introducing dimensionless variables.

Using the straightforward separation of variables as $U = \mathscr{R}(r)\Theta(\theta)$ we multiply Eq. (5.88) by r^2 and find $\mathscr{R}$ determined by:

$$(5.90) \qquad \frac{\partial}{\partial r}\left(r^2 \frac{\partial \mathscr{R}}{\partial r}\right) = \beta^2 \mathscr{R}$$

where β^2 is the separation constant. This equation is put into a simple form by the substitution of independent variable

$$(5.91) \qquad r = e^s$$

which yields:

$$(5.92) \qquad \frac{d^2\mathscr{R}}{ds^2} + \frac{d\mathscr{R}}{ds} - \beta^2\mathscr{R} = 0$$

and if we take $\beta^2 = n(n + 1)$ the solution is found to be,

$$(5.93) \qquad \mathscr{R} = A\,e^{ns} + B\,e^{-(n+1)s} = Ar^n + B\frac{1}{r^{n+1}}$$

Thus we now assume $\mathscr{R}$ as having this form and find upon using the product form $\mathscr{R}\Theta$ in Eq. (5.87) that Θ is given as the solution of:

$$(5.94) \qquad \frac{1}{\sin\theta}\frac{\partial}{\partial\theta}\left(\sin\theta\frac{\partial\Theta}{\partial\theta}\right) + n(n + 1)\Theta = 0$$

and here if we let $x = \cos\theta$ this takes the form,

$$(5.95) \qquad (1 - x^2)\frac{d^2\Theta}{dx^2} - 2x\frac{d\Theta}{dx} + n(n + 1)\Theta = 0$$

which is *Legendre's* differential equation. Consequently Θ can have the form

$$(5.96) \qquad \Theta = CP_n(\cos\theta) + DQ_n(\cos\theta)$$

However since the Q_n functions are not finite at $\theta = 0$ or $\theta = \pi$ we must take the constant $D = 0$.

Finally then we can express the general solution of Laplace's equation in spherical coordinates, for *axial symmetry*, as:

$$(5.97) \qquad U = \sum_{n=0}^{\infty}\left(A_n r^n + B_n\frac{1}{r^{n+1}}\right)P_n(\cos\theta)$$

However this does not yet fit the boundary conditions of our problem.

First of all we note that for ∇U to remain *finite* as $r \to \infty$ we must have $A_n = 0$ for $n > 1$. Then,

$$(5.98) \qquad U = A_0 + A_1 rP_1(\cos\theta) + \sum_{n=0}^{\infty}B_n\frac{P_n(\cos\theta)}{r^{n+1}}$$

Also since $P_1(x) = x$ we see the second term here as of the form $A_1 r\cos\theta = A_1 z$ so that the condition in Eq. (5.88) is fulfilled if $A_0 = 0$ and $A_1 = v$.

Next when we examine $\partial U/\partial r = 0$ on the surface $r = R$, we have,

(5.99)
$$0 = vP_1(\cos \theta) - \sum_{n=0}^{\infty} \frac{(n + 1)B_n}{R^{n+2}} P_n(\cos \theta)$$

or with the notation again $x = \cos \theta$ and rearrangement

(5.100)
$$vP_1(x) = \sum_{n=0}^{\infty} \frac{(n + 1)B_n}{R^{n+2}} P_n(x)$$

Here then we use the orthogonality properties of $P_n(x)$. Multiply both sides by say $P_m(x)$ and integrate from -1 to $+1$ on x. Then we see all $B_n = 0$ except B_1 and we obtain

(5.101)
$$B_1 = \frac{vR^3}{2}$$

so that our final solution is:

(5.102)
$$U = vrP_1(\cos \theta) + \frac{vR^3}{2} \frac{P_1(\cos \theta)}{r^2}$$

or more simply,

(5.103)
$$U = v \cos \theta \left(r + \frac{R^3}{2r^2} \right)$$

from the definition of $P_1(\cos \theta)$.

Example (5)

Waves on the surface of a sphere. The differential equation for waves on the surface of a sphere is,

(5.104)
$$\frac{1}{\sin \theta} \frac{\partial}{\partial \theta} \left(\sin \theta \frac{\partial U}{\partial \theta} \right) + \frac{1}{\sin^2 \theta} \frac{\partial^2 V}{\partial \phi^2} = \frac{R^2}{c^2} \frac{\partial^2 U}{\partial t^2}$$

where R is the radius of the sphere and c is the phase velocity. Here we seek a general expression for the frequency spectrum of *steady* waves on a sphere of given radius and having a fixed phase velocity c for such waves.

Since we seek sinusoidal behavior in time† we take U to have the complex time dependence $e^{i\omega t}$;

(5.105)
$$U = V e^{i\omega t}$$

then

(5.106)
$$\frac{1}{\sin \theta} \frac{\partial}{\partial \theta} \left(\sin \theta \frac{\partial V}{\partial \theta} \right) + \frac{1}{\sin^2 \theta} \frac{\partial^2 V}{\partial \phi^2} + \frac{\omega^2 R^2}{c^2} V = 0$$

Then since V must obviously have *periodicity* 2π in the angular coordinate ϕ, in order to be a single valued function of ϕ on the sphere, we take $V = \Theta(\theta) e^{im\phi}$ where m is some integer or zero. Hence we have,

(5.107)
$$\frac{1}{\sin \theta} \frac{\partial}{\partial \theta} \left(\sin \theta \frac{\partial \Theta}{\partial \theta} \right) - \frac{m^2 \Theta}{\sin^2 \theta} + \frac{\omega^2 R^2}{c^2} \Theta = 0$$

† Steady waves.

determining the $\Theta(\theta)$ function. Here if we let

(5.108)
$$\frac{\omega^2 R^2}{c^2} = n(n + 1), \qquad n = 0, 1, 2, \ldots$$

and make the substitution $x = \cos \theta$, this appears as:

(5.109)
$$(1 - x^2)\frac{d^2\Theta}{dx^2} - 2x\frac{d\Theta}{dx} + \left[n(n + 1) - \frac{m^2}{1 - x^2} \right]\Theta = 0$$

which is the differential equation for the *associated Legendre functions*. Thus we know that *well behaved* solutions $\Theta = P_n^m(\cos \theta)$ exist for Θ if n is fixed as zero or an integer as stated in Eq. (5.108) above. Consequently we have directly the angular frequencies given by Eq. (5.108). The general representation for such steady surface waves is thus:

(5.110)
$$U = \sum_{n=0}^{\infty} \sum_{m=0}^{n} A_{nm}\, e^{i\left[m\phi + \frac{c}{R}\sqrt{n(n + 1)}\,t \right]} P_n^m(\cos \theta)$$

where the A_{nm} are *complex* constants. Note that here we have taken note of the fact that $P_n^m(\cos \theta) \equiv 0$ if $m > n$. Here either the real or imaginary parts of the solution could be made to apply to a physical problem by proper choice of the A_{nm}. Given some *initial amplitude, U,* and velocity, $\partial U/\partial t$, functions we would use the orthogonality properties to obtain these.

Example (6)

Potential problems of *axial symmetry* in which *the potential on the axis is known.*

In many electrostatic problems one can calculate the potential on an axis of symmetry by some elementary means, then one seeks the potential everywhere. For example consider the uniformly charged wire in the form of a circle of radius R. By the elementary application of Coulomb's law we have, with λ being the charge per unit length of wire and ε_0 a constant,

(5.111)
$$V = \frac{\lambda R}{2\varepsilon_0\sqrt{R^2 + z^2}}$$

at any point on the axis of the circle a distance Z from the plane of the circle, as indicated in Fig. 5-1.

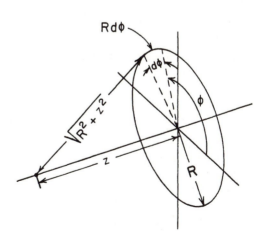

FIGURE 5.1. Geometry for Computing the Field on the Axis of a Uniformly Charged Circular Wire.

Now obviously we have *axial symmetry* so V satisfies,

(5.112)
$$\frac{1}{r^2}\frac{\partial}{\partial r}\left(r^2\frac{\partial V}{\partial r}\right) + \frac{1}{r^2\sin\theta}\frac{\partial}{\partial\theta}\left(\sin\theta\frac{\partial V}{\partial\theta}\right) = 0$$

everywhere, *except* on the circle of charge itself; there V must be singular (see Chapter X).
    We put the problem in dimensionless form as

(5.113)
$$\overline{U} = \frac{2V\varepsilon_0}{\lambda R}, \quad \rho = \frac{r}{R}, \quad \xi = \frac{r}{R}\cos\theta = \rho\cos\theta = \frac{z}{R}$$

then have the equation:

(5.114)
$$\frac{1}{\rho^2}\frac{\partial}{\partial\rho}\left(\rho^2\frac{\partial^2\overline{U}}{\partial\rho}\right) + \frac{1}{\rho^2\sin\theta}\frac{\partial}{\partial\theta}\left(\sin\theta\frac{\partial\overline{U}}{\partial\theta}\right) = 0$$

with

(5.115)
$$\begin{cases} \overline{U} \to 0 & \text{as } \rho \to \infty \\[2mm] \overline{U} = \dfrac{1}{\sqrt{1+\rho^2}} & \text{as } \cos\theta = \pm1, \quad \text{(on axis)} \end{cases}$$

Now we have already seen that the separation of variables applied to Eq. (5.114) gives

(5.116)
$$\overline{U} = \sum_{n=0}^{\infty}\left(A_n\rho^n + \frac{B_n}{\rho^{n+1}}\right)P_n(\cos\theta)$$

as a well behaved solution, so we consider this as a possible solution.
    First consider the *far* solution as $\rho \gg 1$, then in order to fit the first part of Eq. (5.115) we *must* put all $A_n = 0$ for $n > 0$. Then we compare this to the second part of Eq. (5.115). We put $\cos\theta = 1$ and have the requirement, [noting that $P_n(1) = 1$],

(5.117)
$$\frac{1}{\sqrt{1+\rho^2}} = A_0 + \sum_{n=0}^{\infty}\frac{B_n}{\rho^{n+1}}, \quad \rho \gg 1$$

Since $\rho > 1$ we can factor ρ^2 out of the radical on the left and use a binomial expansion to write

(5.118)
$$\frac{1}{\rho}\left(1 - \frac{1}{2}\frac{1}{\rho^2} + \frac{1\cdot3}{2\cdot4}\frac{1}{\rho^4} - \frac{1\cdot3\cdot5}{2\cdot4\cdot6}\frac{1}{\rho^6} - \cdots\right) = A_0 + \sum_{n=0}^{\infty}\frac{B_n}{\rho^{n+1}}$$

Thus comparing coefficients term by term,

(5.119)
$$\begin{cases} A_0 = 0 \quad B_n = 0, \quad n \text{ odd} \\[2mm] B_0 = 1 \quad B_{2m} = \dfrac{(-1)^m(2m+1)!}{[2(m!)]^2} \quad m = 0, 1, \ldots \end{cases}$$

where $n = 2m$ is *even*.
    Thus we have for $\rho > 1$

(5.120)
$$\overline{U} = \sum_{m=0}^{\infty}\frac{B_{2m}P_{2m}(\cos\theta)}{\rho^{2m+2}}$$

with the B_{2m} as given above. (Note that $\overline{U}$ is an even function of θ.)

For $\rho < 1$ the binomial expansion in Eq. (5.118) is *not* valid but in this case we can write, for $\rho < 1$,

$$(5.121) \qquad \left(1 - \tfrac{1}{2}\rho^2 + \frac{1\cdot3}{2\cdot4} - \frac{1\cdot3\cdot5}{2\cdot4\cdot6}\rho^6 + \ldots\right) = \sum_{n=0}^{\infty}\left(A_n\rho^n + \frac{B_n}{\rho^{n+1}}\right)$$

again by putting $\cos\theta = 1$ (on the axis), using the required conditions in Eq. (5.115) and the binomial expansion of $(1 + \rho^2)^{-\frac{1}{2}}$ for $\rho < 1$ as above. Here we compare coefficients term by term as before and see *all* $B_n = 0$, and the $A_n = 0$ for *odd* n; for even $n = 2m$ we have,

$$(5.122) \qquad A_0 = 1, \qquad A_{2m} = (-1)^m \frac{1\cdot3\cdot5\ldots(2m+1)!}{[2(m!)]^2}, \qquad m = 0, 1, \ldots$$

and hence *for $\rho < 1$*

$$(5.123) \qquad \bar{U} = \sum_{m=0}^{\infty} A_{2m}\rho^{2m}P_{2m}(\cos\theta)$$

with the A_{2m} given above. (Note again $\bar{U}$ is even in θ, or $z = r\cos\theta$.)

Example (7)

Coherent scattering problems; plane wave expansions. There exists a great variety of physical problems in which some sort of plane wave is incident on an "object" of some kind and is *scattered* by the object *without change* in frequency of the wave. The scalar wave equation in three dimensions is the appropriate equation for the simplest class of problems and if we assume a *sinusoidal* time dependence $Ue^{i\omega t}$ with angular frequency ω this yields the Helmholtz equation,

$$(5.124) \qquad \nabla^2 U + \frac{\omega^2}{c^2}U = 0$$

for the space dependent part of the wavefunction.

Since the equation is *linear* we seek a solution in the general form, $(k = \omega/c)$

$$(5.125) \qquad U = A e^{-ikz} + W(r, \theta, \phi)$$

where the first term represents the incident plane wave with wave number $k = 2\pi/\lambda$, where λ equals the wavelength, moving in the $+z$ direction, and the function $W(r, \theta, \phi)$ represents the "scattered wave." Here we have expressed W in terms of spherical coordinates. Actually the *choice of coordinate system is dictated by the shape* of the *scattering object*, but for *approximate far* solutions these are always used.

For a spherical object we use spherical coordinates. Carrying through the straightforward separation of variables, or eigenfunction, procedure we find that a general form for W (with no *zero* separation constants imposed) is:

$$(5.126) \qquad W(r, \theta, \phi) = \sum_{n=0}^{\infty}\sum_{m=0}^{n} A_{mn} e^{im\phi}P_n^m(\cos\theta)\frac{H_{n+\frac{1}{2}}^{(2)}[(\omega/c)r]}{\sqrt{r}}$$

Here we have excluded the functions, Q_n^m, since these are singular on isolated lines in the domain. Also note that to achieve an *outgoing* wave at large r we have selected $H_{n+\frac{1}{2}}^{(2)}$ instead of $H_{n+\frac{1}{2}}^{(1)}$. Also note that here these functions occur only in the *half-integer orders*.

The next step in the solution is to expand the plane wave, now written as $Ae^{-ikr\cos\theta}$ in the *same* set of functions as appear in W. For example in the present case we can employ the expansion given in Eq. (4.81) of Chapter Four, i.e.,

$$(5.127) \qquad e^{-ikz} = e^{ikr\cos\theta} = \sum_{n=0}^{\infty} (2n+1)i^n P_n(\cos\theta)\left(\frac{\pi}{2kr}\right)^{\frac{1}{2}} J_{n+\frac{1}{2}}(kr)$$

Thus we now have Eq. (5.125) in suitable form to fit the boundary conditions on the *surface* of the scattering object. That is, say in acoustics for example, we may require

$$(5.128) \qquad \frac{\partial U}{\partial r} = 0, \qquad (r = R)$$

on the surface $r = R$ of a *spherical* scattering object. If the object were say a *cylinder* instead of a sphere we would have carried through the whole analysis using *cylindrical coordinates* instead of spherical coordinates.

In any event the procedure is the same and fitting the boundary conditions on the surface of the scatterer determines the unspecified constants, A_{nm}, in the function W. It is quite evident in the case of the sphere above for example that *axial symmetry* exists and only the A_{0n} are nonzero.

EXECUTE PROBLEM SET (5-3)

Example (8)

Heat loss from a cylindrical, fluid-filled pipe. This illustrates the solution of a boundary value problem in an *inhomogeneous domain*.

We assume no variation of temperature along the pipe; the fluid is stationary. The fluid has conductivity, specific heat, and density, k_1, c_1, and ρ_1, respectively, while those of the pipe material are k_2, c_2, and ρ_2, respectively. Let the temperature at any point in the fluid be $T_1(r, t)$, and that at any point in the pipe wall be $T_2(r, t)$ at time t. Here we also note that due to the axial symmetry there is no angular dependance in either T_1 or T_2. We use r as the radial coordinate from the axis of the pipe.

Thus we have:

$$(5.129) \qquad k_1 \frac{\partial}{\partial r}\left(r\frac{\partial T_1}{\partial r}\right) = C_1\rho_1\frac{\partial T_1}{\partial t}, \qquad 0 < r < a$$

and

$$(5.130) \qquad k_2 \frac{\partial}{\partial r}\left(r\frac{\partial T_2}{\partial r}\right) = C_2\rho_2\frac{\partial T_2}{\partial t}, \qquad a < r < b$$

as the partial differential equations appropriate to the two domains. The boundary conditions are:

$$(5.131) \qquad \left.\begin{array}{r} k_1\dfrac{\partial T_1}{\partial r} = k_2\dfrac{\partial T_2}{\partial r} \\[2mm] T_1 = T_2 \end{array}\right\} \quad \text{at } r = a, \qquad t > 0$$

$$(5.132) \qquad -k_2\frac{\partial T_2}{\partial r} = h(T_2 - T_0), \qquad \text{at } r = b, \qquad t > 0$$

and the initial condition is here assumed to be some general *axially symmetric* form,

$$(5.133) \qquad \left. \begin{array}{l} T_1 = f_1(r) \\ T_2 = f_2(r) \end{array} \right\} t = 0$$

where f_1 and f_2 must be consistent with Eq. (5.131).

From the physical point of view Eq. (5.131) says that the temperature is a *single valued* function at $r = a$, and that heat conducted *to* the interface from one side is conducted *from* the interface on the other side at the *same rate*, i.e., heat does not accumulate in a "mathematical surface" $r = a$. The condition in Eq. (5.132) is "Newton's law of cooling" and says that the heat is radiated from the surface $r = b$ at a rate, $h(T_2 - T_0)$, equal to the rate at which it reaches the surface by conduction [T_0 equals ambient temperature of the surrounding air.]

With the definitions:

$$\rho = \frac{r}{a}, \qquad \beta = \frac{b}{a}, \qquad \tau = \frac{k_1 t}{C_1 \rho_1 a^2}, \qquad \gamma^2 = \frac{C_2 \rho_2 k_1}{C_1 \rho_1 k_2}$$

$$(5.134) \qquad U_1 = \frac{T_1 - T_0}{T_0}, \qquad U_2 = \frac{T_2 - T_0}{T_0}, \qquad \frac{k_2}{k_1} = k, \qquad \delta = \frac{ha}{k_2}$$

$$F_1(\rho) = \frac{f_1(r) - T_0}{T_0}, \qquad F_2(\rho) = \frac{f_2(r) - T_0}{T_0}$$

the problem appears in the dimensionless form:

$$(5.135) \qquad \frac{1}{\rho} \frac{\partial}{\partial \rho}\left(\rho \frac{\partial U_1}{\partial \rho}\right) = \frac{\partial U_1}{\partial \tau}, \qquad 0 < \rho < 1, \qquad \tau > 0$$

$$(5.136) \qquad \frac{1}{\rho} \frac{\partial}{\partial \rho}\left(\rho \frac{\partial U_2}{\partial \rho}\right) = \gamma^2 \frac{\partial U_2}{\partial \tau}, \qquad 1 < \rho < \beta, \qquad \tau > 0$$

$$(5.137) \qquad \left. \begin{array}{l} \dfrac{\partial U_1}{\partial \rho} = k \dfrac{\partial U_2}{\partial \rho} \\[2mm] U_1 = U_2 \end{array} \right\} \rho = 1, \qquad \tau > 0$$

$$(5.138) \qquad \frac{\partial U_2}{\partial \rho} + \delta U_2 = 0, \qquad \rho = \beta, \qquad \tau > 0$$

$$(5.139) \qquad \left. \begin{array}{l} U_1 = F_1(\rho) \\ U_2 = F_2(\rho) \end{array} \right\} \tau = 0$$

and $F_1(\rho)$ and $F_2(\rho)$ satisfy Eq. (5.137) also.

To solve a problem of this type we first construct a formal solution to *each* of the differential equations *but* we cannot determine all parameters in either solution separately. The significance of this becomes evident as follows.

By the usual separation of variables or eigenfunction procedure, we find that U_1 can be formed from

$$(5.140) \qquad \left\{ \begin{array}{ll} A J_0(\lambda \rho) + B Y_0(\lambda \rho), & e^{-\lambda^2 \tau} \qquad \lambda \neq 0 \\ A' + B' \ln \rho & \lambda = 0 \end{array} \right.$$

where λ is the separation constant. Immediately we see that both B and B' must be zero because our solution must be *finite* at $\rho = 0$. Thus we write

(5.141)
$$U_1 = \sum_{n=1}^{\infty} A_n J_0(\lambda_n \rho)\, e^{-\lambda_n^2 \tau}$$

where we tacitly assume that a spectrum $\lambda_1, \lambda_2, \ldots$ etc. for λ can be determined. Note that $\lambda_n = 0$ is included in this sum as a *possibility*.

Turning now to the function U_2 we find in the same way,

(5.142)
$$\begin{cases} \bar{A} J_0(\lambda' \gamma \rho) + \bar{B} Y_0(\lambda' \gamma \rho), & e^{-\gamma^2 \lambda'^2 \tau} & \lambda' \neq 0 \\ \bar{A}_0 + \bar{B}_0 \ln \rho, & & \lambda' = 0 \end{cases}$$

Even a cursory examination of the number of unspecified constants in these solutions reveals that we have more parameters than should be required to satisfy all of the given conditions. Thus now make the tacit assumption, (actually just a redefinition of λ').

(5.143)
$$\lambda_n = \gamma \lambda'_n$$

Next note that if we substitute our solutions for U_1 and U_2 into Eq. (5.137) these can hold for *all values of* τ only if corresponding coefficients of the exponentials, $e^{-\lambda_n^2 \tau}$, are equal in the two series appearing on each side of these equations. Thus we require $\bar{A}_0 = 0$ and $\bar{B}_0 = 0$, and:

(5.144)
$$\begin{cases} A_n J_0(\lambda_n) = \bar{A}_n J_0(\lambda_n) + \bar{B}_n Y_0(\lambda_n) \\ A_n J_1(\lambda_n) = k[\bar{A}_n J_1(\lambda_n) + \bar{B}_n Y_1(\lambda_n)] \end{cases}$$

From these we find $\bar{A}_n$ and $\bar{B}_n$ in terms of A_n.

The spectrum of λ_n is found when we substitute our solution for U_2 into the condition in Eq. (5.138) at $\rho = \beta$. Noting that $\bar{A}_n$ and $\bar{B}_n$ are already expressed as proportional to A_n, i.e., from Eq. (5.144)

(5.145)
$$\bar{A}_n = A_n \frac{k J_0(\lambda_n) Y_1(\lambda_n) - J_1(\lambda_n) Y_0(\lambda_n)}{k[J_0(\lambda_n) Y_1(\lambda_n) - J_1(\lambda_n) Y_0(\lambda_n)]}$$

and

(5.146)
$$\bar{B}_n = A_n \frac{J_0(\lambda_n) J_1(\lambda_n)(1 - k)}{k[J_0(\lambda_n) Y_1(\lambda_n) - J_1(\lambda_n) Y_0(\lambda_n)]}$$

or

(5.147)
$$\bar{A}_n = A_n S(\lambda_n, k)$$

and,

(5.148)
$$\bar{B}_n = A_n S'(\lambda_n, k)$$

where S and S' are the indicated functions above.

With these substituted into U_2 and then U_2 substituted into Eq. (5.138) we see that to be true for *all* τ the series terms must vanish term by term. Thus the λ_n are the roots of:

(5.149)
$$\begin{cases} \frac{\partial}{\partial \rho}[S(\lambda_n, k) J_0(\lambda_n \rho) + S'(\lambda_n, k) Y_0(\lambda_n \rho)] \\ + \delta[S(\lambda_n, k) J_0(\lambda_n \rho) + S'(\lambda_n, k) Y_0(\lambda_n \rho)] \end{cases} = 0, \quad \text{at } \rho = \beta$$

which has an infinity of solutions.

Finally the constants A_n are determined by using the orthogonality property of the various Bessel functions involved, i.e., the $J_0(\lambda_n\rho)$ in U_1 and the linear combination† of $SJ_0(\lambda_n\rho) + S'Y_0(\lambda_n\rho)$ in U_2, as we fit U_1 and U_2 to f_1 and f_2 after putting $\tau = 0$, in Eq. (5.139).

EXECUTE PROBLEM SET (5-4)

GENERAL FORMAT FOR SOLUTION OF LINEAR HOMOGENEOUS BOUNDARY VALUE PROBLEMS

The linear homogeneous boundary value problems appear in the general form:

$$(5.150) \qquad OU = (O_x + O_\xi + \ldots)U = 0$$

where O is a *linear* operator‡ of some sort such as, $\nabla^2 - \partial/\partial t$, for example. If, as indicated here, the operator can be written as a *sum* of linear operators of some kind then U can as a rule be constructed as a *product* of functions V_x, V_ξ etc. such that V_x for example is an eigenfunction of the operator O_x,

$$(5.151) \qquad O_x V_x = a_x V_x$$

where a_x is the *eigenvalue* of O_x corresponding to the *eigenfunction* V_x. Thus Eq. (5.150) above then specifies a *constraint* on the eigenvalues of the operators,

$$(5.152) \qquad O(V_x V_\xi \ldots) = (a_x + a_\xi + \ldots)(V_x V_\xi \ldots) = 0$$

or,

$$(5.153) \qquad (a_x + a_\xi + \ldots) = 0$$

i.e., as in Example (1), Eq. (5.20).

In general we find that if the *domain* of the solution is *finite* in a coordinate, say x, then the boundary condition will determine a *discrete spectrum* for the admissible values of the eigenvalue, a_x. If the domain is unbounded in the coordinate x then a *continuous* spectrum is generally admitted for a_x. Cases of the latter type have *not* been considered explicitly in this chapter but will be in connection with *integral transforms* in subsequent chapters.

Since the differential equation, Eq. (5.150) is *linear* we can form a solution by *adding* many such product solutions, as just described, with some unspecified multiplying constant on each product solution in the sum. Finally these constants are adjusted to fit the remainder of the boundary conditions *not* used in fixing the eigenvalue spectra. The latter procedure generally makes use of the orthogonality properties of some of the eigenfunctions.

This brief outline of the methods employed in this chapter points up one very important fact. Since the *form* of the operator O in Eq. (5.150) above depends upon the choice of coordinate systems we *must* choose a coordinate system in which O is separable as indicated in order to use this procedure. We also must try to adjust our choice of coordinate systems to meet the symmetry of the given physical problem in order to simplify fitting the boundary conditions. In many cases these objectives are in direct conflict. In such cases we *must* choose the separable route if we wish to use the methods of this chapter.

A great number of the important classical problems of physics and engineering can be reduced to solving the *Helmholtz equation*, i.e., if we put a time dependence $e^{i\omega t}$ in the wave equation, or $e^{-\lambda t}$ in the diffusion equation, etc.

---

† See Problems (1) and (2) of Chapter Four.
‡ O is a linear operator if: $O(AU_1 + BU_2) = A(OU_1) + B(OU_2)$ where A and B are constants.

The Helmholtz equation is *separable* in eleven different curvilinear, orthogonal coordinate systems. However the resulting ordinary differential equations correspond in most cases to functions which have not been so extensively studied and tabulated as have the trigonometric, hyperbolic, Bessel, and Legendre functions. For this reason we usually attempt to formulate most problems in rectangular, cylindrical, or spherical coordinates. The student should consult the references at the end of Chapter Four for more extensive information on the other special functions; the small volume by Magnus and Oberhettinger is especially compact and at the same time rather complete.

TAKING ADVANTAGE OF THE LINEARITY OF A PROBLEM

We call attention to an important characteristic of *linear* boundary value problems which has not been stressed in the problems thus far, namely that the solution of a given *linear* boundary value problem can be expressed as the *sum of the solutions of two or more distinct boundary value problems*. We have, in *every* example, seen a solution of the differential equation formed as a linear combination of solutions, *each* of which in general satisfied the boundary conditions of the problem. The procedure suggested here is a further generalization of this concept.

For example, suppose we are solving a linear partial differential equation, for a function U,

$$(5.154) \qquad\qquad OU = 0$$

in a region R, subject to two boundary conditions,

$$(5.155) \qquad\qquad \begin{cases} U = F_1, & \text{on } S_1 \\ U = F_2, & \text{on } S_2 \end{cases}$$

where S_1 and S_2 are disjoint and make up the boundary of the region R. This solution can obviously be constructed as

$$(5.156) \qquad\qquad U = U_1 + U_2$$

where U_1 and U_2 are solutions of the *two distinct boundary value problems*,

$$(5.157) \qquad\qquad OU_1 = 0, \quad \text{in } R$$

with,

$$(5.158) \qquad\qquad \begin{cases} U_1 = F_1, & \text{on } S_1 \\ U_1 = 0, & \text{on } S_2 \end{cases}$$

and

$$(5.159) \qquad\qquad OU_2 = 0, \quad \text{in } R$$

with,

$$(5.160) \qquad\qquad \begin{cases} U_2 = 0, & \text{on } S_1 \\ U_2 = F_2, & \text{on } S_2 \end{cases}$$

These procedures prove useful throughout the remainder of the text. The student should examine the problems of this chapter to see how this could be applied.

PROBLEMS

Set (5-1)

(1) Steady temperature in a rectangular plate.

$$\frac{\partial^2 T}{\partial x^2} + \frac{\partial^2 T}{\partial y^2} = 0, \qquad \begin{array}{l} 0 < x < a \\ 0 < y < b \end{array}$$

with

$$T(0, y) = 0, \qquad T(a, y) = 0$$

$$T(x, b) = 0, \qquad T(x, 0) = F(x)$$

Do the general case here, then the special case for $F(x) = Ax$, $A = $ constant.

(2) Vibration of a rectangular membrane,

$$\frac{\partial^2 Z}{\partial x^2} + \frac{\partial^2 Z}{\partial y^2} = \frac{1}{c^2} \frac{\partial^2 Z}{\partial t^2} \qquad \begin{array}{l} 0 < x < a \\ 0 < y < b \\ t > 0 \end{array}$$

with,

$$Z(0, y, t) = Z(x, 0, t) = Z(a, y, t) = Z(x, b, t) = 0$$

and,

$$Z(x, y, 0) = F(x, y)$$

$$\frac{\partial Z(x, y, 0)}{\partial t} = V(x, y)$$

Do the general case, then the special case of $V = 0$, $F = Axy(a - x)(b - y)$, $A = $ constant.

(3) Vibration of a loaded string. For a string with a load proportional to the distance from one end,

$$\frac{\partial^2 y}{\partial x^2} + Ax = \frac{1}{c^2} \frac{\partial^2 y}{\partial t^2}, \qquad \begin{array}{l} 0 < x < L \\ t > 0 \end{array}$$

$$A = \text{constant}$$

where the term Ax is the load term. The boundary and initial conditions are,

$$y(0, t) = y(L, t) = 0$$

and

$$y(x, 0) = F(x), \qquad \frac{\partial y(x, 0)}{\partial t} = 0$$

Hint: Let

$$y(x, t) = Y(x, t) + \psi(x)$$

then show that

$$\frac{d^2 \psi}{dx^2} = -Ax$$

and

$$\psi(0) = \psi(L) = 0$$

a homogeneous problem in $Y(x, t)$ is obtained. Solve the general case then the special case of $F(x) = Bx(L - x)$, $B = $ constant.

(4) Linear flow of heat in a wire with radiation losses to the surroundings:

$$k\frac{\partial^2 T}{\partial x^2} - h(T - T_0) = C\rho\frac{\partial T}{\partial t}, \qquad \begin{array}{l} 0 < x < L \\ t > 0 \end{array}$$

where T_0 equals the constant ambient temperature of the surroundings, and,

$$T(0, t) = T_0 = \text{constant}$$

$$T(L, t) = T_L = \text{constant}$$

$$T(x, 0) = F(x)$$

Here h is the "radiation constant." Solve the general problem then the special case of $T_0 = T_L = 0$ and $F(x) = B = $ constant.

Set (5-2)

(1) Vibration of a circular membrane:

$$\frac{1}{r}\frac{\partial}{\partial r}\left(r\frac{\partial Z}{\partial r}\right) + \frac{1}{r^2}\frac{\partial^2 Z}{\partial \theta^2} = \frac{1}{c^2}\frac{\partial^2 Z}{\partial t^2}$$

$$0 \le r < R, \qquad 0 \le \theta < 2\pi, \qquad t > 0$$

with,

$$Z(R, \theta, t) = 0$$

and,

$$Z(r, \theta, 0) = F(r, \theta)$$

$$\frac{\partial Z}{\partial t}(r, \theta, 0) = V(r, \theta)$$

Solve the general problem, then the special case of $V = 0$, $F = A(R^2 - r^2)$. Compare the lowest frequency of the latter solution to that in the special case of the rectangular membrane of Problem (2) in Set (5-1) above, i.e., for $a = b = \sqrt{2R}$ in Problem (2) above.

(2) Clamped circular *disk*. Here we have the *fourth-order* equation,

$$\left[\frac{1}{r}\frac{\partial}{\partial r}\left(r\frac{\partial}{\partial r}\right) + \frac{1}{r^2}\frac{\partial^2}{\partial \theta^2}\right]^2 U = \frac{-1}{c^2}\frac{\partial^2 U}{\partial t^2}$$

$$0 \le r < R, \qquad 0 < \theta < 2\pi, \qquad t > 0$$

with

$$U(R, \theta, t) = \frac{\partial U}{\partial r}(R, \theta, t) = 0$$

and

$$U(r, \theta, 0) = F(r, \theta)$$

$$\frac{\partial U(r, \theta, 0)}{\partial t} = V(r, \theta)$$

Hint: Since U must be single valued and periodic with period 2π in θ and we seek periodic solutions in time, put $U = \mathcal{R}(r)(A \sin n\theta + B \cos n\theta) \cdot (C \sin \omega t + D \cos \omega t)$, and show that

$$\left\{\frac{d^2}{dr^2} + \frac{1}{r}\frac{d}{dr} - \frac{n^2}{r^2}\right\}^2 \mathcal{R} + \frac{\omega^2}{c^2}\mathcal{R} = 0$$

and this factors as,

$$\left\{\frac{d^2}{dr^2} + \frac{1}{r}\frac{d}{dr} + \left(\frac{\omega}{c} - \frac{n^2}{r^2}\right)\right\}\left\{\frac{d^2}{dr^2} + \frac{1}{r}\frac{d}{dr} - \left(\frac{\omega}{c} + \frac{n^2}{r^2}\right)\right\}\mathcal{R} = 0$$

Then show that this is equivalent to the *two* second-order equations

$$\frac{d^2\mathcal{R}_1}{dr^2} + \frac{1}{r}\frac{d\mathcal{R}_1}{dr} + \left(\frac{\omega}{c} - \frac{n^2}{r^2}\right)\mathcal{R}_1 = 0$$

$$\frac{d^2\mathcal{R}_2}{dr^2} + \frac{1}{r}\frac{d\mathcal{R}_2}{dr} - \left(\frac{\omega}{c} + \frac{n^2}{r^2}\right)\mathcal{R}_2 = 0$$

where $\mathcal{R} = \mathcal{R}_1 + \mathcal{R}_2$, and hence has a linear combination of Bessel functions as a solution. Just show how to construct the general solution for arbitrary initial conditions.

(3) Vibration of the annular segment of a thin membrane, $a < r < b, 0 < \theta < \pi/4, t > 0$:

$$\frac{1}{r}\frac{\partial}{\partial r}\left(r\frac{\partial Z}{\partial r}\right) + \frac{1}{r^2}\frac{\partial^2 Z}{\partial \theta^2} = \frac{1}{c^2}\frac{\partial^2 Z}{\partial t^2}$$

with,

$$0 = Z(a, \theta, t) = Z(b, \theta, t) = Z(r, 0, t) = Z(r, \pi/4, t)$$

and,

$$Z(r, \theta, 0) = F(r, \theta)$$

$$\frac{\partial Z(r, \theta, 0)}{\partial t} = V(r, \theta)$$

Just show how to construct the general solution for arbitrary initial conditions.

(4) Steady heat conduction in cylindrical solid segment of length L, i.e.,

$$\frac{1}{r}\frac{\partial}{\partial r}\left(r\frac{\partial T}{\partial r}\right) + \frac{1}{r^2}\frac{\partial^2 T}{\partial \theta^2} + \frac{\partial^2 T}{\partial z^2} = 0$$

in the domain,

$$0 < r < R$$

$$0 < z < L$$

$$0 < \theta < \pi/2$$

with,

$$T(r, 0, z) = T(R, \frac{\pi}{2}, z) = 0$$

$$T(r, \theta, z) = T(r, \theta, L) = 0$$

$$T(r, \theta, 0) = T_1$$

(5) Radiation from the surfaces of a thin circular disk heated uniformly on its edge at a constant rate:

$$k\frac{1}{r}\frac{\partial}{\partial r}\left(r\frac{\partial T}{\partial r}\right) - 2h(T - T_0) = C\rho\frac{\partial T}{\partial t}$$

$$0 < r < R, \qquad t > 0$$

Here h is the "radiation constant" and T_0 is the uniform ambient temperature of the surroundings. We have,

$$T(r, 0) = T_0$$

and

$$\frac{\partial T}{\partial r}(R, t) = \frac{Q}{2\pi R\varepsilon k} = \text{constant}$$

where ε equals the thickness of the disk and Q is the quantity of heat per unit time supplied to the disk. Construct the general solution for $T(r, 0) = F(r)$, then that for the special initial condition above.

Set (5-3)

(1) Steady acoustic waves in an infinite medium generated by a vibrating sphere:

$$\frac{1}{r^2}\frac{\partial}{\partial r}\left(r^2\frac{\partial U}{\partial r}\right) + \frac{1}{r^2 \sin^2 \theta}\frac{\partial^2 U}{\partial \phi^2} + \frac{1}{r^2 \sin \theta}\frac{\partial}{\partial \theta}\left(\sin \theta \frac{\partial U}{\partial \theta}\right) = \frac{1}{c^2}\frac{\partial^2 U}{\partial t^2}$$

in the domain

$$r > R$$

$$t > 0$$

with the boundary condition on the sphere, $r = R$,

$$\frac{\partial U}{\partial r} = \frac{\partial R}{\partial r} = f(\theta, \phi) e^{i\omega t}$$

Here the normal velocity of the fluid at the surface of the sphere is equal to the normal velocity of the surface of the sphere and is represented approximately by this relationship. (Use the asymptotic condition of outgoing waves at large r.) Formulate the general solution and then evaluate for the special case $f(\theta, \phi) = A \cos \theta$, $A = \text{constant}$.

(2) Potential due to a spherical shell of electric charge of radius R, with charge density $\sigma(\theta, \phi)$:

$$0 = \frac{1}{r^2}\frac{\partial}{\partial r}\left(r^2\frac{\partial V}{\partial r}\right) + \frac{1}{r^2 \sin^2 \theta}\frac{\partial^2 V}{\partial \phi^2} + \frac{1}{r^2 \sin \theta}\frac{\partial}{\partial \theta}\left(\sin \theta \frac{\partial V}{\partial \theta}\right)$$

with $V = V_1(r,\theta,\phi)$ for $r < R$ and $V = V_2(r,\theta,\phi)$ for $r > R$ the boundary conditions are

$$V_1(R,\theta,\phi) = V_2(R,\theta,\phi) \text{ and } \frac{\partial V_1}{\partial r}(R,\theta,\phi) - \frac{\partial V_2}{\partial r}(R,\theta,\phi) = \frac{\sigma(\theta,\phi)}{\epsilon}$$

with $V_2(r,\theta,\phi) \to 0$ as $r \to \infty$.

(3) For the special case of the above problem in which the charge distribution is *axially* symmetric, $\sigma = \sigma(\theta)$, exhibit the form of the solution. For $\sigma(\theta) = A \cos^2 \theta$ evaluate the constants in this solution using the orthogonality properties of the functions in the solution.

(4) For the charge distribution, $\sigma(\theta) = A \cos^2 \theta$ of Problem (3) above compute the potential *on the axis* using "Coulomb's law" as:

$$V(z) = \frac{1}{4\pi\epsilon} \int_0^{2\pi} \int_0^{\pi} \frac{\sigma(\theta')R^2 \sin \theta' \, d\theta' \, d\phi}{\sqrt{R^2 + z^2 - 2Rz \cos \theta'}}$$

Then use the method of Example (6) of this chapter to construct $V(r, \theta, \phi)$ everywhere. Compare this to your solution of Problem (3) above.

(5) Find the temperature at any point in a sphere at time t after being dropped into a uniform heat bath:

$$\nabla^2 T = \frac{C\rho}{k} \frac{\partial T}{\partial t}$$

with

$$T(R, \theta, \phi, t) = T_B = \text{constant}$$

$$T(r, \theta, \phi, 0) = F(r, \theta, \phi)$$

consider the special case in which $F(r, \theta, \phi) = T_0 = \text{constant}$, *after* you have exhibited the general solution.

(6) For the scattering of acoustic waves from the surface of a sphere *complete* the solution for scattering of the plane wave by the sphere as initiated in *Example 7* of this chapter, i.e., evaluate the constants A_{mn} of Eq. (5.126). Then evaluate the quantity, for *large r* (asymptotic form),

$$\sigma = r^2 \frac{W^*W}{A^2}$$

where W is as defined in Eq. (5.125) and A is the amplitude of the incident plane wave. This quantity, σ, is the differential scattering cross section of the sphere, i.e., the fraction of the incident energy scattered into the solid angle,† $\sin \theta \, d\theta \, d\phi = d\Omega$.

(7) Consider steady waves from a *point source* incident on a sphere and set up and solve the acoustic scattering problems, i.e., the incident wave is described by:

$$U = \frac{Ae^{i(\omega t - kr')}}{4\pi r'}$$

where r' is the distance from the point source.

† See Appendix (VIII) for definition of solid angle.

Set (5-4)

(1) Show that orthogonality conditions can indeed be applied to complete the solution of Example Problem (8) of this chapter.

(2) Solve the problem of *steady state* heat conduction in the infinite slab composed of two layers, i.e.,

$$\frac{\partial^2 T_1}{\partial x^2} = 0, \qquad 0 < x < L_1$$

$$\frac{\partial^2 T_2}{\partial x^2} = 0, \qquad L_1 < x < L$$

with

$$T_1(0) = T_a$$

$$\left\{ \begin{array}{c} T_1 = T_2 \\ k_1 \dfrac{\partial T_1}{\partial x} = k_2 \dfrac{\partial T_2}{\partial x} \end{array} \right\} \quad \text{at } x = L_1$$

$$T_2(L) = T_b$$

(3) Taking the solution of Problem (2) above as the *initial* temperature distribution in the slab suppose *both* surfaces $x = 0$ and $x = L$ suddenly given the temperature T_a. (Assume the specific heats and densities of the two layers to be C_1, ρ_1 and C_2, ρ_2.) Find $T(x, t)$ throughout the slab.

REFERENCES

H. Bateman, *Partial Differential Equations of Mathematical Physics* (Dover Publications, Inc., New York, 1964).

R. V. Churchill, *Fourier Series and Boundary Value Problems* (McGraw-Hill Book Company, Inc., New York, 1941).

R. Courant and D. Hilbert, *Methods of Mathematical Physics*, Vol. II (Interscience Publishers, Inc., New York, 1962).

J. W. Dettman, *Mathematical Methods in Physics and Engineering* (McGraw-Hill Book Company, Inc., New York, 1962).

G. Goertzel and N. Tralli, *Some Mathematical Methods of Physics* (McGraw-Hill Book Company, Inc., New York, 1960).

J. Irving and N. Millineux, *Mathematics in Physics and Engineering* (Academic Press Inc., New York, 1959).

H. Jefferys and B. Jefferys, *Methods of Mathematical Physics* (Cambridge University Press, New York, 1956).

O. D. Kellog, *Foundations of Potential Theory* (Dover Publications, Inc., New York, 1953).

E. A. Kraut, *Fundamentals of Mathematical Physics* (McGraw-Hill Book Company, Inc., New York, 1967).

N. N. Lebedev, I. P. Skalskaya, and Y. S. Uflyand, *Problems of Mathematical Physics*, (translated by R. A. Silverman), Prentice-Hall Inc., Englewood Cliffs, N.J. (1965).

P. M. Morse and H. Feshbach, *Methods of Theoretical Physics* (McGraw-Hill Book Company, Inc., New York, 1953).

A. Sommerfeld, *Partial Differential Equations in Physics* (Academic Press Inc., New York, 1949).

CHAPTER SIX

Elementary applications of the Laplace transform

In this chapter we introduce the *integral transform* of Laplace and develop the method of application which makes use of a *table* of inverse transforms. In Chapter Eight we develop the more sophisticated aspects of Laplace transforms in terms of the complex inversion integral and in Chapter Nine we treat transform methods in general. The rather important domain of application of the Laplace transform to boundary value problems is introduced in the present chapter. We also illustrate the great utility of this tool in other areas.

DEFINITION OF THE LAPLACE TRANSFORM AND DEVELOPMENT OF ITS PROPERTIES

Given any function $F(t)$ we can perform the integration,

$$(6.1) \qquad f(p) = \int_0^\infty e^{-pt} F(t)\, dt$$

provided that the integral exists. It can be shown that this integral does exist if $F(t)$ is *sectionally continuous* and of exponential order as $t \to \infty$, i.e., if $F(t)$ satisfies these criteria then there exists some domain of p for which the integral exists.†

---

† *Sectionally continuous* means the domain of t can be subdivided into intervals in each of which $F(t)$ is continuous. *Exponential order* means there exists some value, γ, such that $\lim\limits_{t \to \infty} |F(t)| e^{-\gamma t}$ (real part $\gamma > 0$) is bounded.

The function $f(p)$ so constructed is called the *Laplace transform* of $F(t)$. Simply by selecting various functions $F(t)$ and carrying out the indicated integration we can construct a table of *transform pairs, functions, $F(t)$, and their *transforms, $f(p)$.*

Also we can develop some *general properties* of the pairs, $F(t)$ and $f(p)$, directly from the definition. Let us introduce the notation

(6.2)
$$f(p) = L\{F(t)\}_t$$

to represent the operation in Eq. (6.1), and also let,

(6.3)
$$F(t) = L^{-1}\{f(p)\}_p$$

denote the *inverse transform*. As stated above we do not develop in this chapter an explicit procedure for constructing the L^{-1} operator but obviously it *exists*. Here we say $F(t)$ is the function whose Laplace transform is $f(p)$.

It follows from the definition above that:

(6.4)
$$L\{F_1(t) + F_2(t)\} = L\{F_1(t)\} + L\{F_2(t)\}$$

and that,

(6.5)
$$L\{AF_1(t)\} = AL\{F_1(t)\}$$

where A is any constant, and $F_1(t)$ and $F_2(t)$ are any functions whose individual transforms exist.

Also observe that if we integrate once by parts in Eq. (6.1) we obtain,

(6.6)
$$-\frac{e^{-pt}}{p}F(t)\Big|_0^\infty + \frac{1}{p}\int_0^\infty e^{-pt}\frac{dF(t)}{dt}\,dt = f(p)$$

or, since $F(t)$ must be of exponential order as $t \to \infty$ and sectionally continuous, (and finite) at $t = 0$, this appears as,

(6.7)
$$\int_0^\infty e^{-pt}\frac{dF(t)}{dt}\,dt = pf(p) - F(0)$$

But the left member here is by our definition just $L\{dF/dt\}$, so,

(6.8)
$$L\left\{\frac{dF(t)}{dt}\right\} = pf(p) - F(0)$$

where $f(p)$ still denotes $L\{F(t)\}$.

By similar processes one can establish a great variety of general properties of the Laplace transform. A rather extensive sampling of these general results is listed in the first part of Table 6.1. The second part of this table consists of a collection of functions, $F(t)$, which occur rather frequently in physical problems, and their Laplace transforms, $f(p)$. The first problem set of this chapter is designed to familiarize the student with this table, particularly some of the *general* properties of the Laplace transform.

EXECUTE PROBLEM SET (6-1)

Table 6-1.   Laplace Transforms.

	$F(t)$	$f(p)$
(1)	$F(t)$	$\int_0^\infty e^{-pt} F(t)\, dt$
(2)	$AF(t) + BG(t)$	$Af(p) + Bg(p)$
(3)	$\dfrac{dF(t)}{dt}$	$pf(p) - F(0)$
(4)	$\dfrac{d^{(n)} F(t)}{dt^{(n)}}$	$p^n f(p) - p^{n-1} F(0) - p^{n-2} F'(0)$ $- p^{n-3} F''(0) \ldots - F^{(n-1)}(0)$
(5)	$\displaystyle\int_0^t F(\tau)\, d\tau$	$\dfrac{1}{p} f(p)$
(6)	$\displaystyle\int_0^t \int_0^\tau F(\lambda)\, d\lambda\, d\tau$	$\dfrac{1}{p^2} f(p)$
(7)	$\displaystyle\int_0^t F_1(t - \tau) F_2(\tau)\, d\tau$	$f_1(p) f_2(p)$
(8)	$tF(t)$	$-\dfrac{df(p)}{dp}$
(9)	$t^n F(t)$	$(-1)^n f^{(n)}(p)$
(10)	$\dfrac{1}{t} F(t)$	$\displaystyle\int_p^\infty f(p')\, dp'$
(11)	$e^{at} F(t)$	$f(p - a)$
(12)	$\begin{cases} F(t - b), & t > b \\ \quad 0, & t < b \end{cases}$	$e^{-bp} f(p)$
(13)	$\dfrac{1}{c} F\!\left(\dfrac{t}{c}\right)$	$f(cp)$
(14)	(Periodic function) $F(t + T) = F(t)$	$\dfrac{\int_0^T e^{-pt} F(t)\, dt}{1 - e^{-pT}}$
(15)	$F(t + T) = -F(t)$	$\dfrac{\int_0^T e^{-pt} F(t)\, dt}{1 + e^{-pT}}$
(16)	1	$\dfrac{1}{p}$
(17)	t	$\dfrac{1}{p^2}$

TABLE 6-1.   **Laplace Transforms—(cont.).**

	$F(t)$	$f(p)$
(18)	t^n	$\dfrac{n!}{p^{n+1}}$
(19)	$\dfrac{1}{\sqrt{t}}$	$\sqrt{\dfrac{\pi}{p}}$
(20)	$\sqrt{t}$	$\dfrac{1}{2}\sqrt{\pi}\,p^{-\frac{3}{2}}$
(21)	$t^{n-\frac{1}{2}}$	$\dfrac{\sqrt{\pi}}{2^n}\dfrac{1\cdot 3\cdot 5\ldots(2n-1)}{p^{n+\frac{1}{2}}}$
(22)	e^{at}	$\dfrac{1}{p-a}$
(23)	te^{at}	$\dfrac{1}{(p-a)^2}$
(24)	$t^{n-1}e^{at}$	$\dfrac{(n-1)!}{(p-a)^n}$
(25)	$\dfrac{e^{at}-e^{bt}}{a-b}$	$\dfrac{1}{(p-a)(p-b)}$
(26)	$\sin at$	$\dfrac{a}{p^2+a^2}$
(27)	$\cos at$	$\dfrac{p}{p^2+a^2}$
(28)	$t\sin at$	$\dfrac{2ap}{(p^2+a^2)^2}$
(29)	$e^{at}\sin bt$	$\dfrac{b}{(p-a)^2+b^2}$
(30)	$t\cos at$	$\dfrac{p^2-a^2}{(p^2+a^2)^2}$
(31)	$e^{at}\cos bt$	$\dfrac{p-a}{(p-a)^2+b^2}$
(32)	$t^n\sin at$	$\dfrac{2^n an!p^n}{(p^2+a^2)^{n+1}}$
(33)	$e^{a^2t}\,\mathrm{erfc}\,(a\sqrt{t})$	$\dfrac{1}{\sqrt{p}(\sqrt{p}+a)}$

TABLE 6-1.   Laplace Transforms—(cont.).

	$F(t)$	$f(p)$
(34)	$e^{-at} \operatorname{erf}\left(\sqrt{b - a}\sqrt{t}\right)$	$\dfrac{\sqrt{b - a}}{(p + a)\sqrt{p + b}}$
(35)	$e^{-\frac{a+b}{2}t} I_0\left(\dfrac{a - b}{2}t\right)$	$\dfrac{1}{\sqrt{p + a}\sqrt{p + b}}$
(36)	$\dfrac{e^{-at}}{t} I_1(at)$	$\dfrac{\sqrt{p + 2a} - \sqrt{p}}{\sqrt{p + 2a} + \sqrt{p}}$
(37)	$J_0(at)$	$\dfrac{1}{\sqrt{p^2 + a^2}}$
(38)	$\dfrac{1}{t} J_k(at)$	$\dfrac{1}{ka^k}(\sqrt{p^2 + a^2} - p)^k, \quad k > 0$
(39)	$I_v(at)$	$\dfrac{1}{a^v} \dfrac{(p - \sqrt{p^2 - a^2})^v}{\sqrt{p^2 - a^2}}, \quad v > -1$
(40)	$J_0(2\sqrt{kt})$	$\dfrac{1}{p} e^{-\frac{k}{p}}$
(41)	$\dfrac{\cos 2\sqrt{kt}}{\sqrt{\pi t}}$	$\dfrac{1}{\sqrt{p}} e^{\frac{k}{p}}$
(42)	$\dfrac{\sin 2\sqrt{kt}}{\sqrt{\pi k}}$	$\dfrac{1}{p^{3/2}} e^{-\frac{k}{p}}$
(43)	$\dfrac{k}{2\sqrt{\pi t^3}} e^{-\frac{k^2}{4t}}$	$e^{-k\sqrt{p}}, \quad k > 0$
(44)	$\operatorname{erfc}\left(\dfrac{k}{2\sqrt{t}}\right)$	$\dfrac{1}{p} e^{-k\sqrt{p}} \quad k \geq 0$
(45)	$\dfrac{1}{\sqrt{\pi t}} e^{-\frac{k^2}{4t}}$	$\dfrac{1}{\sqrt{p}} e^{-k\sqrt{p}} \quad k \geq 0$
(46)	$-\operatorname{Ei}\left(-\dfrac{a}{t}\right)$	$\dfrac{2}{p} K_0(2\sqrt{ap})$
(47)	$\operatorname{erf}\left(\dfrac{t}{2k}\right)$	$\dfrac{1}{p} e^{k^2 p^2} \operatorname{erfc}(k\sqrt{p}), \quad k > 0$
(48)	$\dfrac{1}{2t} e^{-\frac{k^2}{4t}}$	$K_0(kp)$

APPLICATION OF THE LAPLACE TRANSFORM TO BOUNDARY VALUE PROBLEMS IN OPEN DOMAINS

At the end of Chapter Five we summarized the general procedures for solving linear homogeneous boundary value problems. There we pointed out that the eigenvalues of the eigenfunctions used to form the product solutions had *discrete spectra* for *bounded domains* but had a *continuous spectrum* for an infinite, or *unbounded domain*. Thus in *bounded* domains we were able to construct solutions in terms of *sums* over the spectra of these eigenvalues. Here we see that the Laplace transform offers a means for treating such problems in *open domains*. We develop the method by examples.

Example (1)

The long uniform string subject to the gravitational force. The differential equation describing the vertical displacement, $Y(x, t)$, of a point located at a horizontal distance, x, from one *fixed* end, at time t, is:

$$(6.9) \qquad \rho \frac{\partial^2 Y}{\partial t^2} = - \rho g + T \frac{\partial^2 Y}{\partial x^2}$$

where ρ is the mass per unit length of the string, T is the tension in the string, here assumed uniform and constant, and g is the acceleration of gravity.

The string is treated as semi-infinite in length, $0 \le x \le \infty$, and initially at rest on a horizontal support. If we suddenly remove the support, keeping the end $x = 0$ tied at its original position but let the end at infinity slide on a vertical axis which causes no vertical force on the string we have the boundary value problem:

$$(6.10) \qquad \frac{\partial^2 Y}{\partial t^2} = c^2 \frac{\partial^2 Y}{\partial x^2} - g, \qquad t > 0, 0 \le x < \infty$$

$$(6.11) \qquad \begin{cases} Y(x, 0) = 0 \\ \dfrac{\partial Y(x, 0)}{\partial t} = 0 \\ Y(0, t) = 0 \\ \lim_{x \to \infty} \dfrac{\partial Y}{\partial x}(x, t) = 0 \end{cases}$$

where $c^2 = T/\rho$.

We multiply *every* equation above by e^{-pt} and integrate on t from zero to infinity. Since this operation is interchangeable with differentiation with respect to x we obtain:

$$(6.12) \qquad p^2 f(x, p) = c^2 \frac{d^2 f}{dx^2}(x, p) - \frac{g}{p}, \qquad 0 \le x < \infty$$

with the boundary conditions,

$$(6.13) \qquad \begin{cases} f(0, p) = 0 \\ \lim_{x \to \infty} \dfrac{df}{dx}(x, p) = 0 \end{cases}$$

where $f(x, p)$ now denotes $L\{Y(x, t)\}_t$. Observe that the first two conditions in Eq. (6.11) were incorporated in the Laplace transform of Y''. Also note that where as our original differential equation was in *two* independent variables it now involves only the *one*, x. Thus we have an *ordinary* differential equation for solution.

Rearranging Eq. (6.12) it appears as,

(6.14)
$$\frac{d^2}{dx^2}\left(f + \frac{g}{p^3}\right) - \frac{p^2}{c^2}\left(f + \frac{g}{p^3}\right) = 0$$

and then the solution is immediately obvious as:

(6.15)
$$f + \frac{g}{p^3} = Ae^{-\frac{p}{c}x} + Be^{+\frac{p}{c}x}$$

Then to fit the last condition in Eq. (6.13) we must put $B = 0$ and to fit the first condition in Eq. (6.13) we must put $A = g/p^3$. Thus we have as the solution for $f(x, p)$,

(6.16)
$$f(x, p) = -\frac{g}{p^3} + e^{-\frac{x}{c}p}\frac{g}{p^3}$$

Now we employ Table 6.1 to construct the inverse transform,

(6.17)
$$Y(x, t) = L^{-1}\{f(x, p)\}$$

Note Entry Number (18) shows that g/p^3 has the inverse $gt^2/2$. Also note that Entry Number (12) indicates that in the second term of $f(x, p)$ above we have two cases to consider, thus,

(6.18)
$$L^{-1}\left\{e^{-\frac{x}{c}p}\frac{g}{p^3}\right\} = \begin{cases} \frac{1}{2}g\left(t - \frac{x}{c}\right)^2, & t > \frac{x}{c} \\ 0, & t < \frac{x}{c} \end{cases}$$

Consequently the final solution is:

(6.19)
$$Y(x, t) = \begin{cases} -\frac{1}{2}gt^2, & x > ct \\ \dfrac{gx}{c^2}\left(\dfrac{x}{2} - ct\right), & x < ct \end{cases}$$

The graph of this solution at some time $t > 0$ has the general appearance shown in Fig. 6.1. Thus the major portion of the string, $x > ct$ moves as a freely falling body starting from rest at $Y = 0$. The "message" that the end at $x = 0$ is tied to $Y = 0$ has traveled out on the string a distance $x = ct$ during the time t, so that the portion of the string between $0 < x < ct$ has a shape dependent on this end condition, while that part *beyond* $x = ct$ does not. This emphasizes the importance of the phase velocity c in wave-type phenomena.

The solution of a problem of this type would be quite impossible in terms of Fourier series as used in the last chapter, although as we will see later the Fourier integral could be adapted to the problem.

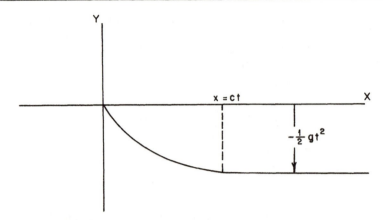

FIGURE 6.1. Semi-Infinite String with End Fixed Falling in a Uniform Gravity Field.

Example (2)

We consider heat conduction in a uniform semi-infinite rod with insulated surface and uniform initial temperature. Two different end conditions will be examined.

In dimensionless form the mathematical statement of the problem appears as:

(6.20)
$$\frac{\partial^2 U}{\partial \xi^2} = \frac{\partial U}{\partial \tau}, \qquad \begin{matrix} \xi > 0 \\ \tau > 0 \end{matrix}$$

(6.21)
$$\begin{cases} U(\xi, 0) = 0 \\ \lim_{\xi \to \infty} \dfrac{\partial U}{\partial \xi} = 0, \qquad \tau \geq 0 \end{cases}$$

Taking the Laplace transform with respect to τ as before we obtain, $(u = L\{U\})$

(6.22)
$$\frac{d^2 u}{d\xi^2} = pu$$

(6.23)
$$\begin{cases} \lim_{\xi \to \infty} \dfrac{du}{d\xi} = 0 \\ \text{Plus end condition on } u(\xi, p) \text{ at } \xi = 0 \end{cases}$$

Obviously the only solution of Eq. (6.22) satisfying the first condition of Eq. (6.23) is

(6.24)
$$u = Ae^{-\sqrt{p}\xi}$$

Thus we have a *general* form for $u(\xi, p)$ to which various end conditions at $\xi = 0$ can be applied.

One of the most general end conditions we could specify would be:

(6.25)
$$\alpha U(0, t) + \beta \frac{\partial U}{\partial \xi}(0, t) = F(t)$$

where α and β are given constants and $F(t)$ is a *specified function of t*. Taking the transform of this we get,

$$(6.26) \qquad \left(\alpha u + \beta \frac{du}{d\xi}\right)_{\xi=0} = f(p)$$

where $f(p)$ is the transform of $F(t)$. Hence, using Eq. (6.24) we have:

$$(6.27) \qquad (\alpha - \sqrt{p}\beta)A = f(p)$$

and therefore,

$$(6.28) \qquad u(\xi, p) = \frac{f(p)e^{-\sqrt{p}\xi}}{\alpha - \beta\sqrt{p}}$$

Here we have two distinct physical situations which arise as we put either $\alpha = 0$, or $\beta = 0$. For $\beta = 0$ the temperature at the end is a *specified function of time*, while for $\alpha = 0$, the heat flux into the end is a specified function of time. We will consider $\beta = 0$ and proceed to invert the transform for this special case.

We have, for $\beta = 0$,

$$(6.29) \qquad u = \frac{f(p)}{\alpha}e^{-\sqrt{p}\xi}$$

This has the form of the *product* of two transforms, one being $f(p)/\alpha$, the other being $e^{-\sqrt{p}\xi}$ which appears in our table as Entry Number (43). Now the product of two transforms has the general inverse as shown in Entry Number (7) of our table. Thus we see that,

$$(6.30) \qquad U(\xi, \tau) = \frac{1}{\alpha}\int_0^\tau F(\tau - \tau')\frac{\xi}{2\sqrt{\pi\tau'^3}}e^{-\frac{\xi^2}{4\tau'}}\,d\tau'$$

is the general solution for *any* specified end temperature. For the special case of $F = F_0$ a constant, this can be shown to reduce to:

$$(6.31) \qquad U(\xi, \tau) = \frac{F_0}{\alpha}\,\text{erfc}\left(\frac{\xi}{2\sqrt{\tau}}\right) = \frac{F_0}{\alpha}\left[1 - \text{erf}\left(\frac{\xi}{2\sqrt{\tau}}\right)\right]$$

which is an extremely convenient form since the error function is a tabulated function.

Example (3)

Potential Problems of Axial Symmetry in a Half-Space. Actually this example covers a rather broad class of axially symmetric boundary value problems in the half-space $z > 0$, but we make the demonstration specific by using Laplace's equation here.

Consider

$$(6.32) \qquad \frac{1}{r}\frac{\partial}{\partial r}\left(r\frac{\partial V}{\partial r}\right) + \frac{\partial^2 V}{\partial z^2} = 0$$

subject to the general condition that

$$(6.33) \qquad V \to 0, \qquad r \to \infty, \qquad \text{or} \qquad |z| \to \infty$$

where we also require $V(r, z)$ to be symmetric about $z = 0$, i.e.,

$$(6.34) \qquad V(r, z) = V(r, -z)$$

Thus we need consider only $z > 0$. Direct separation of variables yields solutions of the form

$$(6.35) \qquad V \approx J_0(\beta r)e^{-\beta z}$$

as possible forms with β being the separation constant. Here we have discarded the $Y_0(\beta r)$ since these are not finite at $r = 0$, and we have also discarded the $e^{\beta z}$, for $z > 0$ since these do not vanish at infinity, with β assumed positive.

Now the functions $J_0(\beta r)$ vanish at infinity so we have here precisely the situation discussed at the end of Chapter Five. In an *open* domain there are no conditions to select a *discrete* spectrum for the eigenvalue, β. Therefore we must admit all possible $\beta > 0$ and write our general solution as:

$$(6.36) \qquad V(r, z) = \int_0^\infty F(\beta)J_0(\beta r)e^{-\beta z}\, d\beta$$

where the function $F(\beta)$ remains to be determined.

The method of determining $F(\beta)$ here is to note that *on the axis, $r = 0$*, we have,

$$(6.37) \qquad V(0, z) = \int_0^\infty F(\beta)e^{-\beta z}\, d\beta$$

so that the potential on the axis, $V(0, z)$ is just the Laplace transform of the unknown function $F(\beta)$.

We have already seen in Chapter Five that in many simple cases we can compute the potential,

$$(6.38) \qquad V(0, z) = f(z)$$

on the axis by elementary integrals. Thus if $f(z)$ is so determined and we find it in our table of Laplace transforms then we immediately find $F(\beta)$.

Using Example (6) of Chapter Five we had V given on the axis as,

$$(6.39) \qquad f(z) = \frac{\lambda R}{2\varepsilon_0\sqrt{R^2 + z^2}}$$

Thus comparing this to Entry Number (37) in our table of transforms we see that

$$(6.40) \qquad F(\beta) = \frac{R\lambda}{2\varepsilon_0}J_0(\beta R)$$

Thus we have,

$$(6.41) \qquad V(r, z) = \frac{R\lambda}{2\varepsilon_0}\int_0^\infty J_0(\beta r)J_0(\beta R)e^{-\beta z}\, d\beta$$

as the potential of the uniformly charged ring of radius R and charge density λ per unit length. This holds for $z > 0$. For $z < 0$ we just use $|z|$, or $z \to -z$.

Quite obviously the method described here has limitations but occasionally is very convenient and it can be extended to equations other than that of Laplace, i.e., the heat conduction or wave equation for example.

EXECUTE PROBLEM SET (6-2)

BOUNDARY VALUE PROBLEMS WITH TIME-DEPENDENT BOUNDARY CONDITIONS

In our Example (2) just above we saw that the end condition was given *explicitly* as a function of time. This is a major distinction from all the problems considered in Chapter Five.

It is quite obvious why one can apply all the methods of Chapter Five now to problems with time-dependent boundary conditions. When we apply the Laplace transform over the time variable t to the partial differential equation and *all* of the boundary conditions we are left with a boundary value problem only in the space coordinates.

Thus all we have to do is solve this problem by our previous techniques and then *invert* the transform of the solution we have constructed, i.e., just as in Example (2) above.

However in nearly *every* case this procedure leads to severe difficulties at the last step, namely inverting the transformed solution to obtain the solution of the original problem in the time domain. Later in Chapters Eight and Nine we will develop a more general technique for this inversion process which makes this general procedure much easier.

These remarks should not be taken to mean that there are not a great number of problems which *can* be carried through just as our Example (2) above.

There is one *general* class of time-dependent boundary value problem which can be solved in a general form in terms of a corresponding problem with stationary boundary conditions. This is shown as follows.

DUHAMEL'S FORMULA

Consider the boundary value problem consisting of:

$$(6.42) \qquad O_{x,y,z}U = \frac{\partial U}{\partial t}, \qquad \text{in } R, \qquad t > 0$$

with

$$(6.43) \qquad U(x, y, z, 0) = 0$$

and

$$(6.44) \qquad \alpha U + \beta\frac{\partial U}{\partial n} = F(t), \qquad \text{on } S$$

$$(6.45) \qquad \alpha' U + \beta'\frac{\partial U}{\partial n} = 0, \qquad \text{on } S'$$

where S and S' are distinct portions of the boundary of the domain R. S and/or S' need not be simply connected segments of the boundary, i.e., we may have the general sort of situation indication in Fig. 6.2. Here $O_{x,y,z}$ is *any* sort of linear operator acting *only* on the space coordinates and the coefficients α and β may be functions of position, but *not time*, along the boundary segments.

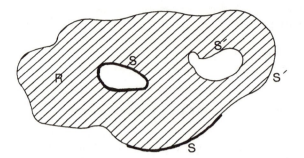

FIGURE 6.2. Domain Definition for Duhamel's Formula.

Now we apply the Laplace transform to this boundary value problem to obtain,

(6.46)
$$O_{x,y,z}u = pu, \quad \text{in } R$$

with

(6.47)
$$\alpha u + \beta\frac{\partial u}{\partial n} = f(p), \quad \text{on } S$$

and

(6.48)
$$\alpha u + \beta\frac{\partial u}{\partial n} = 0, \quad \text{on } S'$$

Here $u = L\{U\}, f = L\{F\}$, and it is important to note that $F(t)$ *must be a function of t only.*

Next we divide *every* equation of this new boundary value problem by the factor $pf(p)$, noting that since this is a *constant* with respect to x, y, z it can be taken inside all operators $O_{x,y,z}$ etc. Thus we obtain

(6.49)
$$O_{x,y,z}u_1 = pu_1, \quad \text{in } R$$

with

(6.50)
$$\alpha u_1 + \beta\frac{\partial u_1}{\partial n} = \frac{1}{p}, \quad \text{on } S$$

and

(6.51)
$$\alpha u_1 + \beta\frac{\partial u_1}{\partial n} = 0, \quad \text{on } S'$$

where we have defined the new symbol

(6.52)
$$u_1(x, y, z, p) = \frac{u(x, y, z, p)}{pf(p)}$$

Comparing this new problem for u_1 with that for u we see that u_1 is just the transform of the solution of the original problem in $U(x, y, z, t)$ with the *single exception* of having $F(t)$ on S replaced by $F(t) = 1$ on S, i.e., u_1 is the transform of the problem in U, but with a *constant homogeneous* boundary condition $F = 1$ on S. Call this solution $U_1(x, y, z, t)$. It can obviously be obtained in most instances by the methods of Chapter Five.

But now look at Eq. (6.52). This can be rearranged to read

(6.53) $$u(x, y, z, p) = [pu_1(x, y, z, p)]f(p)$$

and using Entry Numbers (3) and (7) of our table of transforms can be inverted to read:

(6.54) $$U(x, y, z, t) = \int_0^t F(t - \lambda)\frac{\partial U_1(x, y, z, \lambda)}{\partial \lambda}d\lambda$$

where λ is a dummy variable of integration. This is *Duhamel's Formula*.

Thus if we have already solved some boundary value problem such as given in Eqs. (6.42), (6.43), (6.44), and (6.45) *but with* $F = 1$, then we have $U_1(x, y, z, t)$. We simply put this into the above integral and by a *single integration* write down the solution for the variable boundary condition.

This result should be compared to our result, Eq. (6.30), in our Example (2) above. If in Eq. (6.31) we put $F_0/\alpha = 1$ then this expression for U is just what we have defined here as $U_1(x, y, z, t)$. Differentiation with respect to t and substitution into Duhamel's formula above then yields Eq. (6.30).

It is important in applying Duhamel's formula that all conditions of the problem have the proper form. The one most often overlooked is the initial condition, $U = 0$, is required. Generally by suitable *change of variables* one can put the problem into a suitable form.

EXECUTE PROBLEM SET (6-3)

STEP FUNCTIONS AND DELTA FUNCTIONS†

One of the most useful features of the Laplace transform is the simplification it can introduce into problems involving *discontinuous* functions. As already noted it is necessary only that $F(t)$ be *sectionally* continuous in order that its Laplace transform should exist. Here we define some very general, and extremely useful, sectionally continuous functions. We also define here the linear or *Dirac*, delta function. Delta functions will be treated more completely in our discussion of Green's Functions.

The *unit-step function* is defined as:

(6.55) $$H(x, x') = \begin{cases} 0, & x < x' \\ 1, & x > x' \end{cases}$$

Thus $H(x, x')$ is sectionally continuous but is undefined at $x = x'$. The graph of the function consists of the two horizontal lines, $y = 0$ to the left of $x = x'$ and the line $y = 1$ to the right of $x = x'$ on the x, y plane.

The unit-step function is particularly useful in simplifying notation in problem descriptions involving *changes* in some conditions of the problem. For example consider the simple problem of an R, L, C electrical circuit in which we have an applied electromotive force on the terminals which is defined as follows:

(6.56) $$V = \begin{cases} 0, & t < 0 \\ f_1(t), & 0 < t < t_1 \\ f_2(t), & t > t_1 \end{cases}$$

---

† The modern rigorous treatment of "Delta functions" is in terms of distributions. The student should consult some of the references at the end of this chapter for a more complete and rigorous treatment.

where $f_1(t)$ and $f_2(t)$ might be any sort of continuous functions. This can be represented in terms of unit-step functions as:

(6.57) $$V = [H(t, 0) - H(t, t_1)]f_1(t) + H(t, t_1)f_2(t)$$

Note that here we have formed the quantity in brackets which has the interesting feature of being unity for $0 < t < t_1$ and zero outside this interval, i.e., as shown in Fig. 6.3.

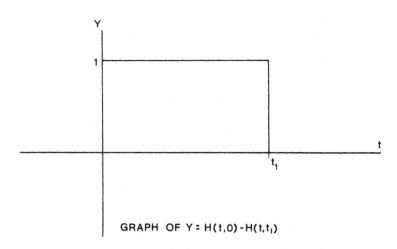

GRAPH OF Y = H(t,0)-H(t,t₁)

FIGURE 6.3.   Illustrating Combination of Unit Step Functions.

Quite obviously we can proceed to build other interesting discontinuous functions on the function $H(x, x')$. For example we can integrate the function

(6.58) $$y(t, t_1) = H(t, 0) - H(t, t_1)$$

shown in Fig. 6.3 to form the function

(6.59) $$s(t, t_1) = \int_{-\infty}^{t} y(t, t_1)\, dt$$

as shown in Fig. 6.4. Some exercises in constructing such functions are provided in the next problem set for the student.

By far one of the most useful "discontinuous" functions is the Dirac delta function. To see the connection of this function, which we denote as $\delta(x - x')$, with the unit-step function defined above consider the following integral.

(6.60) $$I(x') = \int_{a}^{b} H(x, x')\frac{dF(x)}{dx}\, dx$$

where $F(x)$ is some continuous function having a first derivative in the interval a to b.

From the definition of $H(x, x')$ above we see that *if* the *critical point* $x = x'$ is in the interval of integration, $a < x' < b$ then,

(6.61) $$I(x') = \int_{x'}^{b} \frac{dF(x)}{dx}\, dx = F(b) - F(x')$$

Thus now write this for $I(x')$ and then return to consider Eq. (6.60) by performing an

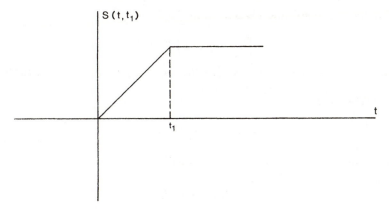

FIGURE 6.4.   **The Function $S(t, t_1)$ Defined as the Integral of the Function Shown in Figure 6.3.**

integration by parts, thus:

$$(6.62) \qquad F(b) - F(x') = H(x, x')F(x)\Big|_a^b - \int_a^b F(x)\frac{dH(x, x')}{dx}\,dx$$

and, from the definition of $H(x, x')$ and the fact that $a < x' < b$ the integrated terms appear as:

$$(6.63) \qquad H(x, x')F(x)\Big|_a^b = F(b)$$

and therefore,

$$(6.64) \qquad F(x') = \int_a^b F(x)\frac{dH(x, x')}{dx}\,dx, \qquad a < x' < b$$

Furthermore we can show that the integral on the right remains the same if we interchange x and x' in $H(x, x')$. This *formal* integration by parts then suggests the definition of a new function.

We define the "function"

$$(6.65) \qquad \delta(x - x') = \frac{dH(x, x')}{dx} = \begin{cases} 0, & x \neq x' \\ \infty, & x = x' \end{cases}$$

subject to the two conditions

$$(6.66) \qquad \delta(x - x') = \delta(x' - x)$$

and

$$(6.67) \qquad F(x') = \int_a^b F(x)\delta(x - x')\,dx, \qquad a < x' < b$$

as the Dirac delta function. Here $F(x)$ is any function having a first derivative on $a < x < b$. In the following discussion we see that a more rigorous definition of the delta function as the limit of a distribution† requires the limits of integration to be specified.

---

† See Avner Friedman, "Generalized Functions and Partial Differential Equations", Prentice-Hall Inc., Englewood Cliffs, N.J. (1963) for rigorous treatment of distributions.

Because of the above definition it is obvious that

(6.68)
$$\int_{-\infty}^{+\infty} \delta(x - x')\, dx = 1$$

For this reason it is often convenient to view the Dirac delta function from another point of view, which by the way will be explored again in Chapter Nine.

Consider some function such as,

(6.69)
$$\Phi_\sigma = \frac{1}{\sqrt{2\pi\sigma}} e^{-\frac{(x - x')^2}{2\sigma}}$$

which has the property of being a distribution such that,

(6.70)
$$\int_{-\infty}^{+\infty} \Phi_\sigma(x - x')\, dx = 1$$

Note that this integral is *not* unity for any other limits. Now examine the integral,

(6.71)
$$I(x') = \int_{-\infty}^{+\infty} F(x)\Phi_\sigma(x - x')\, dx$$

By the mean-value theorem of integral calculus one can show that,

(6.72)
$$I(x') = F(x^*) \int_{-\infty}^{+\infty} \Phi_\sigma(x - x')\, dx'$$

or,

(6.73)
$$I(x') = F(x^*), \qquad -\infty < x^* < +\infty$$

in view of Eq. (6.70) above.

The graph of $\Phi_\sigma(x - x')$ appears as in Fig. 6.5. Here Φ_σ is shown for two values of σ. As σ *decreases* the function $\Phi_\sigma(x - x')$ becomes very narrow, width $\approx 2\sigma$, and very high, height $\approx 1/\sqrt{2\pi\sigma}$ at $x = x'$. In the limit $\sigma \to 0$ then we obtain essentially a vertical line at $x = x'$. In this limiting process we can show that the point x^* in Eqs. (6.72) and (6.73) above approaches x'. Thus the limit

(6.74)
$$\delta(x - x') = \lim_{\sigma \to 0} \frac{1}{\sqrt{2\pi\sigma}} e^{-\frac{(x - x')^2}{2\sigma}}$$

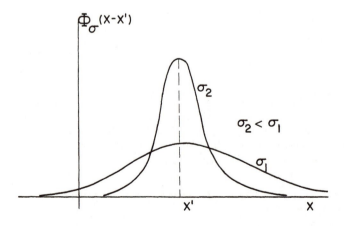

FIGURE 6.5. Graph of $\phi_\sigma(x - x')$ for Two Values of the Parameter σ.

may be taken as the definition of the Dirac delta function. In this case the limits of integration in Eq. (6.67) should be $-\infty$ to $+\infty$. We especially use this definition in visualizing *impulse* problems as shown in the following example.

Example (4)

The simple pendulum struck by a hammer. Newton's second law applied to the simple pendulum of length l with mass m, and subject to a horizontal force $F(t)$, gives:

$$(6.75) \qquad m\frac{d^2x}{dt^2} = -\frac{g}{l}mx + F(t)$$

where we assume the horizontal displacement x from the equilibrium point, $x = 0$, to be small.

Now suppose the applied force $F(t)$ is in the form of horizontal blows from a hammer at times $t_1, t_2, \ldots t_n$. We can view the force of the hammer as being some large force F acting over a short period of time. In fact if we let the axes in Fig. 6.5 above be F and t axes we can view F versus t for a blow of the hammer as of this form.

However the blow is essentially instantaneous so we let $\sigma \to 0$ in this picture but we add on a multiplier "I_i" for the strength of the impulse imparted by the hammer. Thus for the sequence of n hammer blows

$$(6.76) \qquad F(t) = \sum_{i=1}^{n} I_i \delta(t - t_i)$$

where $\delta(t - t_i)$ is just the delta function defined above.

We integrate Eq. (6.75) with $F(t)$ given by Eq. (6.76) simply by applying the Laplace transform, noting that

$$(6.77) \qquad \int_0^\infty e^{-pt}\delta(t - t')\,dt = e^{-pt'}$$

is the Laplace transform of the delta function. Here then:

$$(6.78) \qquad m[p^2u - px(0) - \dot{x}(0)] = -\frac{gm}{l}u + \sum_{i=1}^{n} I_i e^{-pt_i}$$

where $u = L\{x(t)\}$ is our notation. Rearranging gives:

$$(6.79) \qquad u = \frac{px(0) + \dot{x}(0)}{p^2 + \omega^2} + \sum_{i=1}^{n} \frac{I_i}{m}\frac{e^{-pt_i}}{p^2 + \omega^2}$$

where we have used $\omega^2 = g/l$ as the square of the natural angular frequency and of course $x(0)$ and $\dot{x}(0)$ are the initial position and velocity.

We now use the table of transforms to invert this equation to obtain $x(t)$. The first term is readily found as Entry Numbers (27) and (26). The terms in the sum are found as Entry Number (26) subject to the "shift" conditions of the exponential as given in Entry Number (12). We now take care of the zero condition for $t < t_i$ by using our unit-step function, so the solution appears as:

$$(6.80) \qquad x(t) = x(0)\cos \omega t + \frac{\dot{x}(0)}{\omega}\sin \omega t$$

$$+ \sum_{i=1}^{n} \frac{I_i}{m\omega}H(t, t_i)\sin \omega(t - t_i)$$

Thus each blow of the hammer adds a new sinusoidal term which did not exist prior to the blow.

This example illustrates the power and utility of the combination of step and delta functions which are very amenable to treatment by the Laplace transform.

EXECUTE PROBLEM SET (6-4)

LINEAR SUPERPOSITION AND THE CONVOLUTION INTEGRAL

The convolution integral appears as Entry Number (7) of our table of transforms,

$$(6.81) \qquad\qquad I = \int_0^t F_1(t - \tau)F_2(\tau)\,d\tau$$

and has already been indicated as being of great importance in many physical problems. We saw it appearing in Duhamel's formula in particular. This integral has a particularly interesting *physical* interpretation which takes on a different appearance in various contexts but is actually the same in these various contexts. A good understanding of this physical meaning is of great assistance in the solution of all sorts of mathematical problems.

Consider a "thing" which we act on with a "strength," A, at time, τ, and let the "effect" of this action after the elapsed time, $(t - \tau)$, following our action be $f(t - \tau)$. We will say that our thing behaves *linearly* if two *conditions* are fulfilled. These are: first that

$$(6.82) \qquad\qquad f(t - \tau) = A(\tau)F(t - \tau)$$

that is, the "effect" at any time following our action is *proportional* to the "strength" of the action; second is that the effects of distinct actions are *additive*. Thus action $A(\tau)$ at τ produces $A(\tau)F(t - \tau)$ as effect, and $A(\tau')$ produces $A(\tau')F(t - \tau')$ if each were the *only* action, and we require that the *resultant* effect at time t be,

$$A(\tau)F(t - \tau) + A(\tau')F(t - \tau')$$

for the "thing" to be linear.

Quite obviously we generalize by letting there be a continuous "action" $A(\tau) \to F_2(\tau)\,d\tau$ applied to the "thing" and denote the "effect" as $F_1(t - \tau)$ for an instantaneous unit action to obtain Eq. (6.81).

Example (5)

Response of linear instruments. We consider the input signal into some kind of linear instrument as $S(t)\,dt$ during the interval dt and suppose the reading of the instrument at any time t is $I(t)$. If the instrument is linear then,

$$(6.83) \qquad\qquad I(t) = \int_0^t R(t - \tau)S(\tau)\,d\tau$$

Here we say the instrument is also *passive* since no reading exists prior to the initiation of the input at $t = 0$. We call R the *response function* of the instrument.

The meaning of $R(t)$ becomes clear if we consider an input to the instrument in the form of a delta function, i.e., let

$$(6.84) \qquad\qquad S(t) = \delta(t - t_0)$$

then we get

(6.85) $I(t) = H(t, t_0)R(t - t_0)$

So that if the delta function is put into the instrument at $t = 0$ we see $R(t)$ as the reading of the instrument.

Another meaning, or way of getting at $R(t)$, is to consider a unit step function input to the instrument. Thus with

(6.86) $S(t) = H(t, 0)$

we get

(6.87) $I(t) = \int_0^t R(\tau)\, d\tau$

as the output, or reading, of the instrument and hence $R(t)$ is given as $dI(t)/dt$.

INVERSION OF THE CONVOLUTION

Of course what one would like to know in many cases is precisely the form of the input signal, $S(t)$, whereas what one obtains from the instrument is the output $I(t)$. Generally the form of $R(t)$ can be found experimentally by feeding in either a delta function (approximately) or a step function and recording $I(t)$. In the latter case the dI/dt yields $R(t)$.

To obtain $S(t)$ given $I(t)$ and the form of $R(t)$ we take the Laplace transform of Eq. (6.83) to obtain,

(6.88) $\mathscr{I}(p) = \mathscr{R}(p)\mathscr{S}(p)$

where $\mathscr{I}$, $\mathscr{R}$, and $\mathscr{S}$ are the respective transforms of I, R, and S. Then,

(6.89) $\mathscr{S}(p) = \dfrac{\mathscr{I}(p)}{\mathscr{R}(p)}$

and in principle the inverse transform would yield $S(t)$. However this cannot usually be accomplished with just our table of transforms, or for that matter in many cases even with the inversion integral treated in Chapter Nine, because as a rule $I(t)$ is in the form of recorded *tabular data*, not a given function of t. However methods do exist for this problem and are treated in Chapter Nine.

Example (6)

Population problems. This problem occurs in many contexts, human population, bacterial colonies, nuclear reactions, etc. We present it in the context of the human population problem because this is without question one of the most pressing problems facing the world today.

Statistics show that, on the average, the fraction, $F(t)$, of a group of people surviving to age t is represented by a function roughly as shown in Fig. 6.6. Thus if at some time, say *now*, there are N_0 people born, then we can expect $N_0 F(t)$ to be living at time t in the future.

However we continually add to the population, say $R(t')dt'$, is the number of people added to the population in the interval dt' at time t'. Then at time t there will be $R(t')F(t - t')\, dt'$ of these surviving in the population also. Consequently the number $N(t)$ surviving at time t in

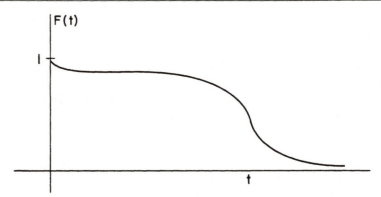

FIGURE 6.6.   Typical Survival Function F(t).

the future is the sum,

$$(6.90) \qquad N(t) = N_0 F(t) + \int_0^t R(t')F(t - t') \, dt'$$

But still another factor to consider is that the rate R at which people are added is *roughly* proportional to the number living at any time t, i.e.,

$$(6.91) \qquad R(t') = \frac{dN(t')}{dt} \approx \alpha N(t')$$

Thus we have the *integral equation*,

$$(6.92) \qquad N(t) = N_0 F(t) + \int_0^t \alpha N(t')F(t - t') \, dt'$$

describing the population growth. Of course α may itself be a function of t' and the survival function F may change *form* with time due to innovations of medical science and technology (i.e., air pollution).

As is immediately obvious the Laplace transform is the natural tool for the solution of this integral equation. Taking the Laplace transform and rearranging yields (α considered constant),

$$(6.93) \qquad n = \frac{N_0 f}{1 - f\alpha}$$

where lower-case letters denote the transforms of N and F, respectively. Unfortunately for *any realistic* form for $F(t)$ the inversion of Eq. (6.93) to obtain $N(t)$ requires more than our simple table of transforms. However if we assume an exponential survival curve such as

$$(6.94) \qquad F(t) = e^{-\frac{1}{T}t}$$

then

$$(6.95) \qquad f = \frac{1}{p + \dfrac{1}{T}}$$

and Eq. (6.93) appears as:

(6.96)
$$n = \frac{N_0}{p + \left(\frac{1}{T} - \alpha\right)}$$

which has the inverse,

(6.97)
$$N(t) = N_0 e^{-\left(\frac{1}{T} - \alpha\right)t}$$

This has the interesting physical interpretation that if the mean lifetime T, of the typical person is *greater* than the reciprocal, $1/\alpha$, of the per capita reproduction rate, α, then the population must *grow* at an exponential rate; whereas for T less than $1/\alpha$ the population decreases in an exponential fashion.

EXECUTE PROBLEM SET (6-5)

APPLICATION OF THE LAPLACE TRANSFORM TO THE EVALUATION OF INTEGRALS

The Laplace transform frequently can be used to advantage in the evaluation of definite, or sometimes indefinite, integrals. An integral of the form,

(6.98)
$$I = \int_a^x F(x', t)\, dx'$$

where t is some parameter, and $F(x', t)$ is of such a form that $L\{F(x', t)\}$ with respect to t exists and I is uniformly convergent for all $0 < t < \infty$ can be treated in this manner.

Example (7)

Evaluate,

(6.99)
$$I(t) = \int_0^\infty \frac{\cos tx\, dx}{x^2 + b^2}$$

We apply the Laplace transform to $I(t)$ and, since this integral is uniformly convergent in t for $0 < t < \infty$, we can interchange the order of integration on x and t to obtain,

(6.100)
$$L\{I(t)\} = \int_0^\infty \frac{p\, dx}{(p^2 + x^2)(x^2 + b^2)} = f(p)$$

since $p/(p^2 + x^2)$ is $L\{\cos tx\}$. Here we expand the integrand by *partial fractions* to yield:

(6.101)
$$f(p) = \frac{p}{p^2 - b^2} \int_0^\infty \left(\frac{1}{x^2 + b^2} - \frac{1}{x^2 + p^2}\right) dx$$

This integrates to,

(6.102)
$$f(p) = \frac{p}{p^2 - b^2} \left\{ \frac{1}{b} \tan^{-1} \frac{x}{b} \Big|_0^\infty - \frac{1}{p} \tan^{-1} \frac{x}{p} \Big|_0^\infty \right\}$$

or

(6.103)
$$f(p) = \frac{\pi}{2b} \frac{1}{p + b}$$

Hence the inverse found in our table yields,

$$(6.104) \qquad\qquad I(t) = \frac{\pi}{2b} e^{-bt}$$

Here we should remark that sometimes it is possible to apply this method even when a parameter such as t does not explicitly appear in the integral. For example suppose in $I(t)$ above just cos x has appeared in the integrand instead of cos tx. In this event we would just insert t to make it have the above form, evaluate $I(t)$ as we have above and *then* in the final result set $t = 1$.

EXECUTE PROBLEM SET (6-6)

PROBLEMS

Set (6–1)

(1) Use integration by parts to establish Formulas (5) and (6) of Table 6.1.
(2) Demonstrate that the arguments, $t - \tau$ and τ of F_1 and F_2 in the integral of Eq. (7) of Table 6.1 are interchangeable.
(3) Derive Formula (12) of Table 6.1.
(4) Derive Formulas (14) and (15) of Table 6.1.
(5) Derive Formula (7) of Table 6.1.

Set (6–2)

(1) An *infinite* string, under tension, is given an initial displacement and released, with zero initial velocity, in that configuration, i.e.,

$$\frac{\partial^2 Y}{\partial t^2} = c^2 \frac{\partial^2 Y}{\partial x^2}, \qquad -\infty < x < +\infty$$

$$t > 0$$

$$Y(x, 0) = F(x), \qquad \frac{\partial Y}{\partial t}(x, 0) = 0$$

$$\underset{x \to \pm\infty}{\text{limit}}\, Y(x, t) = \text{finite value}$$

show that for $t > 0$,

$$Y(x, t) = \tfrac{1}{2}[F(x + ct) + F(x - ct)]$$

(2) A uniform elastic bar is traveling in a direction parallel to its length with a speed v_0, when its end $x = 0$ hits and "sticks" to a rigid wall. Find the displacement, $Y(x, t)$, of a cross section from its unstrained equilibrium position if the bar is initially unstrained and the bar is so long compared to the period of observation that it acts as having infinite length with the "far" end, $x = L \to \infty$, free of strain, i.e., no force applied. We have,

$$\frac{\partial^2 Y}{\partial t^2} = a^2 \frac{\partial^2 Y}{\partial x^2}, \qquad x > 0, \quad t > 0$$

$$Y(x, 0) = 0, \qquad \frac{\partial Y(x, 0)}{\partial t} = -v_0$$

$$Y(0, t) = 0 \qquad \underset{x \to \infty}{\text{limit}} \frac{\partial Y(x, t)}{\partial x} = 0$$

Here a^2 depends on the geometry and elastic properties of the bar.

(3) Consider an infinitely long rod, insulated on its sides, in which heat is generated at the rate $R(t)$ per unit volume (i.e., as by the passage of an electric current through the rod). For zero initial temperature and the end $x = 0$ kept at zero temperature also, we have, if the "far" end is insulated:

$$k\frac{\partial^2 T}{\partial x^2} = C\rho \frac{\partial T}{\partial t} - R, \qquad \begin{array}{l} x > 0 \\ t > 0 \end{array}$$

with

$$T(x, 0) = T(0, t) = 0$$

$$\underset{x \to \infty}{\text{limit}} \frac{\partial T}{\partial x}(x, t) = 0$$

Show that the solution can be expressed as:

$$T(x, t) = U(t) - \frac{2}{\sqrt{\pi}} \int_{\frac{x}{2\sqrt{at}}}^{\infty} U\left(t - \frac{x^2}{4 a\lambda^2}\right) e^{-\lambda^2} d\lambda$$

where $a = k/C\rho$ and

$$U = \frac{1}{C\rho} \int_0^t R(\tau) d\tau$$

(4) The infinitely long thin wire with heat radiation to the surroundings,

$$k\frac{\partial^2 T}{\partial x^2} = C\rho \frac{\partial T}{\partial t} + h(T - T_0), \qquad \begin{array}{l} x > 0 \\ t > 0 \end{array}$$

where h is the radiation factor and T_0 is the ambient temperature of the surroundings. Consider the case

$$T(x, 0) = T_0$$

$$- kA\frac{\partial T}{\partial x}(0, t) = Q(t)$$

$$\underset{x \to \infty}{\text{limit }} T(x, t) = \text{finite value}$$

(5) Find the potential due to a *disk* of electric charge of radius R with charge density $\sigma(r)$, r being the radial distance from the axis, for the special charge distribution $\sigma_0\left(1 - \frac{r^2}{R^2}\right)$.

Hint: Use the solution for the charged ring. The potential of the disk is the "sum" of the potentials of a family of rings of appropriate charge.

(6) Construct the *asymptotic* form of the above solution for large distances from the disk. Also consider the *extreme* case in which $R \to 0$ with $\pi R^2 \sigma_0 = \text{constant}$.

Set (6–3)

(1) Consider linear conduction of heat in the *finite* rod of length, L, with the end $x = 0$ maintained at zero temperature and the initial temperature zero. Let the end $x = L$ have

temperature $F(t)$. Solve by first solving the problem for $F(t) = 1$, by separation of variables, then apply Duhamel's formula. The problem is stated as:

$$\frac{\partial^2 U}{\partial \xi^2} = \frac{\partial U}{\partial \tau}, \qquad \begin{array}{l} \tau > 0 \\ 0 < \xi < 1 \end{array}$$

$$U(\xi, 0) = 0$$

$$U(0, \tau) = 0$$

$$U(1, \tau) = F(\tau)$$

in dimensionless form.

(2) Solve the above problem directly by taking the Laplace transform and showing that

$$u(\xi, p) = f(p)\frac{\sinh \xi\sqrt{p}}{\sinh \sqrt{p}}$$

is the transform of the solution $U(\xi, \tau)$. Here $f(p) = L\{F(\tau)\}$. Show that

$$\frac{\sinh \xi\sqrt{p}}{\sinh \sqrt{p}} = \sum_0^\infty [e^{-(2n+1-\xi)\sqrt{p}} - e^{-(2n+1+\xi)\sqrt{p}}]$$

and then use the table of transforms to invert term by term. Compare this solution to that of Problem (1) above.

Set (6–4)

(1) Evaluate by using a formal integration by parts, assuming $F(t)$ bounded.

$$I(t) = \int_{-\infty}^{+\infty} \frac{d\delta(t' - t)}{dt'} F(t') \, dt'$$

(2) Evaluate

$$I = \int_a^b \delta(x - c)H(x, \alpha)F(x) \, dx,$$

for all possible values of a, b, c and α.

(3) Show that the function defined as

$$\delta(x - x') = \lim_{n \to 0} \frac{e^{\frac{x - x'}{n}}}{n(1 + e^{\frac{x - x'}{n}})^2}$$

has properties like a delta function on $-\infty < x < +\infty$.

Set (6–5)

(1) In example (4) of this chapter, the pendulum, show that for *any* forcing function $F(t)$,

$$x(t) = \frac{1}{m\omega} \int_0^t F(\tau) \sin \omega(t - \tau) \, d\tau$$

if $x(0) = x(0) = 0$. Here $\omega^2 = g/l$. Thus see $x(t)$ as the "response" of a passive linear instrument due to the "action" $F(t)$, and

$$\frac{1}{m\omega} \sin \omega t = R(t)$$

is the "response function" of the instrument. This *form* obviously applies to any simple harmonic oscillator.

(2) Deduce the response function for the simple R, L, C circuit described by

$$L\frac{d^2Q}{dt^2} + R\frac{dQ}{dt} + \frac{Q}{C} = V(t)$$

with $V(t)$ being treated as the input and $R\dfrac{dQ}{dt} = I$ being treated as the output.

(3) *Feedback* consists of introducing all or part of the output from a system back into the system as input. For example a simple oscillator, not passive at time $t = 0$ but executing free oscillations, is described by

$$I(t) = a \sin \omega t + b \int_0^t \sin \omega(t - \tau) S(\tau)\, d\tau$$

where a and b are constants. Find the "output" if the system is given the feedback signal $S(t) = \alpha I(t)$, where, α is constant, i.e., use this for $S(t)$. Under what conditions will we then find

$$I(t) = At$$

with A being a constant?

(4) In the population problem, Eq. (6.90) of the text, what must be the "birth rate", $R(t)$, to assure a *constant population* $N_0 = N(t)$ if the survival function $F(t)$ is the exponential, $\exp - t/T$?

Set (6–6)

(1) Evaluate,

$$I = \int_{-\infty}^{+\infty} \frac{x \sin tx}{a^2 + x^2}\, dx$$

(2) Evaluate,

$$I = \int_0^\infty \frac{\sin tx}{x}\, dx$$

(3) Evaluate,

$$I = \int_0^\infty \frac{\sin x}{\sqrt{x}}\, dx$$

(4) Evaluate,

$$I = \int_0^\infty x \sin \beta x\, e^{-\frac{x^2}{4t}}\, dx$$

REFERENCES

J. A. Aseltine, *Transform Method in Linear System Analysis* (McGraw-Hill Book Company, Inc., New York, 1958).

R. V. Churchill, *Modern Operational Mathematics in Engineering* (McGraw-Hill Book Company, Inc., New York, 1944).

J. Irving and N. Mullineux, *Mathematics in Physics and Engineering* (Academic Press, Inc., New York, 1959).

N. N. Lebedev, I. P. Skalskaya, and Y. S. Uflyand, *Problems of Mathematical Physics*, (translated by R. A. Silverman), Prentice-Hall Inc., Englewood Cliffs, N.J. (1965).

Avner Friedman, *Generalized Functions and Partial Differential Equations*, Prentice-Hall Inc., Englewood Cliffs, N.J. (1963).

CHAPTER SEVEN

Two-dimensional potential problems and conformal mapping; functions of a complex variable

PHYSICAL BACKGROUND

A great many physical problems giving rise to Laplace's equation have their origin in a formulation in terms of some vector function, say $\mathbf{V}(x, y, z)$ subject to the *two* conditions:

$$(7.1) \qquad \nabla \cdot \mathbf{V} = 0$$

and

$$(7.2) \qquad \nabla \times \mathbf{V} = 0$$

For example the first equation here corresponds to the statement that a fluid with velocity $\mathbf{V}$ is incompressible, while the second states that the flow is irrotational, or that the fluid is an ideal fluid free of viscosity. Of course these same or similar equations also arise in many other contexts. Here for the moment we confine our terminology to the fluid-flow example in order to impart a certain visualization to our mathematical discussion.

As we saw in Chapter One, Eq. (7.1) is identically satisfied if $\mathbf{V}$ is the curl of some vector $\mathbf{A}$, thus

$$(7.3) \qquad \nabla \cdot (\nabla \times \mathbf{A}) \equiv 0$$

and this form, $\mathbf{V} = \nabla \times \mathbf{A}$ inserted into Eq. (7.2) yields,

(7.4) $$\nabla \times \nabla \times \mathbf{A} = \nabla(\nabla \cdot \mathbf{A}) - \nabla^2 \mathbf{A}$$

On the other hand Eq. (7.2) is identically satisfied if $\mathbf{V}$ is the gradient of some scalar, say ϕ, thus,

(7.5) $$\nabla \times \nabla \phi \equiv 0$$

and the form $\mathbf{V} = \nabla \phi$ inserted into Eq. (7.1) yields,

(7.6) $$\nabla^2 \phi = 0$$

which is just Laplace's equation.

   Now consider the form these equations take if the vector $\mathbf{V}$ is constrained to have only x and y components. The first thing we note is that with

(7.7) $$\mathbf{V} = \mathbf{1}_x V_x(x, y, z) + \mathbf{1}_y V_y(x, y, z)$$

our Eq. (7.6) becomes,

(7.8) $$\frac{\partial^2 \phi}{\partial x^2} + \frac{\partial^2 \phi}{\partial y^2} = 0$$

even though

(7.9) $$\begin{cases} V_x(x, y, z) = \dfrac{\partial \phi}{\partial x} \\[2mm] V_y(x, y, z) = \dfrac{\partial \phi}{\partial y} \end{cases}$$

and ϕ therefore could depend on z as well as x and y.

   We also note that since $\mathbf{V} = \nabla \times \mathbf{A}$, the vector function $\mathbf{A}$ must be directed along the z axis in order for $\mathbf{V}$ to lie in the xy plane. Thus we put

(7.10) $$\mathbf{A}(x, y, z) = \mathbf{1}_z \psi(x, y)$$

so that the components of $\mathbf{V}$ appear as,

(7.11) $$\begin{cases} V_x(x, y) = \dfrac{\partial \psi}{\partial y} \\[2mm] V_y(x, y) = -\dfrac{\partial \psi}{\partial x} \end{cases}$$

Inserting Eq. (7.10) for $\mathbf{A}$ into Eq. (7.4) yields,

(7.12) $$\frac{\partial^2 \psi}{\partial x^2} + \frac{\partial^2 \psi}{\partial y^2} = 0$$

so that ψ as well as ϕ satisfies Laplace's equation in the *plane*. Furthermore these functions are related by the unique set of relations:

(7.13) $$\begin{cases} \dfrac{\partial \phi}{\partial x} = \dfrac{\partial \psi}{\partial y} \\[2mm] \dfrac{\partial \phi}{\partial y} = -\dfrac{\partial \psi}{\partial x} \end{cases}$$

which we will now show arise in an entirely different context.

ANALYTIC FUNCTIONS OF A COMPLEX VARIABLE AND THE CAUCHY–RIEMANN CONDITIONS

If to the real coordinates x, y of a point in the complex plane we attach the multiplier $i = \sqrt{-1}$ as a multiplier to y and form the linear combination

$$(7.14) \qquad\qquad z = x + iy$$

we call† z a complex variable with real part x and imaginary part y. We can then proceed to develop all of the algebra of such complex variables. Here however we assume the student to be familiar with this and now turn to consider *functions* of a complex variable.

In a function of a complex variable x and y appear in the function *only* in the combination $x + iy$. If we visualize a value of $z = x + iy$ as specifying a *point* in the complex plane then any function $f(z)$ of z can be visualized as the "image" of z in another complex plane, i.e., we separate $f(z)$ into its real and imaginary parts,

$$(7.15) \qquad\qquad \omega = f(z) = u(x, y) + iv(x, y)$$

and see u and v as coordinates of a point in another u-v cartesian plane. Quite obviously this association of points in the two planes is *not* necessarily a one-to-one association.

For example the function

$$(7.16) \qquad\qquad \omega = \ln z$$

is definitely not a single valued mapping. If we write z in polar form‡ as

$$(7.17) \qquad\qquad z = re^{i\theta}$$

we see that z is unchanged upon replacing θ by $\theta + 2n\pi$ where $n = 0, \pm 1, \pm 2, \ldots$ Hence when substituted into Eq. (7.16) we see

$$(7.18) \qquad\qquad \begin{cases} u = \ln r \\ v = \theta, \theta + 2\pi, \theta - 2\pi, \theta + 4\pi, \ldots \end{cases}$$

which is a multiple-point image of z. We call

$$(7.19) \qquad\qquad \omega = \ln r + i\theta$$

the principal value of ω.

A function $f(z)$ is said to be continuous at a point z_1 if $f(z) - f(z_1)$ becomes arbitrarily small as $z \to z_1$ in any manner whatever. When such is the case, and $f(z)$ is also a single-valued function of z in some domain R containing z, then if the limit,

$$\lim_{z \to z_1} \frac{f(z) - f(z_1)}{z - z_1}$$

exists and is independent of the manner in which we let $z \to z_1$ we say the function is differentiable at $z = z_1$. Here z and z_1 must lie in R.

A function which is single valued and differentiable in some region R of the z plane is said to be analytic, or regular, or holomorphic in R. It is just such functions of this class that interests us here.§

---

† Not to be confused with the vertical coordinate z above.
‡ $r = \sqrt{x^2 + y^2}$, $\theta = \tan^{-1} y/x$.
§ See appendix for extension of analytic functions by *analytic continuation.*

CAUCHY–RIEMANN CONDITIONS

Here we indicate in a heuristic fashion the basis for the famed Cauchy–Riemann conditions. We consider the derivative of an analytic function defined as,

$$(7.20) \qquad \frac{df}{dz} = \lim_{\Delta z \to 0} \frac{f(z + \Delta z) - f(z)}{\Delta z}$$

where $\omega = f(z)$ is the analytic function and

$$(7.21) \qquad \Delta z = \Delta x + i\Delta y$$

Since the derivative, if it exists at all, must be independent of the manner in which Δx and Δy approach zero, we first let $\Delta y \to 0$, yielding, to a *first-order approximation*,

$$(7.22) \qquad \frac{df}{dz} = \lim_{\Delta x \to 0} \frac{f(z + \Delta x) - f(z)}{\Delta x}$$

or, since

$$(7.23) \qquad f(z + \Delta x) = u(x + \Delta x, y) + iv(x + \Delta x, y)$$

and

$$(7.24) \qquad f(z) = u(x, y) + iv(x, y)$$

this is:

$$(7.25) \qquad \frac{df}{dz} = \frac{\partial u}{\partial x} + i\frac{\partial v}{\partial x}$$

In the same way if we now let $\Delta x \to 0$ *first* and then let $\Delta y \to 0$ we see Eq. (7.20) as,

$$(7.26) \qquad \frac{df}{dz} = \lim_{i\Delta y \to 0} \frac{f(z + i\Delta y) - f(z)}{i\Delta y}$$

which yields,

$$(7.27) \qquad \frac{df}{dz} = \frac{1}{i}\frac{\partial u}{\partial y} + \frac{\partial v}{\partial y}$$

or

$$(7.28) \qquad \frac{df}{dz} = \frac{\partial v}{\partial y} - i\frac{\partial u}{\partial y}$$

Finally since the result *must* be the same in either limiting process we see that Eq. (7.25) and (7.28) yield upon equating real and imaginary parts,

$$(7.29) \qquad \begin{cases} \dfrac{\partial u}{\partial x} = \dfrac{\partial v}{\partial y} \\[2mm] \dfrac{\partial u}{\partial y} = -\dfrac{\partial v}{\partial x} \end{cases}$$

which are the Cauchy–Riemann conditions.

Comparing Eqs. (7.13) and (7.29) we see that both have *exactly* the same functional structure. Furthermore if we differentiate the first equation of Eq. (7.29) with respect to x and the second with respect to y and add the two equations we will obtain,

$$(7.30) \qquad \frac{\partial^2 u}{\partial x^2} + \frac{\partial^2 u}{\partial y^2} = 0$$

In a similar way we also see that

$$(7.31) \qquad \frac{\partial^2 v}{\partial x^2} + \frac{\partial^2 v}{\partial y^2} = 0$$

so that *both* the real and imaginary parts of an analytic function of a complex variable satisfy Laplace's equation. We call u and v *harmonic* functions of x and y.

COMPLEX POTENTIAL

The above analysis reveals that the functions $\phi(x, y)$, $\psi(x, y)$ discussed at the beginning of this chapter must be the real and imaginary parts of an analytic function of the complex variable z. Thus we define

$$(7.32) \qquad \Omega(z) = \phi(x, y) + i\psi(x, y)$$

as the *complex potential function*. We now examine the physical meaning of ϕ, ψ, and Ω.

First consider the total differential,

$$(7.33) \qquad d\psi = \frac{\partial \psi}{\partial x} dx + \frac{\partial \psi}{\partial y} dy$$

From Eq. (7.11) this can be written as,

$$(7.34) \qquad d\psi = - V_y dx + V_x dy$$

The physical meaning of this is seen in the Fig. 7.1 below. Here we see $- V_y dx$ as the volume

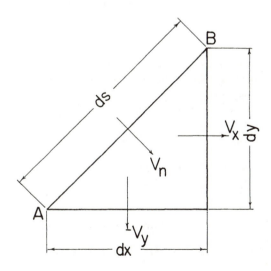

FIGURE 7.1. Geometry for Interpreting the Physical Meaning of the ψ Function.

flow rate *out* of the triangular region of *unit* thickness† across the line element dx and similarly $V_x dy$ as that *out* across dy. Since $\nabla \cdot \mathbf{V} = 0$ this *out flow* must be balanced by an *equal inflow* across the line element ds as indicated. Thus we see $d\psi = \psi_B - \psi_A$ as being equal to the volume flow rate across the line connecting (A) and (B) as indicated (for unit thickness). This physical interpretation of ψ is frequently helpful in solving problems. Thus if we set the function $\psi(x, y)$ equal to a constant, ψ_1, this defines a curve in the x, y plane. Two such curves defined by ψ_1 and ψ_2 define a *tube of flow* in the plane, i.e., we can see from Eq. (7.11) that $\mathbf{V}$ must be parallel to a line of constant ψ at any point in the plane. Similarly from Eq. (7.9) $\mathbf{V}$ must be orthogonal to the lines of constant ϕ.

Here then we have the situation indicated schematically in Fig. 7.2. The ϕ and ψ curves form an *orthogonal* curvilinear net, with the flow as indicated by the arrows.

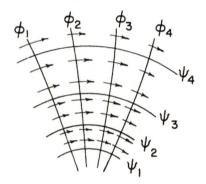

FIGURE 7.2.   Illustrating an Orthogonal Net Formed by Curves of Constant ϕ and Curves of Constant ψ.

EXECUTE PROBLEM SET (7-1)

SOURCES, SINKS, DOUBLETS, AND VORTICES

There are a few potential functions corresponding to special physical conditions which we now list for convenient reference. The first is that for a line source or sink, in the *three dimensional space*, but which is a *point* in the *x-y plane*. This is:

$$(7.35) \qquad \Omega = \frac{q}{2\pi} \ln z + \text{constant}$$

where the constant is a complex constant. Later, in our discussion of *Green's functions* we will see how this can be derived in a rigorous fashion, here we simply note the *physical* meaning of this potential.

We see that the *isopotentials*, ϕ equals a constant, are the circles,

$$(7.36) \qquad \phi = \frac{q}{2\pi} \ln r = \text{constant}$$

and the *streamlines* are

$$(7.37) \qquad \psi = \frac{q}{2\pi} \theta = \text{constant}$$

which are straight lines emanating from the origin.

† Thickness normal to the plane of the paper.

If we form $\nabla\phi$ we get

(7.38)
$$\mathbf{V} = \frac{q}{2\pi r}\mathbf{1}_r$$

where $\mathbf{1}_r$ is a unit radial vector. Thus integrating $\mathbf{V}$ over the surface of a cylinder of radius r about the origin and of unit length we see that q is the volume flow rate per unit length of the line from the source. We call q the strength of the point source in the plane.

With this in mind we let the constant in Eq. (7.35) be zero and we see that between the line $\theta = 0$ and the line $\theta = \pi/2$ for example (one quadrant) exactly one-fourth of q is emitted, i.e.,

(7.39)
$$\psi(\pi/2) - \psi(0) = \tfrac{1}{4}q$$

in accordance with our discussion of Eq. (7.34).

If we simply replace q by $-q$ the line source, or *point source in the plane*, becomes a *point sink in the plane* which absorbs q units of volume per unit time per unit length of the line.

Thus far we have not remarked on the fact that the ϕ or ψ corresponding to this complex potential is a solution of Laplace's equation. As a matter of fact these do satisfy Laplace's equation everywhere *except* at the origin. At $z = 0$ Ω is *not* analytic since its derivative $d\Omega/dz$ does *not* exist. We also note that in order to make Ω analytic elsewhere we must take the *principal value* as we did in Eq. (7.37) above. It is also important to note that *always* if ϕ and ψ satisfy Laplace's equation so does Ω itself.

In fluid dynamics the line integral,

(7.40)
$$\oint_c \mathbf{V}\cdot\mathbf{dr} = \kappa$$

around some closed contour is called the *circulation* on the contour. When this integral is a *constant* if taken about *any* contour in the x-y plane enclosing some point z_0, the point z_0 is said to be a line vortex. The complex potential corresponding to this situation is given by

(7.41)
$$\Omega = -i\frac{\kappa}{2\pi}\ln z,$$

where κ is the circulation of the vortex. Here we have,

(7.42)
$$\begin{cases} \phi = \dfrac{\kappa}{2\pi}\theta \\[2mm] \psi = -\dfrac{\kappa}{2\pi}\ln r \end{cases}$$

and using $\mathbf{V} = \nabla\phi$ in Eq. (7.40) shows that indeed the line integral just has the value κ. Again Laplace's equation is satisfied everywhere *except* the origin.

Quite obviously to represent a source located for example at some point z_1 instead of at the origin we simply translate axes as,

(7.43)
$$\Omega = \frac{q}{2\pi}\ln(z - z_1) + \text{constant}$$

Furthermore, since Laplace's equation is a *linear* equation we are free to *add* solutions so that for example,

(7.44)
$$\Omega = \frac{q}{2\pi}\ln(z - z_1) - \frac{q}{2\pi}\ln(z - z_2) + \text{constant}$$

which corresponds to the concurrent presence of a source of strength q at z_1 and a sink of strength q at z_2. Further generalization is evident. Obviously Ω does *not* satisfy Laplace's equation *at* z_1 or z_2.

METHOD OF IMAGES

If we examine the potential ϕ corresponding to the complex potential Ω in Eq. (7.44) above we will find that if the additive constant is taken as zero then the straight line bisecting the line joining the points z_1 and z_2 corresponds to $\phi = 0$; that is zero potential. Thus if we are given a line source located at a distance l from a plane of zero potential we can write the potential simply by locating an *image line sink* at an equal distance on the opposite side of the plane and then forget the plane exists. Many extensions of this basic idea are possible and will be discussed extensively in our treatment of Green's functions.

CONFORMAL MAPPING

We have already called attention to the fact that a function $\omega = f(z)$ of the complex variable z can be looked upon as a mapping of a point z in the z plane to a point in the ω plane. Now we examine the manner in which a line element dz is mapped into a line element $d\omega$.

We write the derivative of $\omega(z)$,

(7.45)
$$\frac{d\omega}{dz} = f'(z) = ae^{i\gamma}$$

noting that this function has an amplitude (real) a and a phase, γ. Thus we see,

(7.46)
$$|d\omega| = a|dz|$$

and

(7.47)
$$\text{phase}(d\omega) = \text{phase}(dz) + \gamma$$

Thus the line element dz is amplified in length by the factor a and rotated through the angle γ to form the line element $d\omega$ in the ω plane. Of course a and γ are both functions of z, so that the amount of amplification and rotation of dz which occur depend upon the location of the line element in the z plane.

However we note that at any point z we may construct several line elements dz such as indicated in Fig. 7.3. Here we see that since a and γ are the same for both dz_1 and dz_2 both are amplified and rotated by the *same* amount. Hence the *angle* between them *is preserved*. This is what we mean by a *conformal* mapping.

It is quite evident that this conformal characteristic must fail at any point of the z plane for which

(7.48)
$$f'(z) = 0, \quad \text{or} \quad \infty$$

such *singular* points then must always be noted.

Of course we generally have use only for mappings which are one-to-one and have an inverse. Observe that since for the mapping $\omega = f(z) = u + iv$,

(7.49)
$$\begin{cases} du = \dfrac{\partial u}{\partial x}dx + \dfrac{\partial u}{\partial y}dy \\[2mm] dv = \dfrac{\partial v}{\partial x}dx + \dfrac{\partial v}{\partial y}dy \end{cases}$$

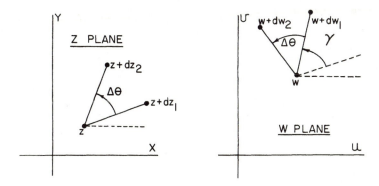

FIGURE 7.3.   **Illustrating the Transformation of Line Elements by a Conformal Mapping.**

we should be able, at any point x, y to solve this set of equations for dx and dy in terms of du and dv. In order that this should be so the determinant

(7.50)
$$\begin{vmatrix} \dfrac{\partial u}{\partial x} & \dfrac{\partial u}{\partial y} \\[2ex] \dfrac{\partial v}{\partial x} & \dfrac{\partial v}{\partial y} \end{vmatrix} = J \neq 0$$

must exist. This is the Jacobian of the transformation and the nonzero requirement assures a unique inverse for the mapping *at the point* at which these derivatives are evaluated. If $f(z)$ is analytic then the Cauchy–Riemann conditions apply and J is seen to reduce to,

(7.51)
$$J = \left(\frac{\partial u}{\partial x}\right)^2 + \left(\frac{\partial v}{\partial x}\right)^2 = |f'(z)|^2$$

Here any point that maps conformally has a unique inverse, i.e., $f'(z) \neq 0, \infty$.

    The value of conformal mapping in the solution of boundary value problems lies in the following fact. If we consider some function $\phi(x, y)$ and employ an analytic function $\omega = f(z) = u + iv$ to express x and y each as functions of u and v we can show that:

(7.52)
$$\frac{\partial^2 \phi}{\partial x^2} + \frac{\partial^2 \phi}{\partial y^2} = |f'(z)|^2 \left(\frac{\partial^2 \phi}{\partial u^2} + \frac{\partial^2 \phi}{\partial v^2}\right)$$

Consequently if the left member is zero, i.e., ϕ satisfies Laplace's equation in some domain R of the x, y plane, and $f'(z) \neq 0, \infty$, then we see that ϕ as a function of u and v also satisfies Laplace's equation in the image of R in the u, v plane.

EXECUTE PROBLEM SET (7-2)

TABLE 7-1.

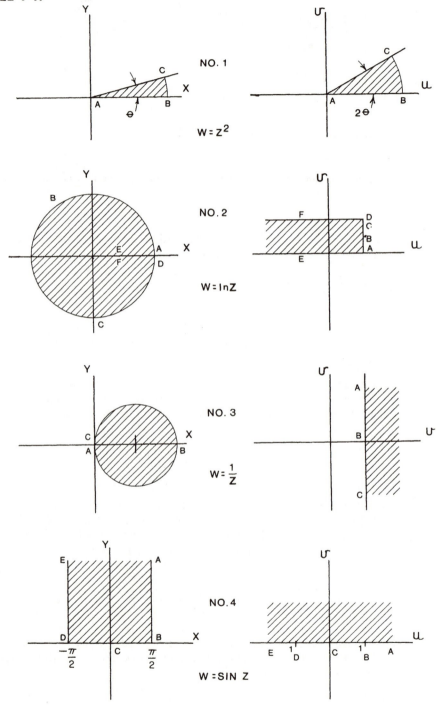

NO. 1

$W = Z^2$

NO. 2

$W = \ln Z$

NO. 3

$W = \dfrac{1}{Z}$

NO. 4

$W = \sin Z$

TABLE 7-1.—(cont.)

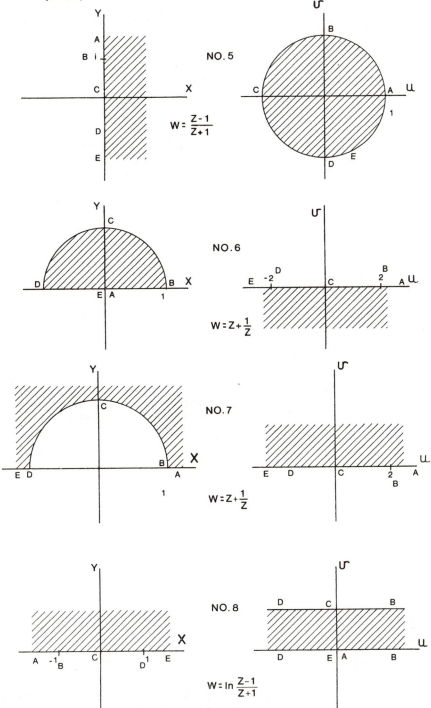

SOLVING PROBLEMS BY CONFORMAL MAPPING

Here we illustrate one approach to the solution of two dimensional potential problems by conformal mapping.

Example (1)

Consider the steady state heat conduction problem,

(7.53)
$$\frac{\partial^2 T}{\partial x^2} + \frac{\partial^2 T}{\partial y^2} = 0, \qquad x > 0, y > 0$$

subject to boundary conditions on the domain indicated below. The mapping function, (see Table 7.1)

(7.54)
$$\omega = \sin^{-1} z = u + iv$$

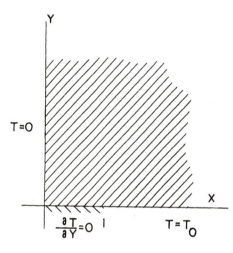

FIGURE 7.4.   Domain and Boundary Conditions for a Heat Conduction Problem in the Z Plane.

transforms the interior of this quadrant of the z plane into the strip in the ω plane. Also the differential equation is just,

(7.55)
$$\frac{\partial^2 T}{\partial u^2} + \frac{\partial^2 T}{\partial v^2} = 0$$

Quite obviously the function,

(7.56)
$$T = \frac{2T_0}{\pi} u$$

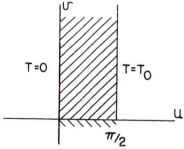

FIGURE 7.5.   The Heat Conduction Problem of Figure 7.4 Mapped to the ω Plane.

satisfies this equation and the indicated boundary conditions. Consequently just *by inspection* it *is* the solution of the problem. However we now must express it back in terms of the original x, y coordinates.

From the above,

(7.57)
$$\sin(u + iv) = x + iy$$

so,

(7.58)
$$\begin{cases} \sin u \cosh v = x \\ \cos u \sinh v = y \end{cases}$$

But since we have the identity,

(7.59)
$$\cosh^2 v - \sinh^2 v = 1$$

we have here,

(7.60)
$$\begin{cases} (\cosh v + \sin u)^2 = (x + 1)^2 + y^2 \\ (\cosh v - \sin u)^2 = (x - 1)^2 + y^2 \end{cases}$$

which yields,

(7.61)
$$2 \sin u = \sqrt{(x + 1)^2 + y^2} - \sqrt{(x - 1)^2 + y^2}$$

Finally then our solution is:

(7.62)
$$T(x, y) = \frac{2T_0}{\pi} \arcsin \tfrac{1}{2}\{[(x + 1)^2 + y^2]^{\frac{1}{2}} - [(x - 1)^2 + y^2]^{\frac{1}{2}}\}$$

It is evident that the essence of this method is to map the domain of the problem into another domain such that the solution is evident by inspection in the new domain. Actually many problems can be solved in this way. Table 7.1 lists a few common mapping functions applied to some simple domains. These are sufficient to treat the following problems.

EXECUTE PROBLEM SET (7-3)

SCHWARTZ–CHRISTOFFEL TRANSFORMATION

While no completely general technique exists for constructing the appropriate transformation to solve any given problem in the manner just discussed above, there are a few methods of great utility. One of these is that for mapping a polygonal line in one plane into the real axis of another plane.

Consider the polygonal line in the z plane as indicated in the sketch below and let this be mapped onto the real axis in the ω plane. To develop this mapping consider,

(7.63)
$$\frac{dz}{d\omega} = A(\omega - a)^c$$

where c is a *real* constant and A is a complex constant. At the point z indicated and its image ω we see:

(7.64)
$$\text{phase}(dz) = \text{phase}(A) + c\,\text{phase}(\omega - a)$$

For ω to the left of the point a, $\omega - a$ is negative on the real axis and hence phase $(\omega - a)$ is

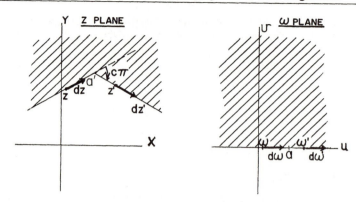

FIGURE 7.6.   Mapping of a Polygonal Line in the Z Plane to the Real Axis of the ω Plane.

π. Thus

(7.65) $$\text{phase } (dz) = \text{phase } (A) + c\pi$$

at this point. On the other hand, if the point ω had been on the right of a, phase $(\omega - a)$ would be zero as at ω', and for the phase of dz, as at z' this would yield

(7.66) $$\text{phase } (dz') = \text{phase } (A)$$

Thus we see that for z anywhere on the first straight line, as at z, the phase of dz is *constant* and given by Eq. (7.65), while for any point on the second straight line it is *constant* and given by Eq. (7.66). Therefore at the vertex a' in the z plane the *change in angle* is:

(7.67) $$\text{phase } (dz') - \text{phase } (dz) = - c\pi$$

or,

(7.68) $$\theta - \theta' = c\pi$$

Here we have an explicit example of the fact that the angle preserving character of mapping with an analytic function fails when the function $z(\omega)$ or $\omega(z)$ has a zero derivative. However it is precisely this feature here which provides the mapping of the polygonal line in the z plane onto the straight line, the real axis, in the ω plane. Quite obviously if the change in angle, $\theta - \theta'$ is given then c is determined by Eq. (7.68). The complex constant A is determined by the angle, θ', of the second straight line and the position of the vertex in the z plane.

Observe that when the polygonal line in the z plane and the real axis in the ω plane are traversed in the *sense* indicated then the region to the left of the polygonal line maps to the *upper* half of the ω plane.

This method can be generalized to a *polygon* of any number of vertices in the form:

(7.69) $$z = A \int (\omega - a_1)^{-\frac{\alpha_1}{\pi}}(\omega - a_2)^{-\frac{\alpha_2}{\pi}} \dots (\omega - a_n)d\omega^{-\frac{\alpha_n}{\pi}} + B$$

where $\alpha_1, \alpha_2, \dots \alpha_n$ are the *exterior* angles of the polygonal line in the z plane and $a_1, a_2, \dots a_n$ are the image points of these vertices on the real axis in the ω plane, i.e., as indicated below. Here we must have for a closed polygon,

(7.70) $$\alpha_1 + \alpha_2 + \dots + \alpha_n = 2\pi$$

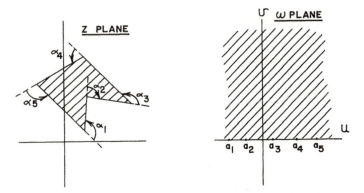

FIGURE 7.7.   Mapping of a Closed Polygon in the Z Plane to the Upper Half Space in the ω Plane.

and we state, without proof, that the locations of any *three* of the image points, $a_1, a_2, \ldots a_n$ may be chosen at will. The others are then fixed by closure and cyclic order conditions.

We also point out the special situation which occurs if we place one of the image points, a_j, at infinity on the u axis. Since this point is "never" reached in our traverse we must delete the corresponding factor, $(\omega - a_j)$, from our expression above. This is illustrated in our Example 5 of this chapter.

The constant A determines the size and orientation of the polygon in the z plane and the constant B determines its location. As a very special and most useful example we here show the mapping of the rectangle into the upper half-plane.

Let the vertices of our rectangle in the z plane be mapped to $-1/\kappa, -1, +1, +1/\kappa$ on the u axis in the ω plane. Here we have chosen *three* of these image points and let the fourth be fixed by symmetry, i.e., we will also require that if $z \to \omega$ then $-z \to -\omega$. Thus since all exterior angles are $\pi/2$ we have:

$$(7.71) \qquad z = A \int_0^\omega \frac{d\omega}{(\omega^2 - 1)^{\frac{1}{2}}\left(\omega^2 - \dfrac{1}{\kappa^2}\right)^{\frac{1}{2}}}$$

since by symmetry we put $B = 0$ and have $z = 0 \to \omega = 0$. Now this is rearranged to read,

$$(7.72) \qquad z = \int_0^\omega \frac{d\omega}{\sqrt{(1 - \omega^2)(1 - \kappa^2\omega^2)}}$$

by factoring out κ^{-1} in the denominator and then putting the constant, $A\kappa = 1$.

Evaluating this integral for the upper limit $\omega = 1$ yields

$$(7.73) \qquad z(\omega = 1) = \int_0^1 \frac{d\omega}{\sqrt{(1 - \omega^2)(1 - \kappa^2\omega^2)}} = K(\kappa)$$

where $K(\kappa)$ is the complete elliptic integral of the first kind of modulus κ and for $\omega = -1$

$$(7.74) \qquad z(\omega = -1) = -K(\kappa)$$

similarly for the upper limit $z = 1/\kappa$ we find

$$(7.75) \qquad z(\omega = 1/\kappa) = K(\kappa) + iK(\kappa^*)$$

where

(7.76) $$\kappa^* = \sqrt{1 - \kappa^2}$$

Thus the rectangle appears in the z plane as in Fig. 7.8 below.

Points in the interior of this rectangle are related to points in the upper half of the ω plane by *incomplete* elliptic integrals which are rather unwieldy to deal with. Even so this mapping does have great utility in many problems since the complete elliptic integrals are readily available in many tables.

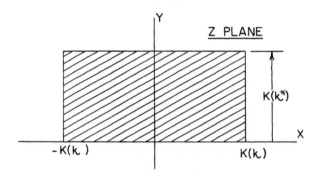

FIGURE 7.8. Rectangle in the Z Plane which is Mapped to the Upper Half Space in the ω Plane.

Example (2)

Capacitance of a slotted cylinder per unit length. We consider a long conducting cylindrical shell of radius, R (having two symmetrically located slots of angular width θ_0) filled with a dielectric of permittivity ε, as shown in Fig. 7.9. We wish to compute the capacitance per unit length of this cylindrical device. From electrostatic theory this is given by,

(7.77) $$C = \frac{q}{\Delta\phi}$$

where q is the charge per unit length on one plate and $\Delta\phi$ is the potential difference between plates. The q is determined as the integral of the charge per unit area, σ, over the area of one

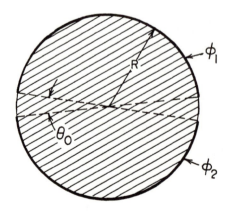

FIGURE 7.9. Cross Section of the Slotted Cylinder Capacitor.

plate in a unit length of the cylinder, and σ is given by:

$$(7.78) \qquad\qquad \sigma = \varepsilon \left(\frac{\partial \phi}{\partial r} \right)_{r=R}$$

i.e., σ is proportional to the normal derivative of ϕ. Thus C is given by:

$$(7.79) \qquad\qquad C = \varepsilon \frac{\displaystyle\int_{\theta_0/2}^{\pi-\theta_0/2} \left(\frac{\partial \phi}{\partial r} \right)_R r\, d\theta}{\phi_2 - \phi_1}$$

If we neglect the edge effect in the slots then we have ϕ satisfying Laplace's equation in the *circular* domain and since there are two lines of symmetry in this domain we represent the boundary conditions as in Fig. 7.10, Here we have rotated the figure 90° and have cut the mid-plane by the x axis which by symmetry must be a "streamline" as must be the arc in the slot.

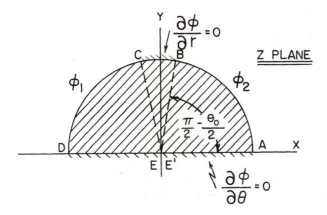

FIGURE 7.10. Boundary Conditions for the Slotted Cylinder Capacitor Using One Half of the Cross Section.

(This is because we neglect the edge effect.) To this we apply a slight modification of the mapping given as Entry Number (6) of our table,

$$(7.80) \qquad\qquad \omega = - \left(\frac{z}{R} + \frac{R}{z} \right)$$

which yields the domain as shown in Fig. 7.11. The letters A, B, C, D, E, E' are used to show the image positions of the various critical points of the original figure. Note that the point E at $0 - \varepsilon$ goes to $+\infty$ while E' at $0 + \varepsilon$ goes to $-\infty$.

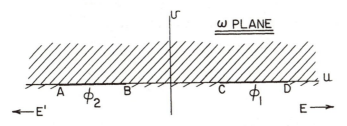

FIGURE 7.11. First Mapping of the Capacitor Problem.

The coordinates of the points B and C are, in the ω plane:

(7.81)
$$\begin{cases} u_B = -2\sin\dfrac{\theta_0}{2} \\[2mm] u_C = +2\sin\dfrac{\theta_0}{2} \end{cases}$$

while those of points A and D are

(7.82)
$$\begin{cases} u_A = -2 \\ u_D = +2 \end{cases}$$

Thus if we now simply define a new transformation as,

(7.83)
$$\omega' = \frac{\omega}{2\sin\dfrac{\theta_0}{2}}$$

the resulting picture will appear as in Fig. 7.12, where

(7.84)
$$\kappa = \sin\frac{\theta_0}{2}$$

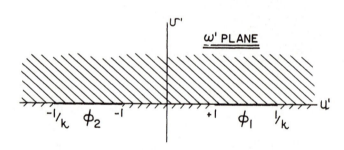

FIGURE 7.12.　Second Mapping of the Capacitor Problem.

Thus we can now apply the mapping of this into the rectangle as developed above; it appears as in Fig. 7.13,† where,

(7.85)
$$\omega'' = \int_0^\omega \frac{d\omega'}{\sqrt{(1-\omega'^2)\left(1-\omega'^2\sin^2\dfrac{\theta_0}{2}\right)}}$$

Obviously the solution to the potential problem here is

(7.86)
$$\phi = \frac{\phi_1 - \phi_2}{2K\left(\sin\dfrac{\theta_0}{2}\right)}\left[u'' + K\left(\sin\dfrac{\theta_0}{2}\right)\right] + \phi_2$$

† Note that $\kappa^* = \sqrt{1 - \sin^2\dfrac{\theta_0}{2}} = \cos\dfrac{\theta_0}{2}$.

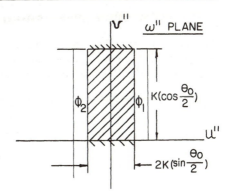

FIGURE 7.13. Final Mapping of the Capacitor Problem.

and hence the normal derivative of ϕ on the left plate is

(7.87)
$$\frac{\partial \phi}{\partial u''} = \frac{\phi_1 - \phi_2}{2K\left(\sin \dfrac{\theta_0}{2}\right)}$$

Then the integral of this over the plate (still taking *unit* length out of the plane of the paper)

(7.88)
$$\int_0^{K\left(\cos \frac{\theta_0}{2}\right)} \frac{\partial \phi}{\partial u''} \, dv'' = (\phi_1 - \phi_2) \frac{K\left(\cos \dfrac{\theta_0}{2}\right)}{2K\left(\sin \dfrac{\theta_0}{2}\right)}$$

But according to Problem (6) of Set 7-2 of this chapter this integral is an invariant and is therefore equivalent to the integral over the corresponding segment of plate in the original domain. Since the integral above corresponds to only *one-half* of one of the original plates we then have, putting this back into the equivalent of Eq. (7.79),

(7.89)
$$C = \varepsilon \frac{K\left(\cos \dfrac{\theta_0}{2}\right)}{K\left(\sin \dfrac{\theta_0}{2}\right)}$$

as the capacitance per unit length of the cylindrical capacitor with slots.

   This example points up several important characteristics of conformal mapping techniques. First a *sequence* of mappings is usually needed to get the problem into a final suitable form. Second, and most important in the present example, is that one can compute many scalar characteristics of a system dependent on the *shape*, such as the capacitance, resistance, etc., in the *final domain*, without expressing the potential, ϕ, ψ, or Ω back in terms of the original variables. Care must be used however to be certain that this invariance applies to the particular quantity of interest, i.e., see the problems of Set (7-2).

EXECUTE PROBLEM SET (7-4)

FLUID FLOW AND AIRFOIL PROBLEMS

The velocity vector $\mathbf{V}$ with components V_x, V_y can be expressed in terms of either $\partial\phi/\partial x$ and $\partial\phi/\partial y$, or $\partial\psi/\partial y$, $-\partial\psi/\partial x$ as we have already noted in Eqs. (7.9) and (7.11). Thus consider the expression $d\Omega/dz$. This can be written as

$$(7.90) \qquad \frac{d\Omega}{dz} = \frac{\partial\Omega}{\partial x} = -i\frac{\partial\Omega}{\partial y}$$

for all points at which Ω is analytic.† Thus we see

$$(7.91) \qquad \frac{d\Omega}{dz} = V_x - iV_y$$

by virtue of our previous equations.

This has great utility in applying conformal mapping to fluid-flow problems. In our earlier examples of applying conformal mapping to steady state heat conduction and other potential problems the quantity of interest was the potential itself, thus having constructed a solution, $\phi(u, v)$ say, in the mapped domain it was then necessary to go through, what in some cases can be rather tedious algebra, to express this back in terms of the original coordinates. Also as we noted in applying the Schwartz–Christoffel transformation above the inverse functions are in many cases quite intractable.

However if only V_x and V_y are of interest we can apply the simple laws of calculus in the following manner. Having mapped with a function

$$(7.92) \qquad \omega = \omega(z) = u + iv$$

to a domain in which we can easily find a solution for Ω, say $\Omega(\omega)$ then we have

$$(7.93) \qquad \frac{d\Omega}{dz} = \frac{d\Omega}{d\omega}\cdot\frac{d\omega}{dz} = V_x - iV_y$$

Here $\Omega(\omega)$ is the easily constructed solution in the mapped domain and $\omega(z)$ is the known mapping function we used. This makes the algebra of obtaining $V_x(x, y)$ and $V_y(x, y)$ much simpler in nearly all cases.

Example (3)

Flow around a cylinder. We consider a cylinder of radius R located with center at the origin in the z plane and suppose an ideal fluid to be flowing around the cylinder with the general direction of flow parallel to the x axis, except of course in the immediate vicinity of the cylinder. Since in this case the x axis is obviously a streamline which must branch and circumvent the cylinder we need only consider the upper half-plane as in Fig. 7.14. The heavy line is a streamline and the points at $z = \pm R$ must obviously be points of zero fluid velocity. These are called *stagnation points.*

We have the differential equation for any of ϕ, ψ, or Ω as Laplace's equation, with the boundary conditions,

$$(7.94) \qquad \begin{cases} \psi = 0 \\[2mm] \dfrac{\partial\phi}{\partial n} = 0 \quad \text{(on the heavy curve in Fig. 7.14)} \end{cases}$$

† That is, by the definition of the *existence* of $d\Omega/dz$ defined at the beginning of this chapter.

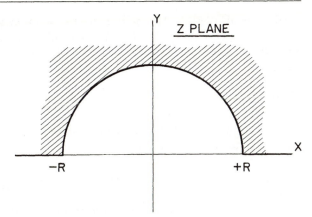

FIGURE 7.14. Fluid Flow Around a Cylinder.

also, at infinite distance from the cylinder we have uniform flow in the $+x$ direction. This can be represented by,

(7.95) $$\text{as} \quad |z| \to \infty, \qquad \frac{\partial \phi}{\partial x} = \frac{\partial \psi}{\partial y} = \frac{d\Omega}{dz} = V_0$$

We map the region *outside* the cylinder in the upper half-plane onto the upper half of the ω plane by the mapping,

(7.96) $$\omega = \frac{z}{R} + \frac{R}{z}$$

which is the appropriate modification of Mapping Number (7) of our table. Thus the u axis is now the streamline which appeared as the heavy line in Fig. 7.14 above. Obviously we have a simple horizontal flow in this plane and therefore in the ω plane

(7.97) $$\Omega = A\omega$$

where A is a constant to be determined.

Looking at the *form* of $\omega(z)$ we see that as $|z| \to \infty$ the z and ω domains differ only by the factor of $1/R$, a constant. Thus

(7.98) $$u \approx \frac{1}{R} x$$

as z is large, thus since *at large z* we have

(7.99) $$\frac{\partial \phi}{\partial x} = \frac{1}{R} \frac{\partial \phi}{\partial u} = V_0$$

we see the constant in Eq. (7.97), must be just RV_0 and hence,

(7.100) $$\Omega = RV_0\omega$$

Here we can easily substitute to obtain

(7.101) $$\Omega = V_0\left(z + \frac{R^2}{z}\right)$$

for the original domain. Also now having this solution, which must apply for the whole z plane outside the circle we note that if the flow is incident not in the direction of the x axis but at some angle θ_0 above the axis we need only multiply z by $e^{i\theta_0}$ to obtain the appropriate potential, potential,

$$(7.102) \qquad \Omega = V_0(ze^{i\theta_0} + \frac{R^2}{ze^{i\theta_0}})$$

We also note now that if the center of our circle is translated to the point z_0 in the z plane then the potential must become,

$$(7.103) \qquad \Omega = V_0\left[(z - z_0) + \frac{R^2}{(z - z_0)}\right]$$

Having this general expression now for the flow of the ideal fluid incident with speed V_0 at angle θ_0 above the x axis on a circle of radius R having center at z_0, we show how this circle is mapped into an airfoil, and hence determine the velocity profile around an airfoil.

Example (4)

Mapping of a circle into an airfoil. An airfoil, as the wing of an airplane, is here approximated as having a uniform cross section at all points of its extension in length. It is a smooth closed curve *except* at one point, the "trailing edge." At this point the tangent to the curve is discontinuous.

We wish to construct a mapping of the region exterior to the circle in the z plane onto the region exterior to our airfoil in the ω plane. Thus the mapping must be conformal for *all* points *exterior* to the circle and all points *on* the circle *except* the one point which maps to the trailing edge of the airfoil. We have this shown pictorally in Fig. 7.15.

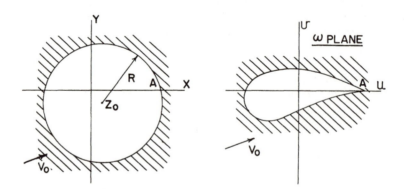

FIGURE 7.15.   Mapping of a Circle Into an Airfoil.

To construct a general mapping of this sort we begin with ω as a general Laurent Series[†] about the origin,

$$(7.104) \qquad \omega = \sum_{n=0}^{\infty} a_n z^n + \sum_{n=1}^{\infty} \frac{b_n}{z^n}$$

But immediately we see that if we want the mapping to be conformal at infinity, $d\omega/dz \neq \infty$ as

[†] This series is discussed in the next chapter.

$z \to \infty$, then we must require all a_n except a_0 and a_1 be zero. Furthermore since a_0 is just a translation of the origin we can put $a_0 = 0$ with no loss of generality.

Now let us suppose for simplicity that the principal part of $\omega(z)$, the second sum above, terminates at N terms, then we have,

$$(7.105) \qquad \omega = a_1 z + \frac{b_1}{z} + \frac{b_z}{z^2} + \ldots + \frac{b_N}{z^N}$$

Then

$$(7.106) \qquad \omega' = \frac{d\omega}{dz} = a_1 - \frac{b_1}{z^2} - \frac{2b_z}{z^3} - \ldots - \frac{Nb_N}{z^{N+1}}$$

and this *vanishes* at points determined by,

$$(7.107) \qquad \omega' = \frac{a_1 z^{N+1} - b_1 z^{N-1} - 2b_2 z^{N-2} - \ldots - Nb_N}{z^{N+1}} = 0$$

or, just when the numerator is zero.

This is a polynomial of $(N + 1)$st order and therefore has $N + 1$ roots. These are singular points of the mapping function and so is the point $z = 0$ at which $\omega' = \infty$. None of these points may be outside the circle but *one* must be *on* the circle. Observe that, since the coefficient of the second highest power of z in the numerator above is zero, this means the *sum* of the roots of the polynomial is zero, i.e., if we let $\beta_1, \beta_2, \ldots \beta_{N+1}$ be the roots of the numerator in Eq. (7.107) then

$$(7.108) \qquad \omega' = \frac{a_1(z - \beta_1)(z - \beta_2)\ldots(z - \beta_{N+1})}{z^{N+1}}$$

and expanding the product gives

$$(7.109) \qquad \beta_1 + \beta_2 + \ldots + \beta_{N+1} = 0$$

to agree with Eq. (7.107). Hence the critical points have the origin as their *centroid*.

Thus we have a general procedure for forming a mapping having the desired features; simply pick a set of critical points β_i with centroid at the origin, such that one, say β_1, is *on* the circle and *all* others are inside the circle.

The simplest mapping of this kind is one having just *two* critical points of $\omega'(z)$, other than $z = 0$ which is *always* included, namely,

$$(7.110) \qquad \omega(z) = a_1 z + \frac{b_1}{z}$$

which is just a generalization of Entry Number (7) of our table. We leave it as an exercise for the student to examine the relationships between the constants a_1, b_1, and the center z_0 and radius R of the circle to the form and orientation of the airfoil.

Of course we desire the velocity components, V_u and V_v, about the airfoil in the ω *plane* while we already have $\Omega(z)$ determined in the z *plane* as Eq. (7.103). These will be given by applying the procedure of Eq. (7.93). Again we leave this to the student as an exercise.

Example (5)

Change of depth in a river bed. Here we illustrate one important technique in applying the Schwartz–Christoffel transformation.

We consider a uniform stream or river having a sudden depth change at some point as illustrated in Fig. 7.16. The line forming the bottom of the stream is a streamline, ψ = constant = 0, and the upper surface is also a streamline, ψ = 1.

We map the interior of this domain to the upper half of the ω plane by the Schwartz–Christoffel transformation as follows. The procedure here is essentially due to Churchill.[†]

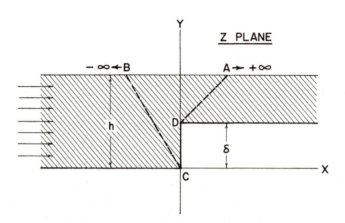

FIGURE 7.16.   Change in Depth in a River.

We view the domain in the z plane as the limiting form of the polygon, indicated by the dotted lines in Fig. 7.16, as the point $A \to +\infty$ and the point $B \to -\infty$. We have four vertices with exterior angles, *in the limit*, $\alpha_A = \pi$, $\alpha_B = \pi$, $\alpha_C = \pi/2$, and $\alpha_D = -\pi/2$. We choose the image points in the ω plane as $a_B = 0$ for the point B, a_C for the point C, and $a_D = 1$ for the point D. We let $a_A \to \infty$, consequently the factor $(\omega - a_A)$ will *not* appear in our expression for the transformation. Thus we have,

(7.111) $$\frac{dz}{d\omega} = A\omega^{-1}(\omega - a_C)^{-\frac{1}{2}}(\omega - 1)^{+\frac{1}{2}}$$

Here it should be noted that we have chosen *three* of the image points so that the remaining one, a_C here, will have to be determined, as will the multiplying constant A. We proceed now to determine these constants.

Observe that in the original domain the fluid flowing from the left must originate somewhere. At any *finite* distance to the left we assume a *uniform* flow toward the right exists, but as we move to an infinite distance there is no loss of generality in assuming *all* of the flow to originate from a point source in the plane. In particular we can let this source be at the point B (which moves to infinity). In the ω plane this point, $\mathscr{A}_B = 0$, is at the origin.

If q is the volume of fluid per unit time flowing in unit width of the stream (unit width out of the plane of the paper) then in the ω plane we have the picture shown below in Fig. 7.17. We see that this appears as a point source in the plane, the whole u axis is a streamline. However the strength of the source is $2q$ because the quantity q is liberated per unit time in the upper half-plane and by symmetry an equal amount must be liberated per unit time in the lower half-plane.

† Churchill, R. V.; *Introduction to Complex Variables and Applications*, McGraw-Hill Book Co., New York (1948).

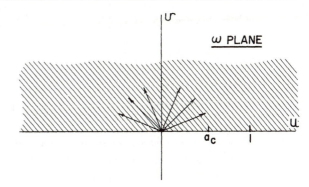

FIGURE 7.17. Mapping of the River Problem.

Also observe that the point A of the original picture ($A \to \infty$) must be an equal sink, but in the ω plane this sink is at infinity and need not be considered.

Thus in the ω plane we see the complex potential as

$$(7.112) \qquad \Omega = \frac{2q}{2\pi} \ln \omega$$

and to find the components of $\mathbf{V}$ in the z plane we write,

$$(7.113) \qquad \frac{d\Omega}{dz} = V_x - iV_y = \frac{d\Omega}{d\omega}\frac{d\omega}{dz} = \frac{q}{\pi}\frac{1}{\omega}\frac{d\omega}{dz}$$

But we have $d\omega/dz$ as the reciprocal of Eq. (7.111), thus

$$(7.114) \qquad V_x - iV_y = \frac{q}{\pi A}(\omega - a_c)^{\frac{1}{2}}(\omega - 1)^{-\frac{1}{2}}$$

Now refer again to our condition on the image of the point B. This point is a real part of $\omega \to \infty$. Thus in the above let $v = 0$ and $u \to +\infty$. We find in the limit

$$(7.115) \qquad V_x = \frac{q}{\pi A}$$

But from physical considerations (definition of q) we see that to the far right of our picture

$$(7.116) \qquad V_x = \frac{q}{h - \delta}$$

Thus we find

$$(7.117) \qquad A = \frac{h - \delta}{\pi}$$

so that the constant A is fixed.

Recalling that $\omega = 0$, the position of the source in the ω plane, corresponds to the limiting position of the point B of our original picture. This must be the region of *uniform* flow in the *deep* part of the channel to the left. Thus now put $\omega = 0$ in Eq. (7.114), using Eq. (7.117) now for A, and obtain,

$$(7.118) \qquad V_x = \frac{q}{h - \delta}\sqrt{a_c}$$

But, again from physical consideration we see that V_x must be in this region,

(7.119)
$$V_x = \frac{q}{h}$$

Hence we see that

(7.120)
$$a_c = \left(\frac{h - \delta}{h}\right)^2$$

Thus all constants of the transformation are determined.

Now we proceed to integrate Eq. (7.111) to express z as an explicit function of ω. We make the substitution,

(7.121)
$$\xi^2 = \frac{\omega - 1}{\omega - a_c}$$

to obtain,

(7.122)
$$-\frac{1}{2}z = A \int \frac{(a_c - 1)\xi^2 \, d\xi}{(1 - \xi^2)(1 - a_c\xi^2)} + K$$

where K is a constant of integration. We then expand the integrand here by partial fractions and the result readily integrates to,

(7.123)
$$z = 2A\left(\frac{1 - a_c}{1 + a_c}\right)\left[\ln \frac{1 + \xi}{1 - \xi} + \frac{1}{\sqrt{a_c}} \ln \frac{1 + \sqrt{a_c}\xi}{1 - \sqrt{a_c}\xi}\right] - 2K$$

Finally putting $z = 0$ and $\omega = a_c$, $\xi \to \infty$, determines the constant K. Then we can use Eq. (7.121) to obtain the final result of z as an explicit function of ω, i.e., since A and a_c are already determined.

In *principle* then we can express Ω as a function of z, however this is algebraically nearly impossible. But this should not be construed as indicating this analysis would not be of practical utility.

EXECUTE PROBLEM SET (7-5)

PROBLEMS

Set (7-1)

(1) If $w = u + iv$ is an analytic function of the complex variable $z = x + iy$, prove that the curves of constant u and constant v are orthogonal at their points of intersection.
(2) Derive the Cauchy–Riemann conditions in plane polar coordinates.

Set (7-2)

(1) Verify Eq. (7.52).
(2) Show that under a conformal mapping a source of strength q is mapped into a source of strength q.
(3) Derive the complex potential for a doublet, i.e., a source q and sink $-q$ separated by an infinitesimal distance Δl, but such that $q\Delta l = m$ is finite.
(4) Show precisely how *a doublet is modified* under a conformal mapping.

(5) Show how a boundary condition, $\phi = \phi(x, y)$ on a curve C, $y = y(x)$, is transformed by a conformal mapping.

(6) Show how a boundary condition, $\partial\phi/\partial n = g(x, y)$ on a boundary curve, $y = y(x)$, is transformed by a conformal mapping. ($\partial\phi/\partial n$ denotes the normal derivative.)
[Hint: if $\mathbf{dr} = \mathbf{1}_x\, dx + \mathbf{1}_y dy$ is a line element of the boundary then,

$$|\nabla\phi \times \mathbf{dr}| = \frac{\partial\phi}{\partial n}\, ds$$

where, $ds = |\mathbf{dr}|$].

(7) Verify that on the plane perpendicularly bisecting the line connecting two equal sources, $\partial\phi/\partial n = 0$.

Set (7-3)

Solve

$$\frac{\partial^2 T}{\partial x^2} + \frac{\partial^2 T}{\partial y^2} = 0$$

in the given domain with the stated boundary conditions; (as follows):

(1) $0 < \sqrt{x^2 + y^2} < R$, $0 < \theta < \theta_m$ (i.e., a "pie slice" of the circle of radius R) with:

$T = 0$ on $\theta = 0$, $0 < r < R$

$T = T_0$ on $\theta = \theta_m$, $0 < r < R$

$\dfrac{\partial T}{\partial r} = 0$ on $r = R$, $0 < \theta < \theta_m$

(2) $x > 0$, $y > 0$ (i.e., the positive quadrant of the plane) with:

$T = 0$ on $y = 0$, $0 < x < \infty$

$T = T_0$ on $x = 0$, $0 < y < \infty$

(3) The semi-infinite strip, $0 < x < L$, $y > 0$, with:

$T = 0$ on $y = 0$, $0 < x < L$

$T = 0$ on $x = L$, $0 < y < \infty$

$T = T_0$ on $x = 0$, $0 < y < \infty$

(4) The semicircular domain of radius R with center at the origin in the upper half-plane defined by $0 < \theta < \pi$, $0 < r < R$, with:

$T = 0$, on $\theta = 0$ and $T = T_0$ on $\theta = \pi$, all r

$T = T_0$ on $r = R$, $\pi/2 < \theta < \pi$

$\dfrac{\partial T}{\partial r} = 0$ on $r = R$, $0 < \theta < \pi/2$

(5) A line heat source (point source in the plane) of strength q, parallel to the axis of a circular cylinder of radius R whose wall is kept at $T = 0$, when the line source is a distance $\delta < R$ from the axis.

(6) A point source, as in Problem (5) above, a distance $\delta < L$ from one wall of a semi-infinite strip of width L, and a distance h from one end, the end and both walls being insulated, $\partial T/\partial n = 0$. (Note that physically there must exist a point *sink* of equal strength at the end at infinity *within* the strip. Thus if under some mapping this infinite end is brought to the *finite* region we must include this sink in our analysis, this occurs in the following problem.)

(7) A point source as in Problem (6) above, a distance $\delta < L$ from one wall of an infinite strip $-\infty < x < \infty$, $0 < y < L$, with both walls insulated. [See comment in Problem (5) above.]

Set (7-4)

(1) Use the Schwartz–Christoffel mapping to derive Mapping Number (4) of our Table 7-1.

(2) Use the Schwartz–Christoffel mapping to derive Mapping Number (8) of our Table 7-1. [Hint: treat the ω plane strip there as the initial z plane, interchange z and ω pictures, then look upon the strip as the *limiting form* of a closed polygon by letting the points marked D and D' be a vertex and those marked B and B' be a vertex then let $D \to -\infty$, $B \to +\infty$, somewhat as we do with points B and A in Fig. 7.16 of the text.]

Set (7-5)

(1) Show that if in the mapping of a circle into an airfoil we let *two* critical points of the mapping be on the circle the airfoil is a straight strip, i.e., let the circle have center at the origin, radius R and let the mapping be

$$\omega = z - \frac{R^2}{z}$$

In this way solve the problem of flow around an infinite strip of width $2R$.

(2) Demonstrate the general dependence of the shape of an airfoil, generated from the circle of radius R, center z_0, by the mapping

$$\omega = z - \frac{K^2}{z}, \; K \; real,$$

on the parameters R, K, and z_0.

(3) Use the Schwartz–Christoffel mapping to solve the problem of a change in depth, δ, in a river of infinite depth.

(4) Use the Schwartz–Christoffel mapping to solve the problem of a thin vertical barrier of height H rising from the flat bed of a river of infinite depth. [Hint: view the barrier as the limiting form of a triangle whose vertex included angle $\to 0$.]

REFERENCES

R. V. Churchill, *Complex Variables and Applications* (McGraw-Hill Book Company, Inc., New York, 1960).

E. T. Copson, *Theory of Functions of a Complex Variable* (Oxford University Press, London, 1935).

J. Irving and N. Mullineaux, *Mathematics in Physics and Engineering* (Academic Press Inc., New York, 1959).

Z. Nehari, *Conformal Mapping* (McGraw-Hill Book Company, Inc., New York, 1952).

P. M. Morse and H. Feshbach, *Methods of Theoretical Physics* (McGraw-Hill Book Company, Inc., New York, 1953).

CHAPTER EIGHT

Calculus of residues

Here we present the essential elements of the calculus of residues with a few illustrative examples. Our primary motivation is to provide the complex inversion integral for the Laplace transform as a tool available for our use. Of course the residue calculus provides us with a powerful means for evaluation of integrals and many other applications.

Proofs are outlined in most instances in only a heuristic fashion, but in sufficient detail to indicate the basis of the fundamental theorems.

INTEGRATION IN THE COMPLEX PLANE

Let $f(z)$ be an analytic function of the complex variable z in a domain R of the z plane. Then let C be any curve joining two points A and B in the domain. This curve is viewed as the limiting form of a polygonal line made up of straight segments of length $|\Delta z_i|$, $i = 1, 2, \ldots n$ such that

$$(8.1) \qquad \lim_{n \to \infty} \sum_{i=1}^{n} |\Delta z_i| = L$$

exists; here *all* $|\Delta z_i| \to 0$ in the limit. At some point ξ_i on each of the line segments let $f(z)$ be evaluated and the sum

$$(8.2) \qquad I_n = \sum_{k=1}^{n} f(\xi_k)\Delta z_k$$

formed. Then as we let all $\Delta z_k \to 0$ the limit of I_n as $n \to \infty$ is defined as the integral of $f(z)$ along C and is written

$$(8.3) \qquad I = \int_C f(z)\, dz$$

It can be shown that the manner in which the points ξ_i are chosen is immaterial.

From this definition it immediately follows that if H is an intermediate point on C between A and B then,

$$(8.4) \qquad \int_A^B c\, f(z)\, dz = \int_A^H c\, f(z)\, dz + \int_H^B c\, f(z)\, dz$$

and quite obviously also,

$$(8.5) \qquad \int_A^B c\, f(z)\, dz = -\int_B^A c\, f(z)\, dz$$

CAUCHY'S THEOREM

From Stokes' theorem of vector calculus we have

$$(8.6) \qquad \oint_C \mathbf{F} \cdot \mathbf{dr} = \iint_A (\nabla \times \mathbf{F}) \cdot \mathbf{dA}$$

where $\mathbf{F}$ is any vector function, continuous in the domain in question, and C is the *closed* curve bounding the area A. (A is *not* necessarily a plane area.) Here we let

$$(8.7) \qquad \mathbf{F} = F_x \mathbf{1}_x + F_y \mathbf{1}_y$$

and choose a plane area with element,

$$(8.8) \qquad \mathbf{dA} = dx\, dy \mathbf{1}_z$$

and bounding line element,

$$(8.9) \qquad \mathbf{dr} = \mathbf{1}_x\, dx + \mathbf{1}_y\, dy$$

Then Eq. (8.6) appears as,

$$(8.10) \qquad \oint_C (F_x\, dx + F_y\, dy) = \iint_A \left(\frac{\partial F_y}{\partial x} - \frac{\partial F_x}{\partial y} \right) dx\, dy$$

Now if we write out Eq. (8.3) with $f = U + iV$ and $dz = dx + i\, dy$, $(i = \sqrt{-1})$, we see,

$$(8.11) \qquad \oint_C f(z)\, dz = \oint_C (U\, dx - V\, dy) + i \oint_C (V\, dx + U\, dy)$$

and in view of Eq. (8.10) above this can be expressed as:

$$(8.12) \qquad \oint_C f(z)\, dz = -\iint \left(\frac{\partial V}{\partial x} + \frac{\partial U}{\partial y} \right) dx\, dy + i \iint \left(\frac{\partial U}{\partial x} - \frac{\partial V}{\partial y} \right) dx\, dy$$

if C is a closed curve, and the line integral is counterclockwise on C.

Thus we see that *if $f(z)$ is an analytic function inside C* the integrals on the right vanish by virtue of the *Cauchy–Riemann* conditions.

Hence *Cauchy's theorem* simply states that,

$$(8.13) \qquad \oint_C f(z)\, dz = 0$$

for $f(z)$ analytic inside and on the closed curve C.

From this it also follows that if C_1 and C_2 are *two* paths connecting a point A to a point B, and $f(z)$ is analytic *between* and *on* the two curves then,

$$(8.14) \qquad \int_A^B {}_{C_1} f(z)\, dz = \int_A^B {}_{C_2} f(z)\, dz$$

i.e., the integral is independent of the path C.

EXTENSION OF CAUCHY'S THEOREM

If a curve C' encloses another Curve C and $f(z)$ is analytic between and on the two curves then it is quite evident that,

$$(8.15) \qquad \oint_{C'} f(z)\, dz = \oint_C f(z)\, dz$$

The proof is left for the student.

CAUCHY'S INTEGRAL

If $f(z)$ is analytic inside and on a closed curve C and z' is any point within C then,

$$(8.16) \qquad f(z') = \frac{1}{2\pi i} \oint_C \frac{f(z)\, dz}{z - z'}$$

To establish this result one places a small circle of radius ε about the point z' at which the *integrand* on the right is *not* analytic. By the extension of Cauchy's integral theorem then,

$$(8.17) \qquad \oint_C \frac{f(z)\, dz}{z - z'} = \oint_\gamma \frac{f(z)\, dz}{z - z'}$$

where γ is this small circle. We add and subtract a term here to get,

$$(8.18) \qquad \oint_C \frac{f(z)\, dz}{z - z'} = \oint_\gamma \frac{f(z) - f(z')}{z - z'}\, dz + \oint_\gamma \frac{f(z')\, dz}{z - z'}$$

In the first integral on the right the integrand just approaches $f'(z')$ as we let $z \to z'$ as the radius of the circle, $\varepsilon \to 0$, and the perimeter $2\pi\varepsilon$ also approaches 0. Since $f'(z')$ is finite this integral vanishes in the limit. In the second integral on the right $f(z')$ may be brought outside and we put $z - z' = \varepsilon e^{i\theta}$ and evaluate the integral to obtain

$$(8.19) \qquad \oint_\gamma \frac{f(z')\, dz}{z - z'} = f(z') \int_0^{2\pi} \frac{\varepsilon e^{i\theta} i\, d\theta}{\varepsilon e^{i\theta}} = 2\pi i f(z')$$

Thus we see that Eq. (8.16) results.

EXTENSION OF CAUCHY'S INTEGRAL

In a fashion very similar to the above analysis one can show that

$$(8.20) \qquad f^{(n)}(z') = \frac{n!}{2\pi i} \oint_C \frac{f(z)\, dz}{(z - z')^{n+1}}$$

where (n) denotes the nth derivative with respect to z and $f(z)$ is again analytic in and on C.

SERIES EXPANSION OF COMPLEX FUNCTIONS

If $f(z)$ is analytic inside and on a circle C with center at $z = a$, and radius R, and z is any point within this circle then,

$$(8.21) \qquad f(z) = \sum_{n=0}^{\infty} \frac{f^{(n)}(a)}{n!}(z - a)^n$$

which is just *Taylor's expansion* about $z = a$. The proof of this expansion can be built on Cauchy's integral.†

Now if we consider two *concentric* circles C_1 and C_2 with center at $z = a$ with radius $R_1 > R_2$ and require $f(z)$ to be analytic on C_1 and C_2 and in the annular domain between the two circles then for any point z in the annular domain,

$$(8.22) \qquad f(z) = \sum_{n=0}^{\infty} a_n(z - a)^n + \sum_{n=1}^{\infty} \frac{b_n}{(z - a)^n}$$

where,

$$(8.23) \qquad a_n = \frac{1}{2\pi i} \int_{C_1} \frac{f(z)\, dz}{(z - a)^{n+1}}, \qquad b_n = \frac{1}{2\pi i} \int_{C_2} \frac{f(z)\, dz}{(z - a)^{1-n}}$$

Equation (8.22) is the *Laurent's expansion* of $f(z)$ in the given domain. Again the proof can be built on Cauchy's integral.‡

We note the *important fact* that if $f(z)$ is analytic *inside* the inner circle, C_2, then the Laurent series reduces to Taylor's series above and then, and *only then* is $a_n = f^{(n)}(a)/n!$.

ZEROS AND SINGULARITIES

If $f(z)$ is analytic inside some circle of radius R about the point $z = a$ then $f(z)$ can be expanded in a Taylor series as in Eq. (8.21). If it should happen that the first $(n - 1)$ coefficients are all zero then the leading term of the series is the term,

$$\frac{f^{(n)}(a)}{n!}(z - a)^n$$

in which case we say $f(z)$ has a *zero of order n* at $z = a$.

Next consider the Laurent expansion of a function $f(z)$ as just described above. The first series portion in *positive* powers of $z - a$ is certainly analytic at $z = a$ but the second series portion in *negative* powers of $z - a$ is *not analytic* at $z = a$. This series is called the *principal part* of $f(z)$ at $z = a$. Quite obviously if the principal part is zero then $f(z)$ is analytic at $z = a$.

† Also see discussion of *analytic continuation* in the appendix.
‡ See appendix.

We distinguish two important cases when the principal part of $f(z)$ is not zero at $z = a$. If the series *terminates* at the term $b_n/(z - a)^n$ then $f(z)$ is said to have a *pole of order n* at $z = a$. A pole is an *isolated singularity*. If the series *does not terminate* then $f(z)$ is said to have an *essential singularity* at $z = a$.

It is quite evident from the above discussion that a simple test for a zero of order n at say $z = a$ is the following: If

$$(8.24) \qquad \lim_{z \to a} \frac{f(z)}{(z - a)^m} = \begin{cases} 0, & m < n \\ \text{finite}, & \neq 0, \quad m = n \\ \infty, & m > n \end{cases}$$

then $f(z)$ has a zero of order n at $z = a$. Similarly if

$$(8.25) \qquad \lim_{z \to a} (z - a)^m f(z) = \begin{cases} \infty, & m < n \\ \text{finite}, & m = n \\ 0, & m > n \end{cases}$$

then $f(z)$ has a pole of order n at $z = a$.

Example (1)

Note that the function $\csc^2 z$ has poles of *second order* at $z = m\pi$, where $m = 0, \pm 1, \pm 2, \ldots$, since:

$$\lim_{z \to m\pi} (z - m\pi)^2 \csc^2 z = \lim_{z \to m\pi} \frac{2(z - m\pi)}{2 \sin z \cos z}$$

$$(8.26) \qquad\qquad = \lim_{z \to m\pi} \frac{2}{2 \cos 2z}$$

$$= 1$$

by successive applications of L'Hopital's rule in evaluating the limit.

The point is that *it is not necessary* to actually construct the Laurent expansion of a function, $f(z)$ in order to determine the nature of its poles and zeroes.

Example (2)

Construct the Laurent series for the function,

$$(8.27) \qquad f(z) = \frac{1}{z^2 - 3z + 2}$$

in the domain $1 < |z - 1| < 2$, i.e., between the circles of radius *one* and radius *two*, respectively, with center $z = 1$.

We first factor the denominator,

$$(8.28) \qquad f(z) = \frac{1}{(z - 2)(z - 1)}$$

Then we use partial fractions to write this as

$$(8.29) \qquad f(z) = \frac{1}{z-2} - \frac{1}{z-1}$$

Now the first term is analytic at $z = 1$ and hence we can just expand it in an ordinary Taylor series about $z = 1$. Thus,

$$(8.30) \qquad f(z) = -\sum_{n=0}^{\infty} (z-1)^n - \frac{1}{z-1}$$

is the required expansion.

This example points up the fact that frequently one can separate a given function into two parts such that one part is a simple analytic function which can be expanded directly by a Taylor's expansion. However we say again that rarely does one need to actually form a Laurent expansion in applied problems.

BRANCH POINTS AND BRANCH LINES

The concept of a branch point and a branch line is best comprehended with a simple example. Consider a function $f(z)$ and examine how this function changes as we allow z to change. In particular let z be taken continuously about some closed curve C in the complex plane. Upon completing the transit of this curve z has returned to its original value *but* the function $f(z)$ *may or may not* have returned to its original value. If the latter situation exists then $f(z)$ is said to have a *branch point* inside C.

As our particular example let

$$(8.31) \qquad f(z) = (z-a)^{\frac{1}{n}}$$

where n is an integer. In this we substitute,

$$(8.32) \qquad z - a = \rho e^{i\theta}$$

Now consider the two curves C and C' indicated in Fig. 8.1. As z traces out the curve C, starting at A, θ increases from zero at A to a maximum value *less than* 2π at B and then *decreases* again to zero. In the second case as z traces out C', starting from A, θ increases *monotonically* from zero to 2π upon returning to A.

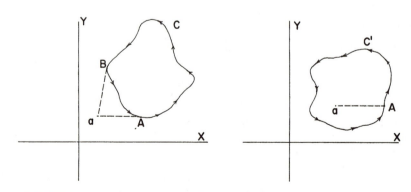

FIGURE 8.1.   Contours Used to Explain the Nature of a Branch Point.

Observe that since

(8.33)
$$f(z) = \rho_n^{\frac{1}{n}} e^{i\frac{\theta}{n}}$$

is a *multivalued function* having the different values, (called *branches*)

(8.34)
$$\begin{cases} f_1(z) = \rho_n^{\frac{1}{n}} e^{i\frac{\theta}{n}} \\ f_2(z) = \rho_n^{\frac{1}{n}} e^{i\frac{\theta + 2\pi}{n}} \\ f_3(z) = \rho_n^{\frac{1}{n}} e^{i\frac{\theta + 4\pi}{n}} \\ \quad \vdots \\ f_n(z) = \rho_n^{\frac{1}{n}} e^{i\frac{\theta + (n-1)2\pi}{n}} \end{cases}$$

we see that $f(z)$ remains say $f_1(z)$ if we start with f_1, so long as $0 < \theta < 2\pi$, but if θ increases beyond 2π, but less than 4π, we jump to the function $f_2(z)$. Thus, in tracing out the curve C above, θ remains bounded, $0 < \theta < 2\pi$, and we maintain the *same branch* of $f(z)$, say $f_1(z)$, all around the contour. But in tracing out curve C' we find upon returning to our starting point we have in essence added 2π to θ so we jump from $f_1(z)$ to $f_2(z)$. If we traversed C' again we would jump to $f_3(z)$ upon returning to A again, etc.

In order to avoid the jump from one *branch* of a multivalued function to another we use the following device. At *any* point on C' we draw a line to the *branch point* a. This we call a *branch line*. Then we indent the curve C' as shown below:

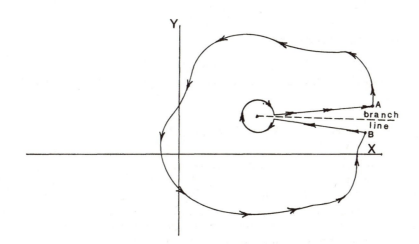

FIGURE 8.2. Illustrating a Branch Cut.

Now we find that if we start, say at A, and trace this new contour, θ increases until we reach point B. Here as we move in along the straight line θ remains fixed, then as we circle a, on the small circle θ *decreases* to its original value and then maintains this value out on the line to A. Thus by introducing the *branch cut†* here we keep θ bounded as $0 < \theta < 2\pi$ and we keep *one* branch of $f(z)$ all the way around the contour.

† The *cut* is assumed to have infinitesimal angular width.

RESIDUES

Let $f(z)$ have a singularity at $z = \alpha$, then its Laurent series appears as

$$(8.35) \quad f(z) = a_0 + a_1(z - \alpha) + a_2(z - \alpha)^2 + \ldots + \frac{b_1}{(z - \alpha)} + \frac{b_2}{(z - \alpha)^2} + \frac{b_3}{(z - \alpha)^3} + \ldots$$

The constant b_1 is called the *residue* of $f(z)$ at α. Observe that if the point α is a pole of order n then the principal part terminates as stated before in the term $b_n/(z - \alpha)^n$. Consequently the *simple test* of a function $f(z)$ to determine the nature of a singularity at $z = \alpha$ is to evaluate the limit,

$$(8.36) \quad \lim_{z \to \alpha} (z - \alpha)^m f(z) = L_m(\alpha)$$

as already pointed out in Eq. (8.25). Certainly if $f(z)$ is regular at α then $L_m(\alpha) = 0$ for any $m = 0, 1, 2, \ldots$. But if $z = \alpha$ is an isolated singularity, a pole of order n then $L_m = \infty$ for $m < n$, $L_m = 0$, $m > n$, and L_m is a nonzero† finite number for $m = n$.

It is readily established that the *residue* at a pole of order n, the constant b_1 above, can be evaluated as:

$$(8.37) \quad b_1 = \frac{1}{(n - 1)!} \lim_{z \to \alpha} \frac{d^{(n-1)}}{dz^{(n-1)}} \{(z - \alpha)^n f(z)\}$$

For a *simple pole*, i.e., of order $n = 1$, this is just,

$$(8.38) \quad b_1 = \lim_{z \to \alpha} (z - \alpha) f(z)$$

CAUCHY RESIDUE THEOREM

This theorem is the key tool in many applications to applied mathematics, we state it as:

"If $f(z)$ is analytic inside and on a closed curve C except at a countable number of poles within C, then:

$$(8.39) \quad \oint_C f(z)\,dz = 2\pi i \sum_i b_1(\alpha_i)$$

where the $b_1(\alpha_i)$ are the residues at *all* the poles of $f(z)$ within C."

The validity of this theorem is visualized with the aid of Fig. 8.3. The dots represent poles at $\alpha_1, \alpha_2, \ldots$. We have made cuts in C joining a small circle of radius ε about each pole to C with parallel straight lines. We examine this as follows. Since there are now *no* singularities in the shaded region $f(z)$ is analytic inside and on this closed contour and by Cauchy's theorem the integral of $f(z)$ about *this* is composed of parts thus:

$$(8.40) \quad 0 = \int_C^* + \int_{\lambda_1} + \int_{\lambda_1'} - \int_{\varepsilon_1} + \int_{\lambda_2} + \int_{\lambda_2'} - \int_{\varepsilon_2} + \int_{\lambda_3} + \int_{\lambda_3'} - \int_{\varepsilon_3} + \ldots$$

Here $\int_C^*$ is the integral around C *except* for the gaps at the cuts,

$$\int_{\lambda_1} \quad \text{and} \quad \int_{\lambda_1'}$$

† Usually.

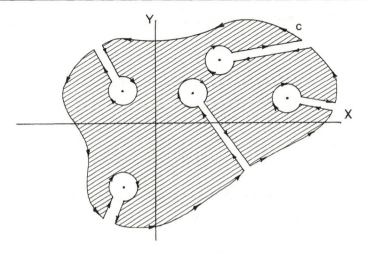

FIGURE 8.3. **Contour Used to Establish Cauchy's Residue Theorem.**

are *in* and *out* on the lines from C to α_1 and

$$-\int_{\varepsilon_1}$$

is about the small circle around α_1. (The minus sign is required because we go around these *clockwise* instead of *counterclockwise*.) The other integrals are similarly defined. The procedure is then to show that which appears quite obvious, namely

$$\int_{\lambda_1} + \int_{\lambda_1'}$$

must add to zero as the width of the cut goes to zero and similarly at other cuts, and that in the limit

$$\int_C^* \qquad \text{is just} \qquad \oint_C$$

Also the integrals around the small circles are examined as follows.

If α_1 is a pole of order n then we can represent $f(z)$ by its Laurent series about α_1 and the principal part terminates in the term of order $(z - \alpha_1)^{-n}$. Now consider the integral of $f(z)$, as the series, term by term, around the small circle of radius ε. A typical term is:

$$(8.41) \qquad \oint_\varepsilon (z - \alpha_1)^m \, dz = \int_0^{2\pi} \varepsilon^m e^{im\theta}(i\varepsilon e^{i\theta} \, d\theta)$$

or

$$(8.42) \qquad \oint (z - \alpha_1)^m \, dz = \frac{\varepsilon^{m+1}}{m+1} e^{i(m+1)\theta} \Big|_0^{2\pi}, \qquad m \neq -1$$

and if $(m + 1)$ is *not* zero this is zero. (Note here we used $z - \alpha_1 = \varepsilon e^{i\theta}$ on the small circle.)

However *if* $m = -1$ this appears as

$$(8.43) \qquad \oint \frac{dz}{z - \alpha_1} = \int_0^{2\pi} \frac{i\varepsilon e^{i\theta} \, d\theta}{\varepsilon e^{i\theta}} = 2\pi i$$

Thus we see that each of the integrals of $f(z)$ about one of the small circles yields

$$(8.44) \qquad \oint_{\varepsilon_i} f(z) \, dz = 2\pi i b_1(\alpha_i)$$

since b_1 is the coefficient of $(z - \alpha_i)^{-1}$ in the Laurent series. Thus as we let all the $\varepsilon_i \to 0$ and the gaps of all the cuts go to zero we see that Eq. (8.40) does indeed yield Eq. (8.39).

EVALUATION OF INTEGRALS; INTEGRATION AROUND THE UNIT CIRCLE

As our first illustration of the use of the calculus of residues we consider evaluation of any integral of the form

$$(8.45) \qquad I = \int_0^{2\pi} F(\sin \theta, \cos \theta) \, d\theta$$

where F is any rational function (quotient of polynomials) in $\sin \theta$, $\cos \theta$.
Substitute

$$(8.46) \qquad \begin{cases} \cos \theta = \dfrac{e^{i\theta} + e^{-i\theta}}{2} = \dfrac{1}{2}\left(z + \dfrac{1}{z}\right) \\[3mm] \sin \theta = \dfrac{e^{i\theta} - e^{-i\theta}}{2i} = \dfrac{1}{2i}\left(z - \dfrac{1}{z}\right) \end{cases}$$

for z on the *unit circle* $r = 1$ ($z = re^{i\theta}$). Then the integral I becomes the integral in the complex plane about the *unit circle*,

$$(8.47) \qquad I = \oint_{|z| = 1} F\left[\frac{1}{2i}\left(z - \frac{1}{z}\right), \frac{1}{2}\left(z + \frac{1}{z}\right)\right] \frac{dz}{iz}$$

where we also have employed,

$$(8.48) \qquad dz = d(1 \cdot e^{i\theta}) = ie^{i\theta} \, d\theta = iz \, d\theta$$

Thus we can apply the residue theorem to evaluate I.

Example (3)

Evaluate the integral

$$(8.49) \qquad I = \int_0^{2\pi} \frac{d\theta}{5 - 4\cos \theta}$$

Putting this in terms of the contour integral about the unit circle as outlined above we get,

$$(8.50) \qquad I = \frac{i}{2} \oint_{|z| = 1} \frac{dz}{z^2 - \frac{5}{2}z + 1}$$

and the denominator factors as

$$(8.51) \qquad z^2 - \tfrac{5}{2}z + 1 = (z - 2)(z - \tfrac{1}{2})$$

so it is obvious that there are simple poles at $z = +2$, $z = +\frac{1}{2}$. Now the pole at $z = +2$ is *not* inside the contour and hence does not concern us. We compute the residue of the pole of the integrand at $z = +\frac{1}{2}$:

(8.52)
$$\text{Residue } (z = +\tfrac{1}{2}) = \lim_{z \to \frac{1}{2}} \frac{(z - \frac{1}{2})}{(z - 2)(z - \frac{1}{2})} = -\frac{2}{3}$$

Then we note that by the residue theorem we can now write Eq. (8.50) as,

(8.53)
$$I = \frac{i}{2}\left\{2\pi i \sum \text{Residues}\right\} = \frac{2\pi}{3}$$

or,

(8.54)
$$\int_0^{2\pi} \frac{d\theta}{5 - 4\cos\theta} = \frac{2\pi}{3}$$

INFINITE INTEGRALS

A certain class of infinite integrals can be evaluated using the calculus of residues as follows. We consider integrals of the type,

(8.55)
$$I = \int_{-\infty}^{+\infty} F(x)\,dx$$

where F, when looked upon as a function of the complex variable z, $F(z)$, has the following properties: it is analytic in the upper half-plane except at a finite number of poles, has only simple poles on the real axis and,

(8.56)
$$|zF(z)| \to 0$$

on the infinite semicircle, center the origin, in the upper half-plane.
For these conditions we consider the integral,

(8.57)
$$I' = \oint_C F(z)\,dz$$

around the contour indicated in Fig. 8-4. Here $\alpha_1, \alpha_2, \ldots$ are simple poles on the axis while

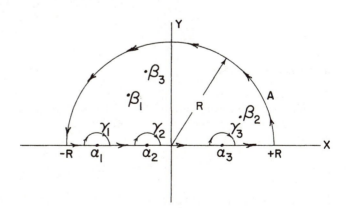

FIGURE 8.4. Contour Used to Evaluate Certain Types of Infinite Integrals.

$\beta_1, \beta_2, \ldots$ are other poles of $F(z)$ in the upper half-plane. We see that by the residue theorem,

$$(8.58) \qquad I' = \int_A + \int_{-R}^R * + \sum_j \int_{\gamma_j} = 2\pi i \sum_k b_1(\beta_k)$$

where A indicates the integral on the arc A, the second integral is along the real axis from $x = -R$ to $x = +R$, and the integrals in the sum are around the indicated arcs, γ_i. Note that in the integral on the axis the asterisk (*) indicates that the "gaps" at each pole on the axis are deleted in the integral. This is called the *principal value* of the integral.

Now because of the condition stipulated in Eq. (8.56) the integral around the arc A will vanish as we let $R \to \infty$. Also if we let the radius of each of the small arcs, γ_i, go to zero we see the integral

$$(8.59) \qquad \int_{-R}^R * \to I = \int_{-\infty}^{+\infty} F(x)\, dx$$

in this multiple limiting process. Hence all that remains to establish a useful means to evaluate I is to examine the integrals on the γ_j about the simple poles on the axis.

Since these are *simple* poles we see by the Laurent's series for $F(z)$, about α_j say, that:

$$(8.60) \qquad F(z) = \frac{b_1(\alpha_j)}{z - \alpha_j} + \sum_{n=0}^{\infty} a_n(z - \alpha_j)^n$$

Thus we put

$$(8.61) \qquad z - \alpha_j = \varepsilon_j\, e^{i\theta}$$

and have

$$\int_{\gamma_j} F(z)\, dz = \int_\pi^0 \frac{b_1(\alpha_j)}{\varepsilon_j\, e^{i\theta}} i\varepsilon_j\, e^{i\theta}\, d\theta + \sum_{n=1}^{\infty} a_{n-1}\varepsilon_j^n \int_\pi^0 e^{in\theta}\, d\theta$$

which reduces to,

$$(8.63) \qquad \int_{\gamma_j} F(z)\, dz = -\pi i b_1(\alpha_j) + \sum_{n=1}^{\infty} a_{n-1}\varepsilon_j^n \frac{[1 - (-1)^n]}{n}$$

and hence as $\varepsilon_j \to 0$ only the term $-\pi_i b_1(\alpha_j)$ survives and $b_1(\alpha_j)$ is just the residue of $F(z)$ at α_j.

Thus we see finally that

$$(8.64) \qquad I = 2\pi i \sum_k b_1(\beta_k) + \pi i \sum_j b_1(\alpha_j)$$

is the value of I.

JORDAN'S LEMMA

An important variation of the above method of evaluating infinite integrals is based on a slight modification of the stipulation that $|zF(z)| \to 0$ on the infinite arc in the upper half-plane. If instead we simply require $|F(z)| \to 0$ on this arc as $R \to \infty$ then we can show that with m real, $m > 0$

$$(8.65) \qquad \int_A F(z)\, e^{imz}\, dz \to 0, \quad \text{as} \quad R \to \infty$$

If $F(z)$ fulfills the other requirements of $F(z)$ above. This is *Jordan's Lemma*.

Example (4)

Evaluate the integral,

$$(8.66) \qquad I = \int_0^\infty \frac{x \sin ax \, dx}{1 + x^2}$$

Here, since the integrand is *even*, we can write this as,

$$(8.67) \qquad I = \tfrac{1}{2} \int_{-\infty}^{+\infty} \frac{x \sin ax}{1 + x^2} \, dx$$

and this is just the *imaginary* part of the integral

$$(8.68) \qquad I'' = \int_{-\infty}^{+\infty} \frac{x e^{iax}}{2(1 + x^2)} \, dx$$

The integrand here is of precisely the type described above, i.e.,

$$(8.69) \qquad F(z) = \frac{z}{2(1 + z^2)}$$

is analytic in the upper half-plane except at the simple pole $z = i$ and $|F(z)| \to 0$ as $|z| \to \infty$ on the infinite arc. Thus if a in the exponent is restricted to *positive* values we can apply the above results directly.

Thus now treating our integral I'' of Eq. (8.68) as we did the integral I of Eq. (8.55) we see that our integrand has no poles on the real axis and only the single simple pole at $z = i$ in the upper half-plane. We evaluate the residue at this pole as,

$$(8.70) \qquad \lim_{z \to i} \frac{z e^{iaz}(z - i)}{2(z^2 + 1)} = \frac{e^{-a}}{4}$$

Then

$$(8.71) \qquad I'' = 2\pi i \left(\frac{e^{-a}}{4} \right) = i \frac{\pi}{2} e^{-a}$$

according to Eq. (8.60).

Retracing our steps now, our original integral, I, of Eq. (8.66) is just the imaginary part of I'' so we have,

$$(8.72) \qquad \int_0^\infty \frac{x \sin ax \, dx}{1 + x^2} = \frac{\pi}{2} e^{-a}, \qquad a > 0$$

ANOTHER TYPE OF INFINITE INTEGRAL; BRANCH-LINE INTEGRALS

Consider a function $F(z)$ having a branch point at α and examine the integral of $F(z)$ about a circle of radius R about this branch point with a cut to the branch point as indicated in Fig. 8.5. As was pointed out earlier with regard to branch points $F(z)$ remains single valued around such a contour as this formed with the circle A, the two lines, λ_+ and λ_-, and the small circle γ about the branch point. If $F(z)$ is otherwise analytic in this contour, except at some isolated poles then

$$(8.73) \qquad \int_A + \int_{\lambda_+} + \int_{\lambda_-} - \int_\gamma = 2\pi i \sum \text{Res}$$

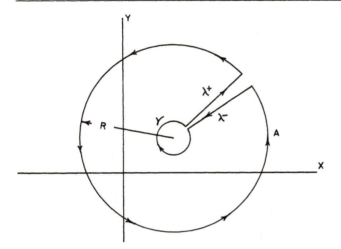

FIGURE 8.5.   Contour With a Branch Cut Used to Evaluate Certain Types of Infinite Integrals.

where the $\sum \text{Res}$ is the sum of all residues of $F(z)$ inside this contour. If $F(z)$ also has the property that $|F(z)z| \to 0$ on the infinite circle, as $R \to \infty$, then we see that the above equation reduces to

$$(8.74) \qquad \int_{\lambda_+} + \int_{\lambda_-} = \int_\gamma + 2\pi i \sum \text{Res}$$

Example (5)

Evaluate

$$(8.75) \qquad I = \int_0^\infty \frac{\sqrt{x}\, dx}{a^2 + x^2}$$

Here the integrand when viewed as a function of z;

$$(8.76) \qquad F(z) = \frac{\sqrt{z}}{a^2 + z^2}$$

is a function having two branches with the branch point at the origin, i.e., this is in the character of $\sqrt{z}$. Thus we evaluate

$$(8.77) \qquad I' = \oint \frac{\sqrt{z}\, dz}{a^2 + z^2}$$

around the contour consisting of the circle of radius R, center the origin, the two straight lines along the cut made on the negative x axis and a small circle about the origin. On this contour all angles are bounded as $-\pi < \theta < \pi$.

Certainly for our integral I' we get no contribution from the integration around the infinite circle. Also, since by inspection of our $F(z)$ the origin is not a pole, we get no contribution from the integration around the small circle.

On the straight line from the infinite circle *in* toward the origin *just above the axis*, we have,

(8.78)
$$\begin{cases} z = re^{i\pi} = -r, & dz = -dr \\ \sqrt{z} = \sqrt{re^{i\frac{\pi}{2}}} = i\sqrt{r} \end{cases}$$

so that

(8.79)
$$\int_{\lambda_+} = \int_\infty^0 \frac{-i\sqrt{r}\,dr}{a^2 + r^2} = i\int_0^\infty \frac{\sqrt{x}\,dx}{a^2 + x^2} = iI$$

simply writing x for r in the last form. Similarly, on the straight line *just below* the negative real axis we are integrating from the origin out to the infinite circle and we have,

(8.80)
$$\begin{cases} z = re^{-i\pi} = -r, & dz = -dr \\ \sqrt{z} = \sqrt{re^{-i\frac{\pi}{2}}} = -i\sqrt{r} \end{cases}$$

and we get in the same way,

(8.81)
$$\int_{\lambda_-} = i\int_0^\infty \frac{\sqrt{x}\,dx}{a^2 + x^2} = iI$$

when we simply write x for r as the last step.

Now our integrand, Eq. (8.76) has two simple poles, one at $+ia$, the other at $-ia$. We evaluate these residues as,

(8.82)
$$\text{Res}(ia) = \lim_{z\to ia} \frac{(z - ia)\sqrt{z}}{z^2 + a^2} = \frac{e^{i\frac{\pi}{4}}\sqrt{a}}{2ia}$$

and,

(8.83)
$$\text{Res}(-ia) = \lim_{z\to -ia} \frac{(z + ia)\sqrt{z}}{z^2 + a^2} = \frac{e^{-i\frac{\pi}{4}}\sqrt{a}}{-2ia}$$

Note that, since all angles are bounded as $-\pi < \theta < \pi$, $-i$ is represented as $e^{-i\pi/2}$. Had we taken the branch cut on the *positive* x axis instead of the negative x axes we would have all angles bounded as $0 < \theta < 2\pi$ and we would then represent $-i$ by $e^{i3\pi/2}$.

We then collect the results of Eqs. (8.79), (8.81), (8.82), and (8.83) and insert them in Eq. (8.74) to obtain,

(8.84)
$$\left\{ 2iI = 2\pi i\, \frac{e^{i\frac{\pi}{4}} - e^{-i\frac{\pi}{4}}}{2i\sqrt{a}} \right\}$$

or

(8.85)
$$I = \int_0^\infty \frac{\sqrt{x}\,dx}{a^2 + x^2} = \frac{\pi}{\sqrt{a}}\sin\frac{\pi}{4} = \frac{\pi}{\sqrt{2a}}$$

EXECUTE PROBLEM SET (8-1)

INVERSION INTEGRAL FOR THE LAPLACE TRANSFORM

In Chapter Eight we introduced the Laplace Transform and developed a few elementary applications. Here we develop the *general* method for inverting this transform. We now view the Laplace transform,

$$(8.86) \qquad f(p) = \int_0^\infty e^{-pt} F(t)\, dt$$

of a function $F(t)$ of the real variable t, as a function of a *complex* variable,

$$(8.87) \qquad p = u + iv$$

Then we can write Eq. (8.86) as,

$$(8.88) \qquad f(p) = \int_0^\infty e^{-ut} \cos vt F(t)\, dt - i \int_0^\infty e^{-ut} \sin vt F(t)\, dt$$

and it is evident that, if $F(t)$ is of *exponential order*, i.e.,

$$(8.89) \qquad \lim_{t \to \infty} e^{-ut} F(t) = 0$$

for any $u > \gamma$, and $F(t)$ is also sectionally continuous, then both the real and imaginary parts of $f(p)$ in Eq. (8.88) exist and will be finite and single valued. It is also verified by differentiation under the integral signs, which can be done because of uniform convergence for $u > \gamma$, that these satisfy the Cauchy–Riemann conditions. Thus $f(p)$ is an *analytic function* of p in the half-space,

$$(8.90) \qquad \text{real}(p) > \gamma > 0$$

Here then recall Cauchy's integral which we here write as

$$(8.91) \qquad f(p) = -\frac{1}{2\pi i} \oint_C \frac{f(z')\, dz'}{p - z'}$$

Note the minus sign in front of the integral and the change of sign in the denominator compensate each other. This does indeed give the value of $f(p)$, if *all* of the contour C is in the half-space defined above, and the point p is inside C. Hence we choose the contour C as shown in Fig. 8.6 and *all possible values of p are provided*, i.e., as $R \to \infty$. Thus

$$(8.92) \qquad f(p) = \frac{1}{2\pi i} \int_{\gamma - i\infty}^{\gamma + i\infty} \frac{f(z')\, dz'}{p - z'}$$

where we have made use of the fact that in view of $f(z')$ being analytic, as defined in Eq. (8.88), the contribution to $f(p)$ on the infinite arc in the right half-plane vanishes as $R \to \infty$. Also we have used the *minus* sign to reverse the *direction* of integration on the vertical line.

Now we take the inverse transform with *respect to p* on both sides thus, *formally at least*, we have,

$$(8.93) \qquad F(t) = \frac{1}{2\pi i} \int_{\gamma - i\infty}^{\gamma + i\infty} f(z') e^{z't}\, dz'$$

since the inverse of $f(p)$ is just $F(t)$, and *inside* the complex integral, the inverse of $(p - z')^{-1}$ is just $e^{z't}$. Several points of mathematical *rigor* are slighted here but this does indicate the basis of the *inversion integral*, Eq. (8.93), for the Laplace transform.

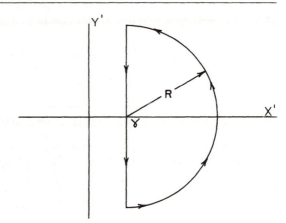

FIGURE 8.6.   Contour Used to Represent $f(p)$ by Cauchy's Integral.

EVALUATION OF THE INVERSION INTEGRAL FOR THE LAPLACE TRANSFORM

While the Laplace transform, $f(z')$, is analytic for $x' > \gamma$ (here $z' = x' + iy'$) it is by no means necessarily analytic for $x' < \gamma$. But it can be shown that the integral of $e^{z't}f(z')$ on the infinite arc in the left half plane vanishes and we use this to advantage. We distinguish two general cases.

Case 1: $f(z)$ is free of branch points

In this case we choose a contour as shown in Fig. 8.7. Then by the residue theorem,

(8.94)
$$\int_{\gamma-iR}^{\gamma+iR} + \int_R = 2\pi i \sum \text{Res}$$

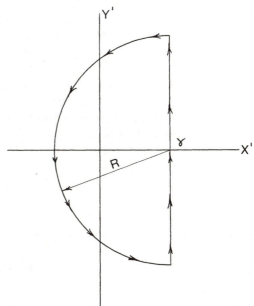

FIGURE 8.7.   Contour Used to Evaluate the Inversion Integral When $f(p)$ has No Branch Points.

where the sum is over the residues of $f(z')$ in this contour. Simply letting $R \to \infty$ then yields,

(8.95)
$$\frac{1}{2\pi i} \int_{\gamma - i\infty}^{\gamma + i\infty} f(z') e^{z't} \, dz' = F(t) = \sum_{x' < \gamma} \text{Res}[f(z') e^{z't}]$$

where the sum is over all residues of $f(z') e^{z't}$.

Case 2: *f(z′)* has a branch point at the origin

In this case we choose a contour as shown in Fig. 8-8. Then by the residue theorem and the fact that our branch line cut has made $f(x')$ single valued on the contour,

(8.96)
$$\int_{\gamma - iR}^{\gamma + iR} + \int_{R} - \int_{C'} + \int_{\lambda_+} + \int_{\lambda_-} = 2\pi i \sum \text{Res}$$

Now as we let $R \to \infty$ *and* the radius, ε, of the small circle C' about the origin go to zero, we obtain

(8.97)
$$\frac{1}{2\pi i} \int_{\gamma - i\infty}^{\gamma + i\infty} f(z') e^{z't} \, dz' = F(t) = \sum_{x' < \gamma} \text{Res}[f(z') e^{z't}] + \frac{1}{2\pi i}(I_0 - I_{\lambda_+} - I_{\lambda_-})$$

where I_{λ_+} and I_{λ_-} are the two branch line integrals and I_0 is the limiting form of the integral around the origin.

SPECIAL CASES

In some special cases, such as those involving multiple branch points, the evaluation of the inversion integral is obtained in terms of *contour integral representations*.

Example (6)

Invert the Laplace transform,

(8.98)
$$f(p) = \frac{\sinh a\sqrt{p}}{p \sinh \sqrt{p}}$$

which arises in the solution of a certain heat conduction problem.

Here since $\sqrt{p}$ appears in the function $f(p)$ we might suspect a branch point at the origin, however if we look at the limiting form of $f(p)$ as $p \to 0$ we see

(8.99)
$$f(p) \approx \frac{1}{p} \frac{a\sqrt{p} - \frac{1}{3!}(a\sqrt{p})^3 + \cdots}{\sqrt{p} - \frac{1}{3!}(\sqrt{p})^3 + \cdots}$$

so that $\sqrt{p}$ divides out, and hence $f(p)$ is single valued. Also we see that

(8.100)
$$\lim_{p \to 0} f(p) \approx \frac{a}{p}$$

and the origin is just a simple pole, *not* a branch point.

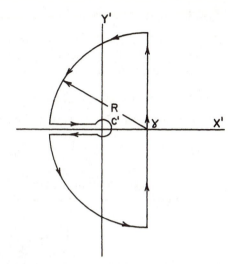

FIGURE 8.8. Contour Used to Evaluate the Inversion Integral When f(p) has a Branch Point at the Origin.

Thus we have *Case 1* and we just need to evaluate the residues of $e^{zt} f(z)$ at the poles of $f(z)$. We note that

(8.101)
$$\sinh \sqrt{z} = 0$$

when

(8.102)
$$z_n = -n^2\pi^2, \qquad n = 0, 1, 2, \ldots$$

but we have already seen above that the origin $z = 0$ ($n = 0$ here) is just a simple pole. It is also evident that all of these z_n, $n \geq 1$, are also simple poles. We have,

(8.103)
$$\text{Residue}(z = z_n) = \lim_{z \to z_n} \frac{(z + n^2\pi^2)\, e^{zt} \sinh a\sqrt{z}}{z \sinh \sqrt{z}}$$

and L'Hopital's rule gives,

(8.104)
$$\text{Residue}(z = z_n) = \frac{(-1)^n 2}{n\pi}\, e^{-n^2\pi^2 t} \sin n\pi a$$

It is also evident from Eq. (8.99) that the residue at the origin is just,

(8.105)
$$\text{Residue}(z = 0) = a$$

so that by Eq. (8.95)

(8.106)
$$F(t) = a + \sum_{n=1}^{\infty} \frac{2}{n\pi}(-1)^n\, e^{-n^2\pi^2 t} \sin n\pi a$$

is the required inverse.

CONTOUR INTEGRAL REPRESENTATIONS OF FUNCTIONS

Our primary interest in contour integral representations arises from their utility in inverting the Laplace transform in certain special cases. Here we consider a single example and refer the student to the references for more elaborate treatment of this and other cases.

Example (7)

Construction of a contour integral representation of the Bessel function $J_0(x)$.

We write Bessel's equation of zero order with x replaced by the complex variable z and $U(z)$ as the dependent variable,

$$(8.107) \qquad z^2 U'' + zU' + z^2 U = 0$$

then seek a solution in terms of an imaginary exponential in z, as an integral,

$$(8.108) \qquad U(z) = \int_C f(\xi) e^{iz\xi} \, d\xi$$

where $f(\xi)$ and the contour, C, are to be determined. Substitution back into the differential equation gives,

$$(8.109) \qquad \int_C (-\xi^2 z^2 + i\xi z + z^2) f(\xi) \, e^{iz\xi} \, d\xi = 0$$

Factoring out iz yields

$$(8.110) \qquad iz \int_C [\xi + iz(\xi^2 - 1)] f(\xi) e^{i\xi z} \, d\xi = 0$$

and if $f(\xi)$ is *chosen* such that,

$$(8.111) \qquad (\xi^2 - 1) \frac{df}{d\xi} = -\xi f$$

then we have the *exact differential form*,

$$(8.112) \qquad iz \int_C \frac{d}{d\xi} \left\{ (\xi^2 - 1) f(\xi) \, e^{i\xi z} \right\} \, d\xi = 0$$

From the form of Eq. (8.111),

$$(8.113) \qquad f(\xi) = \frac{A}{\sqrt{\xi^2 - 1}}$$

where A is a constant.

Now observe that the choice of constraint on $f(\xi)$ in Eq. (8.111) is *unique* in that it reduces the integrand to the *total differential* of a function of ξ. This tells us something about the contour C.

If in Eq. (8.112) we integrate along some contour such that $(\xi^2 - 1) f(\xi) e^{i\xi z}$ has the *same value* at the initial and final point then this equation is satisfied, or if this quantity vanishes at the initial and final points of the contour it is also satisfied.

Our integral for U now appears as,

$$(8.114) \qquad U(z) = A \int_C \frac{e^{i\xi z} \, d\xi}{\sqrt{\xi^2 - 1}}$$

and we have the two choices with regard to the contour described above. Here we also note that there are *two branch points* of this integrand, namely $\xi = \pm 1$. Thus we also need to consider whether the integrand remains single valued on the contour or not.

Now note that if we use a contour such as shown in Fig. 8-9, with the origin as our starting point, then sure enough the quantity,

$$(\xi^2 - 1) f(\xi) e^{i\xi z} = A(\xi^2 - 1)^{\frac{1}{2}} e^{i\xi z}$$

does indeed have the same value at initial and final points of the contour.

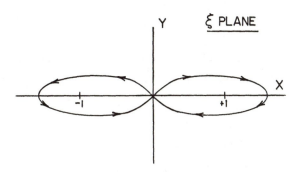

FIGURE 8.9. Contour for Integral Representation of $J_0(Z)$.

As we start at 0 and complete the first loop about the branch point $\xi = +1$ we jump to the *second* branch of the function, presuming we start with the first branch. We stay with this branch as we make the loop about the other branch point at $\xi = -1$, but upon *completing* this loop we again make the jump to another branch. But since the function has but *two* branches we have returned to our *original branch*. Thus our function returns to its starting value as required.

By the extension of Cauchy's integral theorem this contour can be deformed into that shown in Fig. 8-10, and it is this contour we use in the following discussion. In the limit the loops are made infinitely narrow.

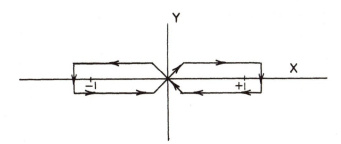

FIGURE 8.10. Modified Contour for Integral Representation of $J_0(Z)$.

Now expand the exponential in our integral, Eq. (8.114) and interchange summation and integration to obtain,

$$(8.115) \qquad U(z) = A \sum_{k=0}^{\infty} \frac{1}{k!} (iz)^k \int_C \xi^k (\xi^2 - 1)^{-\frac{1}{2}} d\xi$$

Since our contour is *symmetric* we note that the integrals here, for *odd k*, must vanish, and hence:

$$(8.116) \qquad U(z) = A \sum_{m=0}^{\infty} \frac{(-1)^m}{(2m)!} z^{2m} \int_C \xi^{2m} (\xi^2 - 1)^{-\frac{1}{2}} d\xi$$

The integrals here, for the chosen contour above, can be shown to yield

(8.117)
$$\int_C \xi^{2m}(\xi^2 - 1)^{-\frac{1}{2}}\, d\xi = \frac{(2m)!}{2^{2m}(m!)^2}\, 2\pi i$$

Thus we have,

(8.118)
$$U(z) = A \sum_{m=0}^{\infty} (-1)^m \frac{1}{(m!)^2} \left(\frac{z}{2}\right)^{2m}$$

which, if we take $A = \dfrac{1}{2\pi i}$, is just $J_0(z)$. Hence we see that the contour integral,   Eq.   (8.114), with suitable contour does represent $J_0(z)$.

EXECUTE PROBLEM SET (8-2)

DISPERSION INTEGRALS

There are a variety of physical problems in which a particular kind of integral relationship arises that is referred to as a dispersion relation. Here we show how one such integral is related to Cauchy's integral and branch-line integrals. In Chapter Nine another viewpoint on dispersion relations is given.

We consider a function, $S(z)$, that is analytic in the whole z plane except for a branch line on the real axis from a point x_0 to ∞. The function is discontinuous across this line. We also require that $|zS(z)| \to 0$ on the infinite circle.

By Cauchy's integral we have,

(8.119)
$$S(z) = \frac{1}{2\pi i} \oint_C \frac{S(z')\, dz'}{z' - z}$$

if we take a contour with a cut around the branch line as shown in Fig. 8-11. Because of the condition imposed on $S(z)$ at infinity we get no contribution on the circle as we let $R \to \infty$

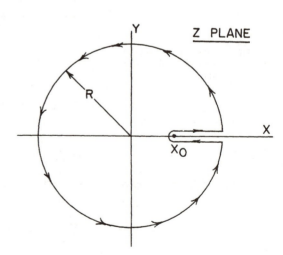

FIGURE 8.11.   Contour Used to Construct a Dispersion Integral.

and then Eq. (8.119) can be written as:

$$(8.120) \qquad S(z) = \frac{1}{2\pi i}\left\{\int_{x_0}^{\infty} \frac{S(x' + i\varepsilon)\,dx'}{x' + i\varepsilon - z} - \int_{x_0}^{\infty} \frac{S(x' - i\varepsilon)\,dx'}{x' - i\varepsilon - z}\right\}$$

where 2ε is the width of the cut about the branch line. It is important to note here that the point z is *not* on the real axis *in* the cut; it is elsewhere in the plane.

Now we can readily show that for *any* analytic function of a complex variable, which takes on *real* values when its argument is real,

$$(8.121) \qquad S^*(z) = S(z^*)$$

where the asterisk here denotes the complex conjugate,

$$(8.122) \qquad z = x + iy, \qquad z^* = x - iy$$

Thus since in the denominators of the integrals in Eq. (8.120) we have

$$(8.123) \qquad \mathrm{Im}(z) \gg \varepsilon$$

as $\varepsilon \to 0$, we neglect ε in both denominators and have,

$$(8.124) \qquad S(z') = \frac{1}{2\pi i}\int_{x_0}^{\infty} \frac{S(x' + i\varepsilon) - S(x' - i\varepsilon)}{x' - z}\,dx'$$

and in view of Eq. (8.121)

$$(8.125) \qquad \lim_{\varepsilon \to 0}[S(x' + i\varepsilon) - S(x' - i\varepsilon)] = 2i\,\mathrm{Im}\,S(x' + i0)$$

where $+i0$ indicates we use the branch of the function on the top of the cut. Then in the limit, as $\varepsilon \to 0$, we have,

$$(8.126) \qquad S(z) = \frac{1}{\pi}\int_{x_0}^{\infty} \frac{\mathrm{Im}[S(x' + i0)]}{x' - z}\,dx'$$

Hence for this special class of functions, $S(z)$, we can represent the value of the function at *any* point of the plane in terms of *just its imaginary part* along the line of discontinuity of the function.

METHOD OF STEEPEST DESCENT OR SADDLE-POINT METHOD

This is a method for the approximate evaluation of integrals of the type,

$$(8.127) \qquad I = \int_C f(z)\,e^{tg(z)}\,dz$$

where C may be open or closed. Here t must be real and positive, and $f(z)$ a "slowly varying" function. This has application in the inversion of the Laplace transform in some of the more complicated cases.

Now the curve C can be continuously deformed into any other curve C' without altering the value of the integral provided no singularities of the integrand lie on or between the two curves. This follows from Cauchy's theorem, or its extension if C is closed.

We assume that on the new curve, C', $g(z)$ is *analytic* and write

$$(8.128) \qquad g(z) = g_1(x, y) + ig_2(x, y)$$

Thus the integrand is now written as

$$f(z)\, e^{ig_2 t}\, e^{tg_1}$$

and the magnitude is determined primarily by the factor e^{tg_1}, particularly if t *is large*. The objective then is to choose a curve C' passing through a point of maximum $g_1(x, y)$, such that g_1 decreases rapidly on either side of this point, i.e., the path of *steepest descent*. Then the main contribution to the integral is in the neighborhood of this point on C'.

Now $g_1(x, y)$ is a maximum if

$$(8.129) \qquad\qquad \frac{\partial g_1}{\partial x} = 0, \qquad \frac{\partial g_1}{\partial y} = 0$$

but $g(z)$ is analytic and hence, by the Cauchy–Riemann conditions,

$$(8.130) \qquad\qquad \frac{\partial g_2}{\partial x} = 0, \qquad \frac{\partial g_2}{\partial y} = 0$$

also. In fact

$$(8.131) \qquad\qquad \frac{dg}{dz}(z_0) = 0$$

defines the point z_0 at which both Eqs. (8.129) and (8.130) hold. However since $g(z)$ is analytic, $g_1(x, y)$ and $g_2(x, y)$ are *both* harmonic functions satisfying Laplaces equation, i.e.,

$$(8.132) \qquad\qquad \frac{\partial^2 g_1}{\partial x^2} + \frac{\partial^2 g_1}{\partial y^2} = 0$$

so that z_0 can be neither a true maximum nor a true minimum; it must be a *saddle point*.†
This is illustrated in Fig. 8-12.

For the moment we suppose g_1 has only *one* saddle point, i.e., Eq. (8.131) has but one solution in the region of interest. Later multiple saddle points can be considered.

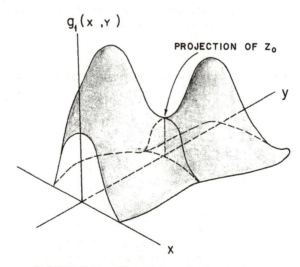

FIGURE 8.12. Illustration of a Saddle Point.

† See Appendix VI.

Although we have defined the point $z = z_0$ through which C' must pass we have yet to specify the *direction*. The *desirable* direction is readily visualized in Fig. (8-12), i.e., g_1 decreases rapidly as we go away from z_0 along C'.

Let $\mathbf{dr}$ be a vector displacement along C' then the change in $g_1(x, y)$ as we move from $\mathbf{r}$ to $\mathbf{r} + \mathbf{dr}$ on C' is:

$$(8.133) \qquad dg_1 = \nabla g_1 \cdot \mathbf{dr} = \left(\cos \alpha \frac{\partial g_1}{\partial x} + \sin \alpha \frac{\partial g_1}{\partial y} \right) ds$$

where α is the angle between $\mathbf{dr}$ and the x axis. We desire now to choose the *direction* of $\mathbf{dr}$, at z_0, such that dg_1 is a maximum for a *given length* ds of $\mathbf{dr}$. Thus we maximize Eq. (8.133) with respect to α.

$$(8.134) \qquad \frac{\partial}{\partial \alpha}(dg_1) = 0$$

or,

$$(8.135) \qquad -\sin \alpha \frac{\partial g_1}{\partial x} + \cos \alpha \frac{\partial g_1}{\partial y} = 0$$

But using the Cauchy–Riemann conditions to express this in terms of g_2 we have,

$$(8.136) \qquad -\sin \alpha \frac{\partial g_2}{\partial y} - \cos \alpha \frac{\partial g_2}{\partial x} = 0$$

but this means,

$$(8.137) \qquad dg_2 = \left(\cos \alpha \frac{\partial g_2}{\partial x} + \sin \alpha \frac{\partial g_2}{\partial y} \right) ds = 0$$

and hence

$$(8.138) \qquad g_2(x, y) = \text{constant}$$

along C' at $z = z_0$ in the direction of steepest descent. Since tg_2 is the phase of the function exp tg, we see here why this method is also sometimes called the method of *stationary phase*.

If we let

$$g_2(x, y) = \phi = \text{constant}$$

be the constant value of g_2 at z_0, which is maintained along C' we have

$$(8.139) \qquad g(z) = g_1(x, y) + i\phi$$

Since we plan to evaluate the integral just over a small segment of C' in the neighborhood of z_0 we expand each side of this in a Taylor series about $z = z_0$.

On the left we get,

$$(8.140) \qquad g(z) = g(z_0) + \tfrac{1}{2}(z - z_0)^2 g''(z_0) + \cdots$$

while on the right, in terms of x and y, we get,

$$(8.141) \qquad g(z) = g_1(x_0, y_0) + \frac{1}{2}\left[\frac{\partial^2 g_1}{\partial x^2} \Delta x^2 + 2 \frac{\partial^2 g_1}{\partial x \partial y} \Delta x \Delta y + \frac{\partial^2 g_1}{\partial y^2} \Delta y^2 \right] + \cdots + i\phi$$

or, with the substitutions,

(8.142)
$$\begin{cases} \Delta x = s \cos \alpha \\ \Delta y = s \sin \alpha \end{cases}$$

(8.143)
$$g(z) = [g_1(x_0, y_0) + i\phi] + \frac{s^2}{2}\left[\cos \alpha \frac{\partial}{\partial x} + \sin \alpha \frac{\partial}{\partial y}\right]^2 g_1 + \cdots$$

Here,

(8.144)
$$\left[\cos \alpha \frac{\partial}{\partial x} + \sin \alpha \frac{\partial}{\partial y}\right]^2 g_1 = \frac{d^2 g_1}{ds^2}$$

which is the second derivative of g_1 with respect to distance s along C' at z_0. Since g_1 is a *relative maximum* along C' at z_0 we see that this second derivative must be *negative*.

Thus comparing Eqs. (8.143) and (8.140), with this result in mind, we see that

$$(z - z_0)^2 g''(z_0)$$

is *real* and negative. Thus write

(8.145)
$$g(z) - g(z_0) = -\frac{1}{2}\frac{u^2}{t} = \frac{1}{2}(z - z_0)^2 g''(z_0)$$

where u is real.

Then, since $f(z)$ is slowly varying we see that

(8.146)
$$I \approx f(z_0)e^{tg(z_0)} \int_{-u_1}^{+u_2} e^{-\frac{1}{2}u^2} \frac{dz}{du} du$$

Furthermore since negligible contributions result from extending the integration to $-\infty, +\infty$ we have,

(8.147)
$$I \approx f(z_0)e^{tg(z_0)} \int_{-\infty}^{+\infty} e^{-\frac{1}{2}u^2} \frac{dz}{du} du$$

From Eq. (8.145), with

(8.148)
$$z - z_0 = se^{i\alpha}$$

we see

(8.149)
$$u^2 = -tg''(z_0)s^2 e^{2i\alpha}$$

or

(8.150)
$$u = |u| = \pm s|tg''(z_0)|^{\frac{1}{2}}$$

Thus

(8.151)
$$\frac{du}{dz} = \frac{du}{ds}\frac{ds}{dz} = \pm e^{-i\alpha}|tg''(z_0)|^{\frac{1}{2}}$$

which is constant with respect to u and can be brought outside the integral in Eq. (8.147). Thus evaluating the remaining real integral of $\exp(-u^2/2)$ from $-\infty$ to $+\infty$ we have finally,

(8.152)
$$I \approx \frac{\sqrt{2\pi}e^{i\alpha}f(z_0)\exp\{tg(z_0)\}}{|tg''(z_0)|^{\frac{1}{2}}}$$

The ambiguity in sign in Eq. (8.151) which is deleted in Eq. (8.152) is always clearly taken care of by the proper sense (direction) along C' being applied.

If $g_1(x, y)$ has more than one saddle point then we apply the same procedure of deforming C into a curve C' through one or more saddle points. It may happen that one of these is much higher than all others and then the others can be neglected. If such is not the case then we get a *sum* of terms of the form of Eq. (8.152), one for each saddle point on C'.

Example (8)

As an illustration of the application of the method of steepest descent consider the *real* integral defining the gamma function (see Chapter Four)

(8.153)
$$\Gamma(\lambda + 1) = \int_0^\infty e^{-x} x^\lambda \, dx$$

Here the curve C in the complex x, y plane is just the real axis from $x = 0$ to ∞.

Make the substitution,

(8.154)
$$z = \frac{x}{\lambda}$$

then

(8.155)
$$\Gamma(\lambda + 1) = \lambda^{\lambda+1} \int_0^\infty z^\lambda e^{-\lambda z} \, dz = \lambda^{\lambda+1} \int_0^\infty e^{\lambda g(z)} \, dz$$

where

(8.156)
$$g(z) = \ln z - z$$

and the last form of the integral in Eq. (8.155) is in the form suitable for our method *if λ is large, real, and positive.*

For the saddle point,

(8.157)
$$\frac{dg(z)}{dz} = \frac{1}{z} - 1 = 0$$

gives $z = 1$, on the real axis, i.e., a point on our original Curve C in this case. But, we must deform C to pass through $z = z_0 = 1$ in the proper *direction*, i.e., we must find α. The path of steepest descent is that for which

(8.158)
$$g_2 = \text{Imag}[g(z)] = \phi = \text{constant}$$

or

(8.159)
$$\tan^{-1} \frac{y}{x} - y = \phi = \text{constant}$$

Taking the derivative with respect to x,

(8.160)
$$\left(1 + \frac{y^2}{x^2}\right)^{-1} \frac{xy' - y}{x^2} - y' = 0$$

and this is to be used to evaluate $y'(1, 0)$. We get $y' = 0$, so $\alpha = 0, \pi$. (Later we see we must use $\alpha = 0$ since $I > 0$.) We also evaluate d^2g/dz^2 at $z = z_0$, $(x = 1, y = 0)$ and get $|g''(z_0)| = 1$.

Thus putting all these results into the formula of Eq. (8.152), with the extra factor $\lambda^{\lambda+1}$ in Eq. (8.152) we find,

(8.161) $\Gamma(\lambda + 1) = \sqrt{2\pi}e^{-\lambda}\lambda^{\lambda+\frac{1}{2}}$

which is the Stirling approximation to the gamma function, i.e., for $\lambda = N$, an integer recall $\Gamma(N + 1) = N!$

EXECUTE PROBLEM SET (8-3)

PROBLEMS

Set (8-1)

(1) Show that the function,

$$f(z) = \frac{\sin z}{J_1(z)}$$

has the infinite set of zeros $z = n\pi, n = \pm 1, \pm 2, \pm 3, \ldots$ and the infinite set of *simple* poles $z = \xi_j, j = 1, 2, \ldots$ where the ξ_j are the *real* roots of $J_1(x) = 0$. Show that the origin is a regular point.

(2) Determine the nature of the singularities, and zeros, of the function

$$f(z) = \frac{\cosh a\sqrt{z}}{J_0(b\sqrt{z})}$$

Is it single valued?

(3) Show that the first four terms of the Laurent series, about $z = 0$, for the function

$$f(z) = \frac{e^z}{z(1 + z^2)} = \frac{1}{z} + (1 - \tfrac{1}{2}z - \tfrac{5}{6}z^2 + \ldots)$$

in the domain $0 < |z| < 1$.

(4) Construct the Laurent's series for

$$f(z) = \frac{z}{(z - 1)(z - 3)}$$

about $z = 1$ which converges for $0 < |z - 1| < 2$.

(5) Use Residue theory to evaluate or verify the following integrals:

(a) $\displaystyle\int_0^\infty \frac{dx}{x^4 + 1} = \frac{\pi\sqrt{2}}{4}$

(b) $\displaystyle\int_0^\infty \frac{J_0(x)\,dx}{a^2 + x^2} =$

(c) $\displaystyle\int_0^{2\pi} \frac{d\theta}{5 + 3\cos\theta} = \frac{\pi}{2}$

(d) $\displaystyle\int_0^{2\pi} \frac{\cos^3 3\theta\,d\theta}{5 - 4\cos 2\theta} =$

(e) $\displaystyle\int_0^\pi \frac{d\theta}{(a + \cos\theta)^2} =$

(f) $\displaystyle\int_{-\infty}^{+\infty} \frac{dx}{x^2 + 2x + 2} =$

(g) $\displaystyle\int_0^\infty \frac{J_0(x)\,dx}{x^2 - a^2} =$

(h) $\displaystyle\int_0^\infty \frac{\sin x}{x}\,dx =$

(i) $\displaystyle\int_0^\infty \frac{(\ln x)^2\,dx}{1 + x^2} =$

(j) $\displaystyle\int_0^\infty \frac{(\ln x)\,dx}{1 + x^3} = \frac{-2\pi^2}{27}$

Set (8-2)

(1) Invert the following Laplace transforms using the complex inversion integral

(a) $\dfrac{2ap}{(p^2 + a^2)^2}$

(b) $e^{-k\sqrt{p}}, \qquad k > 0$

(c) $\dfrac{1}{p}\ln(1 + kp), \qquad k > 0$

(d) $\dfrac{1}{p}\,\dfrac{h\sinh x\sqrt{p} + \sqrt{p}\cosh x\sqrt{p}}{h\sinh\sqrt{p} + \sqrt{p}\cosh\sqrt{p}}$

(2) Solve the heat conduction problem

$$\frac{\partial^2 T}{\partial r^2} + \frac{1}{r}\frac{\partial T}{\partial r} = \frac{\partial T}{\partial t}, \qquad \begin{array}{l} 0 < r < 1 \\ t > 0 \end{array}$$

with

$$T(r, 0) = a$$
$$T(1, t) = 0, \qquad t > 0$$

By taking Laplace transform on t and ultimately using the complex inversion integral.

(3) Solve the problem of transverse vibrations of a bar described by

$$-\frac{\partial^2 Y}{\partial t^2} = a^2\frac{\partial^4 Y}{\partial x^4}$$

where $a^2 = EI/A\rho$. (E is Young's modulus, I is the moment of inertia of the cross-sectional area about the centroid in the vertical direction, EI being the "flexural rigidity," A is the cross-sectional area and ρ the mass per unit volume.) $Y(x, t)$ is the vertical displacement of the bar at the distance x from one end at time t.

Consider the boundary and initial conditions

(a) $Y(x, 0) = \dfrac{\partial Y(x, 0)}{\partial t} = 0$

(b) $Y(0, t) = \dfrac{\partial^2 Y(0, t)}{\partial x^2} = 0$

(c) $Y(L, t) = A \sin \omega t, \qquad \dfrac{\partial^2 Y(L, t)}{\partial x^2} = 0$

These correspond to the bar being at rest horizontally at $t = 0$. The end $x = 0$ is pivoted to swing freely with $x = 0$ fixed (rotational freedom), the end at $x = L$ has a given displacement and is stress free.

Find the resonant frequencies by using the Laplace transform on t and solving by the inversion integral. (Note that *first-order poles do not give rise to resonance* but *second-order poles do give rise to resonance*.)

(4) Verify Eq. (8.117) for the contour given in Fig. 8-10.

(5) Show that the inverse transform of

$$f(p) = \frac{1}{\sqrt{p^2 + a^2}}$$

is $J_0(at)$. [Hint: use the extension of Cauchy's integral theorem to suitably deform an initial contour, taking account of the two branch points.]

Set (8-3)

(1) Obtain the approximate value of

$$I = \int_0^\infty e^{(\alpha\xi - e^\xi)} \, d\xi$$

for α large, real, and positive.

(2) Find an approximate form for

$$f_n(x) = \int_0^\infty \xi^n e^{-x\xi - \frac{\xi^2}{2}} d\xi$$

for n real large and positive.

REFERENCES

R. V. Churchill, *Complex Variables and Applications* (McGraw-Hill Book Company, Inc., New York, 1960).

E. T. Copson, *Theory of Functions of a Complex Variable* (Oxford University Press, London, 1935).

P. Dennery and A. Krzywicki, *Mathematics for Physicists* (Harper and Row, Publishers, Inc., New York, 1967).

J. Irving and N. Mullineux, *Mathematics in Physics and Engineering* (Academic Press, Inc., New York, 1959).

J. Mathews and R. L. Walker, *Mathematical Methods of Physics* (W. A. Benjamin, Inc., New York, 1965).

P. M. Morse and H. Feshbach, *Methods of Theoretical Physics* (McGraw-Hill Book Company, Inc., New York, 1953).

CHAPTER NINE

Integral transforms; the solution of inhomogeneous partial differential equations

In earlier chapters we have developed some techniques suitable for the solution of homogeneous partial differential equations subject to homogeneous boundary conditions. In this chapter we develop the basic concepts of integral transforms from our knowledge of complete sets of orthogonal functions. These transforms are then exhibited as the primary tools for solution of inhomogeneous partial differential equations subject to homogeneous boundary conditions.

DEFINITION OF AN INTEGRAL TRANSFORM PAIR

We consider some *finite* domain $a \leq x \leq b$ of a variable x on which a "class" of functions of the type† $F(x)$ is defined and another given function $K(x, y)$ also defined on the same interval. Here y is some parameter to be specified. We *define*

(9.1) $$f(y) = \int_a^b F(x)K(x, y)\,dx$$

as the *integral transform* of $F(x)$ with kernel, $K(x, y)$. We also suppose that given $f(y)$ there is

† Here the "type" is just those satisfying Direchlet conditions, for other transforms the type class could have other definitions.

some *inversion*, either in the form of a *sum* or *integral* which yields the function $F(x)$, that is, say:

(9.2)
$$F(x) = \int_c^d f(y)K'(x, y)\, dy$$

where $K'(x, y)$ is the kernel of the *inverse* transform. We call the pair of functions $F(x)$, $f(y)$ a *transform pair*.

We have already seen one such transform pair in the *Laplace transform*, i.e.,

(9.3)
$$f(p) = \int_0^\infty F(t)\, e^{-pt}\, dt$$

and

(9.4)
$$F(t) = \frac{1}{2\pi i} \int_{\gamma - i\infty}^{\gamma + i\infty} f(p)e^{pt}\, dp$$

FINITE SINE AND COSINE TRANSFORMS†

While we may not have been aware of it we have already dealt with some transform pairs other than the Laplace pair. In particular the finite sine and cosine transforms.

If $F(x)$ is an *odd* function defined on the interval $0 < x < L$ then we know that it can be represented by its Fourier sine series as,

(9.5)
$$F(x) = \sum_{n=1}^{\infty} f_n \sin \frac{n\pi x}{L}$$

where the coefficients of this series are represented by

(9.6)
$$f_n = \int_0^L F(x) \left\{ \frac{2}{L} \sin \frac{n\pi x}{L} \right\} dx$$

We call f_n the *finite sine transform* of $F(x)$ on the interval $0 \le x \le L$ and we interpret the sum in Eq. (9.5) as the *inversion* of the finite sine transform. Note that here the kernel of the integral transform is just

(9.7)
$$K(x, n) = \frac{2}{L} \sin \frac{n\pi x}{L}$$

so that the parameter of the transform, $y = n$, is a *discrete* parameter. This is a general characteristic of *finite* integral transforms as contrasted to *infinite* integral transforms, as will become more evident later.

If $F(x)$ is an *even* function on the interval $0 \le x \le L$ then it can be represented by its cosine series as

(9.8)
$$F(x) = \sum_{n=0}^{\infty} g_n \cos \frac{n\pi x}{L}$$

† See Problem (7) of Set (9-1) for another viewpoint.

where the g_n are given by

(9.9)
$$\begin{cases} g_0 = \int_0^L F(x) \left\{ \frac{1}{L} \right\} dx, & n = 0 \\[2mm] g_n = \int_0^L F(x) \left\{ \frac{2}{L} \cos \frac{n\pi x}{L} \right\} dx, & n > 0 \end{cases}$$

and we call g_n the finite cosine transform of $F(x)$. Again we view Eq. (9.8) as the inversion of the finite cosine transform.

To illustrate the advantages of integral transforms in the solution of *inhomogeneous partial differential equations* and hence to provide a motivation for further development of such techniques consider the following example.

Example (1)

Diffusion of a substance in a closed tube in which the substance is continually being formed by some chemical process.

The differential equation is:

(9.10)
$$D \frac{\partial^2 C}{\partial x^2} = \frac{\partial C}{\partial t} - F(x, t), \quad 0 < x < L$$

where D is the diffusion coefficient, C is the concentration, and $F(x, t)$ is the rate at which the material is being created in the cross section at x, expressed as quantity per unit volume per unit time.

We have the boundary conditions

(9.11)
$$\frac{\partial C}{\partial x} = 0, \quad \text{at} \quad x = 0, L$$

because the tube is closed and we assume an initial condition of the form

(9.12)
$$C(x, 0) = C_0(x)$$

Now we are going to apply a finite sine transform, or a finite cosine transform, to this problem, which one is quite *immaterial* because our problem is defined only for $x > 0$. But generally we have one clue for a choice. *All* odd functions are zero at the origin.† Thus in view of the boundary conditions Eq. (9.11), we choose the cosine transform at the outset.

Multiply Eq. (9.10) by $(2/L) \cos ax$, where a remains to be chosen and integrate over x from 0 to L, thus:

(9.13)
$$\int_0^L \frac{2}{L} \cos ax \frac{\partial^2 C}{\partial x^2} dx = \frac{1}{D} \frac{\partial}{\partial t} \int_0^L \frac{2}{L} C(x, t) \cos ax \, dx - \frac{1}{D} \int_0^L \frac{2}{L} \cos ax F(x, t) \, dx$$

Note that the x integration is taken inside the partial derivative on t and that we have divided through by D.

Now integrate the left member here by parts *twice* to obtain,

(9.14)
$$\frac{2}{L} \cos ax \frac{\partial C}{\partial x} \Big|_0^L + \frac{2a}{L} \sin ax C \Big|_0^L - a^2 \int_0^L \frac{2}{L} \cos ax C \, dx$$

† At least all *continuous* odd functions.

At the limits 0 and L here the first term vanishes by virtue of the boundary conditions, Eq. (9.11). The second set of integrated terms involving sin ax can also be made to vanish if we now *choose* the unspecified parameter a to be:[†]

$$(9.15) \qquad a = \frac{n\pi}{L}, \quad n = 0, 1, 2, 3, \ldots$$

Then if we define,

$$(9.16) \qquad g_n(t) = \delta_n \int_0^L \frac{2}{L} \cos \frac{n\pi x}{L} C(x, t) \, dx$$

as the finite cosine transform of $C(x, t)$ where $\delta_n = 1$, $n > 0$ and $\delta_n = \frac{1}{2}$, $n = 0$; and in the same way

$$(9.17) \qquad \bar{F}_n(t) = \delta_n \int_0^L \frac{2}{L} \cos \frac{n\pi x}{L} F(x, t) \, dx$$

as the finite cosine transform of $F(x, t)$ we have,

$$(9.18) \qquad -\frac{n^2\pi^2}{L^2} g_n = \frac{1}{D} \frac{dg_n}{dt} - \frac{\bar{F}_n}{D}$$

as the transformed version of our original differential equation. This is now an *ordinary* differential equation for g_n. The *initial condition* on g_n is deduced from the finite cosine transform of Eq. (9.12), i.e.,

$$(9.19) \qquad g_n(0) = \delta_n \int_0^L \frac{2}{L} \cos \frac{n\pi x}{L} C_0(x) \, dx$$

The most general attack we could make on the problem of solving Eq. (9.18) for g_n would be to apply the *Laplace transform* on t. Thus let

$$(9.20) \qquad \begin{cases} u(n, p) = L\{g_n(t)\}_t \\ f(n, p) = L\{\bar{F}_n(t)\}_t \end{cases}$$

and we have, upon taking the Laplace transform of Eq. (9.18),

$$(9.21) \qquad -\frac{n^2\pi^2}{L^2} Du = pu - g_n(0) - f$$

So now

$$(9.22) \qquad u(n, p) = \frac{g_n(0)}{p + \dfrac{n^2\pi^2 D}{L^2}} + \frac{f(n, p)}{p + \dfrac{n^2\pi^2 D^2}{L^2}}$$

and we are then ready to initiate the inversion of *both* the cosine transform *and* the Laplace transform.

We invert the Laplace transform first, noting that, with regard to the parameter p, $g_n(0)$ is constant, while in the last term we have the *product* of two transforms and hence have a

---

[†] Note here sometimes other choices of a are required and also it is sometimes *not* possible to make *all* the integrated terms vanish.

convolution integral upon inversion, thus:

$$(9.23) \qquad g_n(t) = g_n(0)e^{-\frac{n^2\pi^2}{L^2}Dt} + \int_0^t e^{-\frac{n^2\pi^2}{L^2}(t-\tau)} \bar{F}_n(\tau)\,d\tau$$

Finally then we invert the cosine transform simply by multiplying by $\cos(n\pi x/L)$ and forming the sum on n as in Eq. (9.8), thus:

$$(9.24) \qquad C(x,t) = \sum_{n=0}^{\infty} g_n(0)e^{-\frac{n^2\pi^2}{L^2}Dt}\cos\frac{n\pi x}{L} + \sum_{n=0}^{\infty}\cos\frac{n\pi x}{L}\int_0^t e^{-\frac{n^2\pi^2 D}{L^2}(t-\tau)}\,\bar{F}_n(\tau)\,d\tau$$

where $g_n(0)$ and $\bar{F}_n(t)$ are as already defined in Eq. (9.19) and (9.17), respectively.

Observe that not only have these transform methods allowed us to solve the *inhomogeneous* equation, they have permitted us to obtain the general solution for *any* initial distribution $C_0(x)$, and for *any* form of the distributed source function, $F(x, t)$.

FINITE HANKEL TRANSFORMS

The feature of the cosine transform which made it so appropriate for the solution of the example problem above was the way in which the term $\partial^2 C/\partial x^2$ was reduced through the integration by parts in Eq. (9.14), plus the fact that we could so conveniently introduce the boundary condition to fix the parameter a. This feature stems from the fact that $\cos ax$ is an *eigenfunction* of the operator $\partial^2/\partial x^2$.

Had we been faced with an inhomogeneous equation of the form,

$$(9.25) \qquad D\frac{1}{r}\frac{\partial}{\partial r}\left(r\frac{\partial C}{\partial r}\right) = \frac{\partial C}{\partial t} - F(r, t), \quad 0 < r < R$$

the cosine transform could *not* accomplish the same result. However, if here we should multiply the equation through by an eigenfunction of the operator

$$\frac{1}{r}\frac{\partial}{\partial r}\left(r\frac{\partial}{\partial r}\right)$$

with a weight factor of r, i.e., $rJ_0(\beta r)$, we could carry through a similar process. Thus:

$$(9.26) \qquad \int_0^R \frac{1}{r}\frac{\partial}{\partial r}\left(r\frac{\partial C}{\partial r}\right)rJ_0(\beta r)\,dr = r\frac{\partial C}{\partial r}J_0(\beta r)\Big|_0^R - rC\frac{\partial J_0(\beta r)}{\partial r}\Big|_0^R + \int_0^R C(r,t)\left\{\frac{1}{r}\frac{\partial}{\partial r}\left(r\frac{\partial J_0}{\partial r}\right)\right\}r\,dr$$

and since by Bessel's equation,

$$(9.27) \qquad \frac{1}{r}\frac{\partial}{\partial r}\left(r\frac{\partial J_0(\beta r)}{\partial r}\right) = -\beta^2 J_0(\beta r)$$

we obtain

$$(9.28) \qquad \int_0^R \left\{\frac{1}{r}\frac{\partial}{\partial r}\left(r\frac{\partial C}{\partial r}\right)\right\}J_0(\beta r)r\,dr = -\beta^2\int_0^R C(r,t)J_0(\beta r)r\,dr$$

provided that the boundary conditions on C, or $\partial C/\partial r$, at $r = R$, and β, are chosen in such a way that the integrated terms in Eq. (9.26) vanish. Note that all of these vanish at $r = 0$ always.

Here it is evident that if the boundary condition on $C(R, t)$ were $C = 0$ then we would require that β be defined as a root of

$$(9.29) \qquad\qquad J_0(\beta R) = 0$$

while if $\partial C/\partial r = 0$ at R is specified we would choose β as a root of

$$(9.30) \qquad\qquad \frac{\partial J_0(\beta R)}{\partial r} = 0$$

In this manner all integrated terms could be made to vanish and the problem could be treated quite similarly to the above example.

Thus we see the utility and basis for defining the *finite Hankel transform* of a function $F(r)$ as,

$$(9.31) \qquad\qquad h_{nj} = \int_0^R F(r) J_n(\beta_{nj} r) r \, dr$$

with the *inversion*

$$(9.32) \qquad\qquad F(r) = \sum_{j=1}^{\infty} \frac{h_{nj}}{N_{nj}} J_n(\beta_{nj} r)$$

where the β_{nj} are appropriately defined roots of the Bessel functions, usually

$$(9.33) \qquad\qquad J_n(\beta_{nj} R) = 0, \quad \text{or} \quad \frac{\partial J_n(\beta_{nj} R)}{\partial r} = 0$$

as indicated above and N_{nj} is the corresponding *norm*. Here h_{nj} is the finite Hankel transform of *order* n, while in our introductory remarks above we used the example $n = 0$.

We can also define Hankel transforms for *annular* domains such as $a < r < b$ by using say,

$$J_n(a\beta_{nj}) Y_n(\beta_{nj} r) - J_n(\beta_{nj} r) Y_n(a\beta_{nj})$$

in place of $J_n(\beta_{nj} r)$ above. The procedure should be quite obvious and will not be elaborated here.

EXECUTE PROBLEM SET (9-1)

FOURIER INTEGRAL AND FOURIER TRANSFORMS

For $F(x)$ a general function satisfying Dirichlet conditions in the interval $-L < x < L$, we have the Fourier series representation,

$$(9.34) \qquad\qquad F(x) = a_0 + \sum_{n=1}^{\infty} \left(a_n \cos \frac{n\pi x}{L} + b_n \sin \frac{n\pi x}{L} \right)$$

where

$$(9.35) \qquad\qquad a_0 = \frac{1}{2L} \int_{-L}^{L} F(x') \, dx'$$

and

$$(9.36) \quad \begin{cases} a_n = \dfrac{1}{L} \displaystyle\int_{-L}^{L} F(x') \cos \dfrac{n\pi x'}{L} \, dx' \\[4mm] b_n = \dfrac{1}{L} \displaystyle\int_{-L}^{L} F(x') \sin \dfrac{n\pi x'}{L} \, dx' \end{cases}$$

If we insert these into the above expression for $F(x)$ the result is,

$$(9.37) \quad F(x) = \frac{1}{2L} \int_{-L}^{L} F(x') \, dx' + \sum_{n=1}^{\infty} \frac{1}{L} \int_{-L}^{L} F(x') \left(\cos \frac{n\pi x}{L} \cos \frac{n\pi x'}{L} + \sin \frac{n\pi x}{L} \sin \frac{n\pi x'}{L} \right) dx'$$

or, in view of the trigonometric identity,

$$(9.38) \quad \cos \frac{n\pi}{L}(x - x') = \cos \frac{n\pi x}{L} \cos \frac{n\pi x'}{L} + \sin \frac{n\pi x}{L} \sin \frac{n\pi x'}{L}$$

this can be rewritten as,

$$(9.39) \quad F(x) = \frac{1}{2L} \int_{-L}^{L} F(x') \, dx' + \sum_{n=1}^{\infty} \frac{1}{L} \int_{-L}^{L} F(x') \cos \frac{n\pi}{L}(x - x') \, dx'$$

Now the cosine is an *even* function and hence the sum here is *exactly* the same if we sum from $n = -1$ to $n = -\infty$, over the *negative* integers. Also note that the first term here is just one-half the term which would result in the sum if we put $n = 0$. For these reasons this can be written as,

$$(9.40) \quad F(x) = \sum_{n=-\infty}^{+\infty} \frac{1}{2L} \int_{-L}^{L} F(x') \cos \frac{n\pi(x - x')}{L} \, dx'$$

We now examine the limiting form of this expression as we let $L \to \infty$. To do this we define the quantity

$$(9.41) \quad k_n = \frac{n\pi}{L} \quad \text{and} \quad \Delta k = \frac{\pi}{L}$$

and write Eq. (9.40) as,

$$(9.42) \quad F(x) = \frac{1}{2\pi} \sum_{n=-\infty}^{+\infty} \Delta k \int_{-L}^{L} F(x') \cos k_n(x - x') \, dx'$$

Most certainly as we let $L \to \infty$ we have an integral over the infinite domain of x', $-\infty < x' < +\infty$. Also we see that $\Delta k \to 0$ as $L \to \infty$. Now if, as we have already supposed, our series are *uniformly convergent*, even in the limit of $L \to \infty$, we can interchange the integration on x' and the summation on n at will.

Since $\Delta k \to 0$ we have then the sum on n approaching an integral over k in the limit, i.e., as $L \to \infty$, we have,

$$(9.43) \quad F(x) = \frac{1}{2\pi} \int_{-\infty}^{+\infty} \int_{-\infty}^{+\infty} F(x') \cos k(x - x') \, dx' \, dk$$

This is the *Fourier Integral* of $F(x)$. It can be shown† that this does indeed represent $F(x)$, if

† See R. V. Churchill, *Fourier Series and Boundary Value Problems*, (McGraw-Hill Book Company, Inc., New York, 1963) page 89.

$F(x)$ is sectionally continuous in any finite interval and the integral of $|F(x)|^2$ over the domain $-\infty < x < +\infty$ exists.

This integral can be put in a variety of forms. For example, since $\sin k(x - x')$ is an *odd* function of k, the integral, $-\infty < k < +\infty$ over this function must vanish at any fixed $x - x'$. Thus we can add $i \sin k(x - x')$, where $i = \sqrt{-1}$, to the $\cos k(x - x')$ to form,

(9.44)
$$F(x) = \frac{1}{2\pi} \int_{-\infty}^{+\infty} \int_{-\infty}^{+\infty} e^{ik(x-x')} F(x') \, dx' \, dk$$

It is this form of the Fourier integral we most often employ,† and we use it to define the *Fourier transform pair*:

(9.45)
$$\begin{cases} f(k) = \int_{-\infty}^{+\infty} F(x') e^{-ikx'} \, dx' \\[2mm] F(x) = \frac{1}{2\pi} \int_{-\infty}^{+\infty} f(k) e^{ikx} \, dk \end{cases}$$

It is our convention here to put the $1/2\pi$ factor in $F(x)$ as indicated. However other writers follow various practices with regard to this factor.

We call $f(k)$ the Fourier transform of $F(x)$ and $F(x)$ the inverse Fourier transform of $f(k)$. Quite obviously this terminology is interchangeable since the only asymmetry is the sign of the exponent in the integrands and the placement of $1/2\pi$ factor.

Other forms of the Fourier integral and other corresponding transform pairs can be constructed. For example if $F(x)$ *is an odd function* then we can show that Eq. (9.44) reduces to:

(9.46)
$$F(x) = \frac{2}{\pi} \int_0^{\infty} \int_0^{\infty} F(x') \sin kx' \sin kx \, dx' \, dk$$

and we then have the *infinite sine transform pair*:

(9.47)
$$\begin{cases} f_s(k) = \int_0^{\infty} F(x') \sin kx' \, dx' \\[2mm] F(x) = \frac{2}{\pi} \int_0^{\infty} f_s(k) \sin kx \, dk \end{cases}$$

where again we have put the $2/\pi$ factor with $F(x)$.

In precisely the same way we show that *if $F(x)$ is an even function* then we can reduce Eq. (9.44) to

(9.48)
$$F(x) = \frac{2}{\pi} \int_0^{\infty} \int_0^{\infty} F(x') \cos kx' \cos kx \, dx' \, dk$$

and hence we define the *infinite cosine transform pair*:

(9.49)
$$\begin{cases} f_c(k) = \int_0^{\infty} F(x') \cos kx' \, dx' \\[2mm] F(x) = \frac{2}{\pi} \int_0^{\infty} f_c(k) \cos kx \, dk \end{cases}$$

† Here k is *real*, but there are variations of this formalism in which k is complex, see the Wiener–Hopf technique in the Appendix.

These three Fourier transform pairs, Eq. (9.45), (9.47), and (9.49) are the forms we will treat in our subsequent discussions.† However, before we can proceed very far with the application of these transforms there is an item concerning the existence of certain integrals that must be cleared up. We examine this in the following section.

SUMMABILITY OF INTEGRALS

Suppose we apply the infinite cosine transform to the differential equation,

$$\text{(9.50)} \qquad \frac{\partial^2 C}{\partial x^2} = \frac{1}{D}\frac{\partial C}{\partial t} - \frac{1}{D}F(x, t), \quad 0 < x < \infty$$

just as we did in our Example 1 of this chapter, only now we consider an *infinite* domain, $L \to \infty$. The crucial part of that analysis depended upon an integration by parts of the integral like,

$$\text{(9.51)} \qquad I = \int_0^\infty \cos kx \, \frac{\partial^2 C}{\partial x^2} \, dx$$

carrying out the same integration by parts here we get,

$$\text{(9.52)} \qquad I = \cos kx \, \frac{\partial C}{\partial x}\bigg|_0^\infty + k \sin kxC \bigg|_0^\infty - k^2 \int_0^\infty \cos kxC \, dx$$

and we see that we must require $\partial C/\partial x$ and C *both* to vanish as $x \to \infty$ in order to proceed as before. On the other hand there is no difficulty at $x = 0$ if $\partial C/\partial x = 0$ at $x = 0$ is the boundary condition. Then all integrated terms vanish.

Now the remaining integral in Eq. (9.52) will be required to exist at all times. What is the value of this integral for $C = $ constant at $t = 0$? i.e., we sometimes encounter integrals such as,

$$\text{(9.53)} \qquad \mathscr{I} = \int_0^\infty \cos kx \, dx$$

We can *define* a value for such an integral in several ways. For example we may define

$$\text{(9.54)} \qquad \mathscr{I} = \lim_{\lambda \to \infty} \int_0^\lambda \left(1 - \frac{x}{\lambda}\right) \cos kx \, dx$$

and we say the integral is *summable*,‡ order one, $(C, 1)$.

Thus consider again the integration by parts above, but now in terms of the summable form,

$$\text{(9.55)} \qquad I_\lambda = \int_0^\lambda \left(1 - \frac{x}{\lambda}\right)\frac{\partial^2 C}{\partial x^2}\cos kx \, dx$$

We arrive at:

$$\text{(9.56)} \qquad \begin{aligned} I_\lambda = & \left[\left(1 - \frac{x}{\lambda}\right)\frac{\partial C}{\partial x}\cos kx + Ck\left(1 - \frac{x}{\lambda}\right)\sin kx + \frac{C}{\lambda}\cos kx\right]_0^\lambda \\ & - k^2\int_0^\lambda \left(1 - \frac{x}{\lambda}\right)C\cos kx \, dx + \frac{2k}{\lambda}\int_0^\lambda C\sin kx \, dx \end{aligned}$$

---

† It is possible to deduce the Laplace transform pair as a special case of the Fourier transform pair but we do not treat this here.

‡ See references at the end of this chapter for more details.

Now as $\lambda \to \infty$ the last integral tends to zero for C bounded and the quantity in brackets goes to zero as $\lambda \to \infty$, provided C is simply *finite* at infinity. Thus we obtain as the limit, $\lambda \to \infty$,

$$(9.57) \qquad I = \int_0^\infty \frac{\partial^2 C}{\partial x^2} \cos kx \, dx = -k^2 \int_0^\infty C \cos kx \, dx$$

for C simply finite at infinity, and $\partial C / \partial x = 0$ at $x = 0$.

In every case in which such situations arise when applying Fourier transforms we will presume all such integrals to be *summable* as indicated here.

SOME IMPORTANT PROPERTIES OF FOURIER TRANSFORMS

Many of the basic properties of the Fourier transform pair, Eq. (9.45), follow directly from the fact that the transform operation is a *linear* operation, i.e., if we let

$$(9.58) \qquad \mathscr{F}\{F(x)\} = f(k)$$

denote the Fourier transform of $F(x)$ and

$$(9.59) \qquad \mathscr{F}^{-1}\{f(k)\} = F(x)$$

similarly denote the inverse then we can immediately establish the partial list of useful properties shown in Table 9.1.

TABLE 9.1. Fourier Transforms.

	$F(x)$	$f(k)$
	$\dfrac{1}{2\pi} \displaystyle\int_{-\infty}^{+\infty} e^{ikx} f(k) \, dk$	$\displaystyle\int_{-\infty}^{+\infty} e^{-ikx} F(x) \, dx$
(1)	$aF_1(x) + bF_2(x)$	$af_1(k) + bf_2(k)$
(2)	$F(-x)$	$f(-k)$
(3)	$F(ax)$	$\dfrac{1}{a} f\!\left(\dfrac{k}{a}\right)$
(4)	$F(x)e^{-ik_1 x}$	$f(k + k_1)$
(5)	$F(x + x_1)$	$e^{ikx_1} f(k)$
(6)	$\displaystyle\int_{-\infty}^{+\infty} F_1(x')F_2(x - x') \, dx'$	$f_1(k) f_2(k)$
(7)	$F_1(x)F_2(x)$	$\dfrac{1}{2\pi} \displaystyle\int_{-\infty}^{+\infty} f_1(k') f_2(k - k') \, dk'$
(8)[a]	$\delta(x - x_1)$	e^{-ikx_1}
(9)[a]	$\dfrac{1}{2\pi} e^{ik_1 x}$	$\delta(k - k_1)$

[a] $\delta(x - x_1)$ is Dirac delta function.

TABLE 9.1. Fourier Transforms—(cont.).

	$F(x)$	$f(k)$
	$\dfrac{1}{2\pi}\displaystyle\int_{-\infty}^{+\infty} e^{ikx}f(k)\,dk$	$\displaystyle\int_{-\infty}^{+\infty} e^{-ikx}F(x)\,dx$
(10)	$\dfrac{1}{\sqrt{\lvert x\rvert}}$	$\sqrt{\dfrac{2\pi}{\lvert k\rvert}}$
(11)	$\dfrac{\sin ax}{x}$	$\begin{array}{ll}\pi & \lvert k\rvert < a \\ 0 & \lvert k\rvert > a\end{array}$
(12)	$\dfrac{\sin^2 ax}{x^2}$	$\begin{array}{ll}\pi\left(a-\dfrac{k}{2}\right), & \lvert k\rvert < 2a \\[2mm] 0 & \lvert k\rvert > 2a\end{array}$
(13)	$e^{-\lambda x^2}\ \operatorname{Re}(\lambda) > 0$	$\sqrt{\dfrac{\pi}{\lambda}}\,e^{-\frac{k^2}{4\lambda}}\ \begin{array}{l}\operatorname{Re}(\lambda) > 0 \\[2mm] \operatorname{Re}(\sqrt{\lambda}) > 0\end{array}$
(14)	$\dfrac{1}{x^2 + a^2}$	$\dfrac{\pi}{a}\,e^{-a\lvert k\rvert}$
(15)	$\begin{array}{ll}\dfrac{1}{\sqrt{a^2 - x^2}}, & \lvert x\rvert < a \\[4mm] 0 & ,\ \lvert x\rvert > a\end{array}$	$\pi J_0(ak)$
(16)	$\dfrac{e^{-b\sqrt{a^2 + x^2}}}{\sqrt{a^2 + x^2}}$	$2K_0(a\sqrt{b^2 + k^2})$
(17)	$\dfrac{\sin b\sqrt{a^2 + x^2}}{\sqrt{a^2 + x^2}}$	$\begin{array}{ll}0 & \lvert k\rvert > b \\ \pi J_0(a\sqrt{b^2 - k^2}) & \lvert k\rvert < b\end{array}$
(18)	$\dfrac{e^{ib\sqrt{a^2 + x^2}}}{\sqrt{a^2 + x^2}}$	$\pi i H_0^{(1)}(a\sqrt{b^2 - k^2}$
(19)	$\begin{array}{ll}\dfrac{\cosh b\sqrt{a^2 - x^2}}{\sqrt{a^2 - x^2}} & \lvert x\rvert < a \\[4mm] 0 & \lvert x\rvert > a\end{array}$	$\pi J_0(a\sqrt{k^2 - b^2})$
(20)	$\begin{array}{ll}0 & \lvert x\rvert < a \\[2mm] \dfrac{-1}{\sqrt{x^2 - a^2}} & \lvert x\rvert > a\end{array}$	$\pi Y_0(a\lvert k\rvert)$

Items (6) and (7) of this table are the two forms of the *convolution theorem* for the Fourier transform. To illustrate how some of these are arrived at we derive Entry Number (7) of the table.

We require

(9.60)
$$\mathscr{F}\{F_1(x)F_2(x)\} = \int_{-\infty}^{+\infty} e^{-ikx} F_1(x)F_2(x) \, dx$$

Into this we substitute the representations of $F_1(x)$ and $F_2(x)$ in terms of their individual transforms,

(9.61)
$$F_1(x) = \frac{1}{2\pi} \int_{-\infty}^{+\infty} e^{ik'x} f_1(k') \, dk'$$

and

(9.62)
$$F_2(x) = \frac{1}{2\pi} \int_{-\infty}^{+\infty} e^{ik''x} f_2(k'') \, dk''$$

to obtain:

(9.63)
$$\mathscr{F}\{F_1(x)F_2(x)\} = \frac{1}{4\pi^2} \int_{-\infty}^{+\infty} \int_{-\infty}^{+\infty} \int_{-\infty}^{+\infty} f_1(k') f_2(k'') \, dk' \, dk'' \, e^{-i(k-k'-k'')t} \, dt$$

upon interchanging the order of integration. Now the integral over t,

(9.64)
$$\frac{1}{2\pi} \int_{-\infty}^{+\infty} e^{-i(k-k'-k'')t} \, dt = \delta(k - k' - k'')$$

according to Entry Number (9) of our table, which by the way is trivial to establish. Thus

(9.65)
$$\mathscr{F}\{F_1(x)F_2(x)\} = \frac{1}{2\pi} \int_{-\infty}^{+\infty} \int_{-\infty}^{+\infty} f_1(k') f_2(k'') \, \delta(k - k' - k'') \, dk' \, dk''$$

Then integrating on say k' we get

(9.66)
$$\mathscr{F}\{F_1(x)F_2(x)\} = \frac{1}{2\pi} \int_{-\infty}^{+\infty} f_1(k - k'') f_2(k'') \, dk''$$

or, integrating on k''

(9.67)
$$\mathscr{F}\{F_1(x)F_2(x)\} = \frac{1}{2\pi} \int_{-\infty}^{+\infty} f_1(k') f_2(k - k') \, dk'$$

Many other properties of these transforms can be established in similar fashion.

APPLICATIONS OF THE FOURIER TRANSFORM

Fourier transforms have *many* areas of application but two of the most notable are in problems of wave propagation and in the analysis of alternating current circuits consisting of passive elements. Quite obviously since the factor, e^{ikx}, is an eigenfunction of the operator, $\partial^2/\partial x^2$, we find some form of the Fourier transform suitable for application to equations involving this operator if the domain of x is infinite or semi-infinite in the given problem.

Example (2)

The finite string with a prescribed force applied at one point. We use this example to demonstrate *two* things, one is the solution of a wave-propagation problem using Fourier transforms, and the other is the use of the Dirac delta function to describe *an isolated source of excitation.*

For a uniform string of length L having mass ρ per unit length, and under tension T, having applied vertical force of $\bar{F}(x, t)$ per unit length we have:

$$(9.68) \qquad \rho \frac{\partial^2 Y}{\partial t^2} = T \frac{\partial^2 Y}{\partial x^2} + \bar{F}(x, t), \quad 0 < x < L$$

for the vertical displacement $Y(x, t)$. Here since we are considering the ends fixed we have

$$(9.69) \qquad Y(0, t) = Y(L, t) = 0$$

We stated above that the string had an applied force at only *one point*. This means that here $\bar{F}(x, t)$ must be zero everywhere except at that particular point, say $x = x'$. However $\bar{F}$ is a force per unit length so $\bar{F}\,dx$ is the applied force on the length dx. Thus let

$$(9.70) \qquad \lim_{\Delta x \to 0} \int_{x-\Delta x}^{x+\Delta x} \bar{F}(x, t)\,dx = \begin{cases} 0, & x \neq x' \\ F(t), & x = x' \end{cases}$$

where $F(t)$ is the force applied at $x = x'$. This suggests

$$(9.71) \qquad \bar{F}(x, t) = F(t)\,\delta(x - x')$$

as the appropriate form for $\bar{F}(x, t)$. Thus our equation is:

$$(9.72) \qquad \frac{\partial^2 Y}{\partial x^2} = \frac{1}{c^2} \frac{\partial^2 Y}{\partial t^2} - \frac{F(t)}{\rho} \delta(x - x')$$

where $c^2 = (T/\rho)$ as usual and we have made some rearrangements.

Here we do not require any initial conditions in this inhomogeneous equation if we suppose the excitation to be in existance for *all* time $-\infty < t < +\infty$, i.e., *there is no initial instant.*

Thus we apply the complex Fourier transform on the time variable by multiplying each term in Eq. (9.72) by $e^{-i\omega t}$ and integrating on t from $-\infty$ to $+\infty$, defining

$$(9.73) \qquad y(x, \omega) = \int_{-\infty}^{+\infty} e^{-i\omega t}\,Y(x, t)\,dt$$

as the transform of $Y(x, t)$ on t. Note that since the argument, ωt here, of *any* transcendental function must be *dimensionless*, ω has units of reciprocal time. In fact we often write $\omega = 2\pi v$ where v is the "frequency" in reciprocal time units.

Requiring $Y(x, t)$ to be *finite* at $t = \pm\infty$ and using our previous integration by parts procedure on the term in $\partial^2 Y/\partial t^2$, with the *summability* condition imposed, we obtain,

$$(9.74) \qquad \frac{d^2 y}{dx^2} = -\frac{\omega^2}{c^2} y - \frac{1}{\rho} f(\omega)\,\delta(x - x')$$

where we use $f(\omega)$ to denote the above Fourier transform applied to $F(t)$.

To this equation we now apply the *finite sine transform*. Our choice of this transform is dictated by two factors, the domain is finite, hence a finite transform, and the function $Y(x, t)$, and hence $y(x, \omega)$, is zero at the origin, thus calling for an *odd* function of x. Thus multiplying by $(2/L) \sin (n\pi x/L)$ and integrating from $x = 0$ to $x = L$ in Eq. (9.74) we obtain,

$$(9.75) \qquad -\frac{n^2\pi^2}{L^2} \bar{y}_n = -\frac{\omega^2}{c^2} \bar{y}_n - \frac{2}{\rho L} f(\omega) \sin \frac{n\pi x'}{L}$$

Here we used the integration by parts as before on d^2y/dx^2 and we also used the property

$$(9.76) \qquad U(x') = \int_0^L U(x)\, \delta(x - x')\, dx, \quad 0 < x' < L$$

of the delta function. Our notation here is

$$(9.77) \qquad \bar{y}_n = \frac{2}{L}\int_0^L y(x, \omega) \sin \frac{n\pi x}{L}\, dx$$

Rearranging Eq. (9.75) then yields,

$$(9.78) \qquad \bar{y}_n = \frac{-\dfrac{2}{\rho L} f(\omega) \sin \dfrac{n\pi x'}{L}}{\dfrac{\omega^2}{c^2} - \dfrac{n^2\pi^2}{L^2}}$$

so that inversion of the two transforms yields formally:

$$(9.79) \qquad Y(x, t) = \frac{1}{2\pi} \sum_{n=1}^{\infty} \frac{2}{\rho L} \sin \frac{n\pi x}{L} \sin \frac{n\pi x'}{L} \int_{-\infty}^{+\infty} \frac{f(\omega)e^{i\omega t}\, d\omega}{\dfrac{n^2\pi^2}{L^2} - \dfrac{\omega^2}{c^2}}$$

Here the factor $1/2\pi$ times the integral is just the inversion of the Fourier transform but note that the integrand is just the *product* of two Fourier transforms, i.e.,

$$f(\omega) \quad \text{and} \quad \frac{-c^2}{\omega^2 - \dfrac{n^2\pi^2c^2}{L^2}}$$

Hence, according to Entry Number (6) of Table 9.1 the inverse should be in the form of a convolution integral. To write the inverse in this form we need to know what function has the second form above as its Fourier transform. This will be,

$$(9.80) \qquad G(t) = \frac{1}{2\pi}\int_{-\infty}^{+\infty} \frac{-c^2 e^{+i\omega t}\, d\omega}{\omega^2 - \dfrac{n^2\pi^2c^2}{L^2}}$$

which is an integral of the type considered as in Example (4) of Chapter Eight.
Thus we must consider the integral,

$$(9.81) \qquad I = \oint_C \frac{e^{+izt}\, dz}{z^2 - a^2}$$

on an appropriate contour as described there. Here, since there are simple poles at $z = \pm a$ on the *real* axis, we take the contour of the real axis from $-R$ to $+R$ with small arcs about each of these to include the poles *in* the contour, and the semicircle of radius R in the *upper* half-plane for $t > 0$, or the same arc in the lower half-plane for $t < 0$. The two distinct cases are necessary for Jordan's Lemma to apply. By this process we obtain:

$$(9.82) \qquad G(t) = -\frac{Lc^2}{2n\pi} \sin \frac{n\pi}{L} ct \; , t > 0$$

Thus now using the above convolution property we have,

$$(9.83) \qquad Y(x,t) = -\frac{c^2}{\pi\rho} \sum_{n=1}^{\infty} \frac{1}{n} \sin \frac{n\pi x}{L} \sin \frac{n\pi x'}{L} \int_{-\infty}^{t} F(t') \sin \frac{n\pi}{L} c(t - t') \, dt'$$

as the final form of the solution for any applied forcing function $F(t)$ at the point x' on the string.

A particular case of interest having a rather simple form is that in which we apply an impulse to the string as a "blow by a hammer," i.e.,

$$(9.84) \qquad F(t) = E_0 \, \delta(t)$$

where the blow is struck at $t = 0$ and E_0 is the impulse of the blow. We obtain,

$$(9.85) \qquad Y(x,t) = -\frac{c^2 E_0}{\pi\rho} \sum_{n=1}^{\infty} \frac{1}{n} \sin \frac{n\pi x}{L} \sin \frac{n\pi x'}{L} \sin \frac{n\pi}{L} ct \; , \text{ for } t > 0$$

and $Y(x,t) = 0$, for $t < 0_g$ as the solution for this case.

Example (3)

In *the analysis of simple alternating current circuits* consisting of passive elements the equation describing the *output signal, $I(t)$* as a function of the input signal, $S(t)$ can usually be put in the form of a linear differential equation with constant coefficients, i.e.,

$$(9.86) \qquad a_n \frac{d^n I}{dt^n} + a_{n-1} \frac{d^{n-1} I}{dt^{n-1}} + \ldots + a_0 I = S(t)$$

Here the constants $a_0, a_1, \ldots a_n$, as well as n itself will be determined by the nature and arrangement of the circuit elements, that is resistances, inductances, and capacitances. All the a_j are *real*.

Now we could apply the *Laplace* transform to this equation, assuming I and all its derivatives to be zero at time zero, and thereby derive Eq. (6.87) of Example (5), Chapter Six for the *response* of this *linear instrument*, but we leave this to the student as an exercise.

If instead we apply the Fourier transform and impose suitable summability criteria on all integrals such as,

$$\int_{-\infty}^{+\infty} e^{-i\omega t} \frac{d^n I}{dt^n} \, dt$$

we obtain by the repeated integration by parts,

$$\left[\sum_{j=0}^{n} a_j (i\omega)^j \right] \mathscr{I}_F = \mathscr{S}_F$$

Where $\mathscr{I}_F$ and $\mathscr{S}_F$ are the Fourier transforms of the output signal $I(t)$ and the input signal $S(t)$, respectively.

Rearranging and defining,

$$\mathscr{R}_F(\omega) = \frac{1}{\sum\limits_{j=0}^{n} a_j(i\omega)^j}$$

as the Fourier transform of the *response function* of the circuit we have,

(9.89) $$\mathscr{I}_F(\omega) = \mathscr{R}_F(\omega)\mathscr{S}_F(\omega)$$

which is exactly analogous to Eq. (6.88) of Chapter Six. Thus, here taking the inverse Fourier transform we get,

(9.90) $$I(t) = \int_{-\infty}^{+\infty} \bar{R}(t - t')S(t')\,dt'$$

The major difference between the convolution integral here and that in Chapter Six is the lower limit. Here there is *no initial time* at which the circuit, or instrument, is quiescent.

DISPERSION RELATIONS

The convolution relationship between the output $I(t)$ of a linear instrument and the input signal, $S(t)$, given in Eq. (9.90) above was deduced for a particular physical situation. However this form must exist in general. In Chapter Six we discussed linear superposition and the convolution integral, now we present these ideas in a more complete fashion. By doing so we arrive at dispersion relations.

We consider a physical system having an input $S(t)$ and an output $I(t)$. It is assumed that:

   (A) the internal properties of the system are constant in time;
   (B) the system is *causal*, relating output to input in a causal manner and;
   (C) the output is a *linear* functional of the input.

Because of the condition (C) we can write the general expression:

(9.91) $$I(t) = \int_{-\infty}^{+\infty} \bar{R}(t, t')S(t')\,dt'$$

where $\bar{R}(t, t')$ characterizes the output at time t associated with the input at time t'. The *causal* condition (B) requires that:

(9.92) $$\bar{R}(t, t') \equiv 0 \quad \text{for} \quad t < t'$$

implying that an effect cannot exist prior to a cause. The condition (A) implies that $\bar{R}(t, t')$ must be a function of $t - t'$ only and not t explicitly. Thus these three conditions are met by the functional structure:

(9.93) $$I(t) = \int_{-\infty}^{+\infty} \bar{R}(t - t')S(t')\,dt'; \qquad \bar{R}(t - t') \equiv 0 \quad \text{for} \quad t - t' < 0$$

From the Fourier representation:

(9.94) $$\mathscr{R}_F(\omega) = \int_{-\infty}^{+\infty} \bar{R}(\tau)e^{-i\omega\tau}\,d\tau$$

it follows that since $\bar{R}(\tau) \equiv 0$ for $\tau < 0$ then $\mathscr{R}_F(\omega)$ must be an analytic function of ω regular in the lower half of the ω plane, i.e., since only positive τ contributes to the integral. To see this write ω as $\alpha + i\beta$ and have:

$$(9.95) \qquad \mathscr{R}_F(\omega) = \int_0^\infty \bar{R}(\tau) \cos \alpha\tau \, e^{\beta\tau} \, d\tau$$

$$- i \int_0^\infty \bar{R}(\tau) \sin \alpha\tau \, e^{\beta\tau} \, d\tau$$

where the lower limit has been written as zero in view of the condition on $\bar{R}(\tau)$. These integrals are the real and imaginary parts of $\mathscr{R}_F(\omega)$ and if bounded and uniformly convergent in α and β satisfy the Cauchy–Reimann conditions. Here we can see that they are so bounded for *negative* β and reasonable conditions on $\bar{R}(\tau)$. Thus the *causality condition*, $\bar{R}(\tau) \equiv 0$ for $\tau < 0$, determines in large measure the form of $\mathscr{R}_F(\omega)$.

Now for $\text{Im}\,\omega < 0$ the value of $\mathscr{R}_F(\omega)$ can be written in terms of Cauchy's integral:

$$(9.96) \qquad \mathscr{R}_F(\omega) = \frac{1}{2\pi i} \int_C \frac{\mathscr{R}_F(\omega')\,d\omega'}{\omega' - \omega}$$

or, for *real* values of ω,

$$(9.97) \qquad \mathscr{R}_F(\omega) = \frac{1}{2\pi i} {}^*\!\!\int_{-\infty}^{+\infty} \frac{\mathscr{R}_F(\alpha')\,d\alpha'}{\alpha' - \omega} + \frac{\pi i \mathscr{R}_F(\omega)}{2\pi i}$$

where we use as the contour C the real axis with a small semicircle indented above the point ω on the real axis and closed by the infinite semicircle in the lower half-plane. The asterisk on the integral above indicates the Cauchy principal value and the last term results from the integral on the small semicircle. Here it is *assumed* that the contribution from the infinite semicircle is zero, i.e., we assume $|\mathscr{R}_F(\omega)| \to 0$ as $|\omega| \to \infty$ in the lower half-plane.

Finally then we see from this that, *for real* ω,

$$(9.98) \qquad \mathscr{R}_F(\omega) = \frac{1}{\pi i} {}^*\!\!\int_{-\infty}^{+\infty} \frac{\mathscr{R}_F(\alpha')\,d\alpha'}{\alpha' - \omega}$$

and hence that,

$$(9.99) \qquad \text{Re}[\mathscr{R}_F(\omega)] = \frac{1}{\pi} {}^*\!\!\int_{-\infty}^{+\infty} \frac{\text{Im}[\mathscr{R}_F(\alpha')]\,d\alpha'}{\alpha' - \omega}$$

and,

$$(9.100) \qquad \text{Im}[\mathscr{R}_F(\omega)] = -\frac{1}{\pi} {}^*\!\!\int_{-\infty}^{+\infty} \frac{\text{Re}[\mathscr{R}_F(\alpha')]\,d\alpha'}{\alpha' - \omega}$$

These two equations constitute *dispersion relations*; they relate the real and imaginary parts of the Fourier transform of the response function of a causal, linear, constant system.† Such relationships have many practical uses. Thus in some cases the real and imaginary parts of $\mathscr{R}_F(\omega)$ have different physical meanings and these relations allow calculation of one when the other is measured.

† The particular form here is that of the Hilbert transform which we do not treat in this text, i.e., $\text{Re}(\mathscr{R}_F)$ and $\text{Im}(\mathscr{R}_F)$ are here Hilbert transforms of each other, a transform pair.

Dispersion relations can also be used to test mathematical models of physical systems to determine whether the models are causal or not. Also these relations can be used directly on data of physical systems to see if the systems behave causally.

Note that according to Eq. (9.98) the function $\mathcal{R}_F(\omega)$ at any point in the lower half-plane is determined entirely by its values on the real axis. This is a dispersion integral of a type somewhat different from that discussed in Chapter Eight.

EXECUTE PROBLEM SET (9-2)

INFINITE HANKEL TRANSFORMS

In earlier chapters we found that we could represent a function as a series in Bessel functions such as,

$$(9.101) \qquad F(r) = \sum_{j=1}^{\infty} A_{nj} J_n(\beta_{nj} r), \qquad 0 < r < R$$

where the A_{nj} are given by,

$$(9.102) \qquad A_{nj} = \frac{1}{N_{nj}} \int_0^R F(r') J_n(\beta_{nj} r') r' \, dr'$$

with the *norm* being N_{nj} and the β_{nj} being appropriate roots of a relation in Bessel functions. Here we wish to consider the situation $R \to \infty$ just as we did in the case of the Fourier series with $L \to \infty$.

For example suppose the β_{nj} are roots of

$$(9.103) \qquad J_n(\beta_{nj} R) = 0$$

then the norm is

$$(9.104) \qquad N_{nj} = \frac{R^2}{2} J_{n+1}^2(\beta_{nj} R)$$

Since we are going to let $R \to \infty$ we can use the simplest *asymptotic* forms† for J_n and J_{n+1} in these two equations, thus:

$$(9.105) \qquad \left(\frac{2}{\pi \beta_{nj} R}\right)^{1/2} \cos\left(\beta_{nj} R - \frac{2n+1}{4}\pi\right) = 0$$

replaces Eq. (9.103) and this has the roots given by

$$(9.106) \qquad \beta_{nj} R - \frac{2n+1}{4}\pi = \frac{2j+1}{2}\pi, \qquad j = 0, 1, 2, \ldots$$

Then since the asymptotic form of Eq. (9.104) is;

$$(9.107) \qquad N_{nj} = \frac{R^2}{2}\left(\frac{2}{\pi \beta_{nj} R}\right) \cos^2\left(\beta_{nj} R - \frac{2n+1}{4}\pi - \frac{\pi}{2}\right)$$

we get, using Eq. (9.106) for the roots, β_{nj},

$$(9.108) \qquad N_{nj} = \frac{R}{\pi \beta_{nj}}$$

† See Eq. (4.6) of Chapter Four.

Also note that in the asymptotic form for $R \to \infty$, we have from Eq. (9.106)

$$(9.109) \qquad \Delta \beta_{nj} = \beta_{nj+1} - \beta_{nj} = \frac{\pi}{R}$$

Thus if we now substitute Eq. (9.102) into Eq. (9.101) and use these asymptotic forms we get:

$$(9.110) \qquad F(r) = \sum_{j=1}^{\infty} J_n(\beta_{nj}r)\beta_{nj} \, \Delta \beta_{nj} \int_0^R F(r')J_n(\beta_{nj}r')r' \, dr'$$

Thus, since from Eq. (9.109) above we see $\Delta \beta_{nj} \to 0$ as $R \to \infty$, we have in the *limit* as $R \to \infty$,

$$(9.111) \qquad F(r) = \int_0^{\infty} \int_0^{\infty} F(r')J_n(\beta r)J_n(\beta r')r' \, dr' \beta \, d\beta$$

That is, the sum on j approaches the integral on β in the limit and we drop the subscripts on β since it is obvious that now it is superfluous. This limiting form is exactly comparable to the Fourier integral already treated. On this basis we define the *Hankel Transform Pair of order n* as:†

$$(9.112) \qquad \begin{cases} f_n(\beta) = \displaystyle\int_0^{\infty} F(r')J_n(\beta r')r' \, dr' \\[2mm] F(r) = \displaystyle\int_0^{\infty} f_n(\beta)J_n(\beta r)\beta \, d\beta \end{cases}$$

This transform pair has particular utility in problems having a symmetry appropriate to cylindrical coordinates and an *infinite domain* in the r coordinate. To illustrate the *class* of problems for which these transforms are especially appropriate we present the following example.

Example (4)

Find the *electrostatic potential distribution in all of space due to a given distribution of charge confined to a finite part of space*, i.e., the charge density, $\rho \to 0$, as $\mathbf{r} \to \infty$.

The appropriate differential equation is Poisson's equation and we write this in cylindrical coordinates as:

$$(9.113) \qquad \frac{1}{r}\frac{\partial}{\partial r}\left(r\frac{\partial V}{\partial r}\right) + \frac{1}{r^2}\frac{\partial^2 V}{\partial \theta^2} + \frac{\partial^2 V}{\partial z^2} = -\frac{\rho}{\varepsilon}$$

where ε is a constant, $V(r, \theta, z)$, is the potential at any point and $\rho(r, \theta, z)$ is the given charge distribution.

Now we proceed to apply *appropriate* integral transforms. First, since the domain of z is $-\infty < z < +\infty$ and e^{-ikz} is an eigenfunction of the operator, $\partial^2/\partial z^2$, we use the Fourier transform on z, carrying out the usual integration by parts. This yields

$$(9.114) \qquad \frac{1}{r}\frac{\partial}{\partial r}\left(r\frac{\partial v}{\partial r}\right) + \frac{1}{r^2}\frac{\partial^2 v}{\partial \theta^2} - k^2 v = -\frac{\bar{\rho}(r, \theta, k)}{\varepsilon}$$

We then select a finite integral transform for θ since its domain is $-\pi < \theta < \pi$. The question arises whether to use the sine or cosine transform; we have seen the first to be appropriate to *odd* functions while the second applies to *even* functions.

† The argument given above is purely heuristic, but can be made rigorous.

Any function can be written as the sum of an even function and an odd function. Thus we see ρ as

$$(9.115) \qquad \rho = \rho_0 + \rho_e$$

where ρ_0 is an *odd* function of θ while ρ_e is an *even* function of θ. We multiply Eq. (9.114) by †$(\delta_n/\pi)\cos n\theta$ and integrate from $-\pi$ to π and obtain, using the usual integration by parts on $\partial^2/\partial\theta^2$, and the fact that v is *also* composed of an odd and even part to obtain,

$$(9.116) \qquad \frac{1}{r}\frac{\partial}{\partial r}\left(r\frac{\partial \bar{v}_e}{\partial r}\right) - \frac{n^2}{r^2}\bar{v}_e - k^2\bar{v}_e = -\frac{\bar{\rho}_e}{\varepsilon}$$

where $\bar{v}_e$ is the finite cosine transform of the *even part* of v, as is $\bar{\rho}_e$ of $\bar{\rho}$. A similar process yields, with the sine transform, an identical equation with the e subscript replaced by an o subscript for the *odd* parts. Thus we only need to consider one part since the other is treated in exactly similar fashion.

Now multiply Eq. (9.116) through by $J_n(\beta r)r$ and integrate from $r = 0$ to $r = \infty$. In the first term we can do two integrations by parts to obtain the form:

$$r\frac{\partial \bar{v}_e}{\partial r}J_n(\beta r)\bigg|_0^\infty - r\bar{v}_e\frac{\partial J_n(\beta r)}{\partial r}\bigg|_0^\infty + \int_0^\infty \bar{v}_e\left\{\frac{1}{r}\frac{\partial}{\partial r}\left(r\frac{\partial J_n}{\partial r}\right)\right\}r\,dr$$

Here the integrated terms must vanish and we then have the form,

$$\int_0^\infty \bar{v}_e\left\{\frac{1}{r}\frac{\partial}{\partial r}\left[r\frac{\partial J_n(\beta r)}{\partial r}\right] - \frac{n^2}{r^2}J_n(\beta r)\right\}r\,dr$$

$$(9.117) \qquad - k^2\int_0^\infty \bar{v}_e J_n(\beta r)r\,dr = -\frac{1}{\varepsilon}\int_0^\infty \bar{\rho}_e J_n(\beta r)r\,dr$$

In this equation we recognize the term in brackets in the first integral as being just $-\beta^2 J_n(\beta r)$ according to Bessel's equation (i.e., Eq. (4.1) of Chapter Four or, in the same form as used here, in Eq. (5.75) of Chapter Five.) Thus we have just,

$$(9.118) \qquad -\beta^2\int_0^\infty \bar{v}_e J_n(\beta r)r\,dr = -\beta^2\bar{u}_e$$

replacing the first integral above. Here we use the symbol $\bar{u}_e$ to denote the Hankel transform of order n of $\bar{v}_e$. Similarly we will now use $\bar{\gamma}_e$ for the Hankel transform of order n of $\bar{\rho}_e$, which appears with the coefficient $-1/\varepsilon$ on the right in Eq. (9.108).

Thus we now have

$$(9.119) \qquad (\beta^2 + k^2)\bar{u}_e = -\frac{1}{\varepsilon}\bar{\gamma}_e$$

and hence

$$(9.120) \qquad \bar{u}_e = \frac{1}{\varepsilon}\frac{\bar{\gamma}_e}{\beta^2 + k^2}$$

† Note: $\delta_n = \tfrac{1}{2}, \qquad n = 0$

$\delta_n = 1, \qquad n > 0$

Then applying the appropriate inverse transforms,

$$(9.121) \qquad V_e = \sum_{n=0}^{\infty} \frac{\cos n\theta}{2\pi\varepsilon} \int_{-\infty}^{+\infty} \int_{0}^{\infty} \frac{\bar{\bar{\jmath}}_e(k, n, \beta)}{\beta^2 + k^2} e^{ikz} J_n(\beta r)\beta \, d\beta \, dk$$

Similarly we can write an exactly equivalent form for the odd part of V, V_0, with $\bar{\bar{\jmath}}_0$ and $\sin n\theta$ replacing $\bar{\bar{\jmath}}_e$ and $\cos n\theta$.

Here we should note that the odd and even parts of a function of θ are formed as,

$$(9.122) \qquad V_o(r, \theta, z) = \tfrac{1}{2}[V(r, \theta, z) - V(r, -\theta, z)]$$

and

$$(9.123) \qquad V_e(r, \theta, z) = \tfrac{1}{2}[V(r, \theta, z) + V(r, -\theta, z)]$$

and obviously V is the sum of V_o and V_e.

Now we have constructed a general form for the solution, but in order to develop a specific case so that the full details of the method are clear let us now choose a *specific* charge distribution.

Consider a thin cylindrical shell of uniform charge density, say λ per unit area, of radius R and length L, $-L/2 < z < L/2$. Let the axis coincide with the z axis. This charge distribution is described by,

$$(9.124) \qquad \rho(r, \theta, z) = \left[H\left(z, \frac{-L}{2}\right) - H\left(z, \frac{+L}{2}\right) \right] \frac{\delta(r - R)}{r} \lambda$$

where $H(z, z')$ is the unit step function and $\delta(r - R)$ is the *Dirac* delta function. That this is the correct form can be seen by multiplying by $2\pi r \, dr \, dz$ and integrating over all space. The correct total charge, $2\pi RL\lambda$, results.

Having this form we now determine the appropriate multiple transforms, $\bar{\bar{\jmath}}_e$ and $\bar{\bar{\jmath}}_o$. Obviously here we have *only* an even function. Thus, since the order in which we take the transforms is immaterial, we apply the finite cosine transform,

$$(9.125) \qquad \frac{\delta_n}{\pi} \int_{-\pi}^{\pi} \rho(r, \theta, z) \cos n\theta \, d\theta = \begin{cases} \rho(r, z), & n = 0 \\ 0, & n > 0 \end{cases}$$

and find that this vanishes except for $n = 0$ since the problem has *axial* symmetry in this case.

Next we apply the Hankel transform, and find, noting that only $n = 0$ need be considered,

$$(9.126) \qquad \int_{0}^{\infty} \rho(r, z) J_0(\beta r) r \, dr = \left[H\left(z, \frac{-L}{2}\right) - H\left(z, \frac{L}{2}\right) \right] \frac{J_0(\beta R)}{2\pi} \lambda$$

because of the property of the delta function.

Finally we apply the Fourier transform on z to obtain, using the above results:

$$(9.127) \qquad \bar{\bar{\jmath}}_e = 0, \qquad n > 0$$

and

$$(9.128) \qquad \bar{\bar{\jmath}}_e = \frac{\lambda}{2\pi} J_0(\beta R) \int_{-\frac{L}{2}}^{+\frac{L}{2}} e^{-ikz} \, dz$$

or,

$$(9.129) \qquad \bar{\bar{\jmath}}_e = \frac{\lambda}{k\pi} J_0(\beta R) \sin \frac{kL}{2}$$

This is then inserted into Eq. (9.121) to yield:

(9.130)
$$V = \int_{-\infty}^{+\infty} \int_0^{\infty} \frac{\lambda}{2\pi^2 \varepsilon} \frac{J_0(\beta R)J_0(\beta r)}{k(k^2 + \beta^2)} e^{ikz} \sin \frac{kL}{2} \beta \, d\beta \, dk$$

as the potential of a thin cylindrical shell of length L, radius R, and uniform surface charge density λ.

Of course one feels a certain amount of dissatisfaction with any solution of a problem in the form of a rather "nasty" multiple integral as this. However, if we wished to carry the analysis of this problem further we might make some studies of limiting cases here, i.e., near the axis for example we could use the appropriate approximate form for $J_0(\beta r)$; or at large distance from the axis, etc.

This example does clearly illustrate when one should choose a particular Hankel transform for a given problem.

EXECUTE PROBLEM SET (9-3)

Z TRANSFORM AND ITS INVERSION

In Chapter Six we discussed the response of linear instruments in terms of a convolution integral and again in the present chapter. Here we use the description given in Chapter Six as the motivation for defining a new transform, the z transform. This was not introduced in Chapter Six because the calculus of residues is required for the formal inversion of this transform and this technique was not available to us there.

Consider the equation giving the output $I(t)$ of a *linear* instrument, quiescent at $t = 0$, having response function $R(t)$ and input signal $S(t)$:†

(9.131)
$$I(t) = \int_0^t R(t - \tau)S(\tau) \, d\tau$$

As we have already pointed out $R(t)$ is the form $I(t)$ will have if the input is a *delta* function at $t = 0$, $S(t) = \delta(t)$. Thus suppose we *approximate* the input signal by a sequence of equally spaced delta functions, each multiplied by the signal, $S(t)$, actually existing at that instant, i.e.,

(9.132)
$$S(t) \approx \sum_{n=0}^{\infty} S(nT) \delta(t - nT) T$$

where T is the spacing between the "sampling" points, $t = nT$. To be sure, this is a highly unrealistic description of $S(t)$ but with this we *do* get a very good approximation to $I(t)$.

Putting Eq. (9.132) for $S(t)$ into Eq. (9.131) we get for $I(t)$ the form,

(9.133)
$$I(t) \approx \sum_{n=0}^{N} R(t - nT)S(nT) T$$

where N is the *largest* value of n satisfying,

(9.134)
$$n \le \frac{t}{T}$$

Thus $I(t)$ is now represented by a sum.

Recall now that the Laplace transform of a convolution integral such as Eq. (9.131) is just the *product* of the transforms of the two functions, $R(t)$ and $S(t)$, in the integrand. Therefore we

---

† We require $R(t) = 0$ for $t < 0$ for *causality*.

examine the Laplace transform of $S(t)$ as approximated in Eq. (9.132). We obtain,

(9.135) $$L\{S(t)\} = \sum_{n=0}^{\infty} S(nT)Te^{-pnT}$$

where p is the transform parameter. Here if we define

(9.136) $$z^{-1} = Te^{-pnT}$$

we see Eq. (9.135) as:

(9.137) $$Z\{S(t)\} = \sum_{n=0}^{\infty} S(nT)z^{-n} = s(z)$$

where we now denote the transform operator as Z instead of L. We call the sum on the right here the *Z transform of the function $S(t)$*.

Now look upon the variable z as a *complex* variable and consider the form,

(9.138) $$z^{m-1}s(z) = (S_0 z^{m-1} + S_1 z^m + \ldots + S_{m-1}) + \left(\frac{S_m}{z} + \frac{S_{m+1}}{z^2} + \ldots\right)$$

which has the appearance of the Laurent series about the origin of some function of z with the *principal part* being the second parenthesis. [Here S_m denotes $S(mT)$.] Thus if we integrate on some contour about the origin,

(9.139) $$\oint_C z^{m-1}s(z)\,dz = 2\pi i S_m$$

by the *residue theorem* (see Chapter Eight), since the coefficient of $1/z$ in the Laurent series is the residue if the principal part terminates. Thus

(9.140) $$S(mT) = \frac{1}{2\pi i} \oint_C z^{m-1}s(z)\,dz$$

is the *inversion integral* for the Z transform.

Thus we have the *Z transform pair*:

(9.141)
$$
\begin{cases}
Z\{S(t)\}_T = \sum_{n=0}^{\infty} S(nT)z^{-n} = s(z) \\[2mm]
Z^{-1}\{s(z)\} = \dfrac{1}{2\pi i} \oint_C z^{n-1}s(z)\,dz = S(t), \qquad t = nT
\end{cases}
$$

One should not be misled by the route we have used to introduce Z transforms. Equations (9.141) stand alone and do not have an explicit connection to the Laplace transform. Quite obviously the appearance of discrete values of $S(t)$ here suggests the greatest applicability of these transforms to equations involving finite differences and discrete data. We illustrate this with an example of great *practical* value.

Example (5)

Discrete representation of linear instruments. Consider the convolution integral of Eq. (9.131) representing the response $I(t)$ of a linear instrument to the input $S(t)$. *Suppose $S(t)$ is zero* *for* $t < 0$ and *approximate* the integral then as the sum,

(9.142)
$$I_n = \sum_{k=0}^{\infty} \bar{R}_{n-k} S'_k$$

where

(9.143)
$$\begin{cases} I_n = I(nT) \\ \bar{R}_{n-k} = \bar{R}(nT - kT), \quad 0 \quad \text{for} \quad k > n, \\ S'_k = TS(kT) \end{cases}$$

and T is supposed sufficiently small so that the sum approximation for the integral is sufficiently good.

Now it can be shown, and we leave this as an exercise for the student, that

(9.144)
$$Z^{-1}\{s_1(z)s_2(z)\} = \sum_{k=0}^{\infty} S_1(kT)S_2(nT - kT)$$

consequently if we take the Z transform of Eq. (9.142) we get,

(9.145)
$$\mathscr{I}(z) = r(z)s'(z)$$

where $\mathscr{I}$, r, and s' are the Z transforms of $I(t)$, $R(t)$, and $S'(t)$, respectively.

Written out this appears as

(9.146)
$$\sum_{n=0}^{\infty} I_n z^{-n} = \sum_{k=0}^{\infty} \bar{R}_k z^{-k} \cdot \sum_{l=0}^{\infty} S'_l z^{-l}$$

and then dividing by the $\bar{R}$ sum,

(9.147)
$$\sum_{l=0}^{\infty} S'_l z^{-l} = \frac{\displaystyle\sum_{n=0}^{\infty} I_n z^{-n}}{\displaystyle\sum_{k=0}^{\infty} \bar{R}_k z^{-k}}$$

Now as a rule the response of an instrument is highly damped so that an impulse at $t = 0$ does not cause $\bar{R}(t)$ to persist different from zero for a very long period, i.e., $\bar{R}_k \approx 0$ for $k > K$ where K is not a very large number. This means the denominator here is a finite polynomial in z^{-1} and the division can be formally carried out to any desired number of terms. Then we can use the fact that two polynomials are equal only when coefficients of corresponding powers are equal to obtain explicit expressions for S'_1, S'_2, etc., in terms of combinations of the I_n and $\bar{R}_n$. Alternatively we can use the formal inversions of Eq. (9.141).

This means that if we have discrete data as output of a linear instrument and also a discrete description of the characteristic response function $\bar{R}(t)$, then we can actually construct, from this output, what the input signal was, but only in the discrete form which is approximate.

OTHER INTEGRAL TRANSFORMS

We have by no means exhausted the subject of integral transforms but here we have presented those which are most generally applicable to physical problems. For others, and more details on those given here the student should consult the references.

EXECUTE PROBLEM SET (9-4)

SUMMARY OF THE GENERAL FORMAT FOR SOLUTION OF INHOMOGENEOUS, LINEAR, PARTIAL DIFFERENTIAL EQUATIONS SUBJECT TO HOMOGENEOUS BOUNDARY CONDITIONS

We consider the linear partial differential equation as the general form,

$$(9.148) \qquad OU(\chi, \eta, \gamma, \tau) = F(\chi, \eta, \gamma, \tau)$$

where we use χ, η, γ, τ to indicate *some* set of variables, i.e., $x, y, z, t, r, \theta, \phi, z$, etc. O is a *linear* differential operator of some sort, such as say, $\nabla^2 - \partial/\partial t$ and F is a given function.

The solution, $U(\chi, \eta, \gamma, \tau)$ is to be obtained in some domain, D, bounded by surface S, and U is subject to *homogeneous* conditions on S, i.e., *neither U nor $\partial U/\partial n$ is a function of position on S* although different homogeneous conditions may apply on various finite segments of S, i.e., one condition for the ends of a cylinder say and another for the curved lateral surface.

If the boundary conditions *are* homogeneous as just described, but are *time dependent*, then we can immediately take the Laplace, or Fourier, transform of the equation *and* all the boundary conditions so that we then treat a time independent problem. In many cases this end is achieved in terms of Duhamel's formula as discussed in Chapter Six.

Now suppose the operator O is separable,

$$(9.149) \qquad O = O_\chi + O_\eta + O_\gamma + O_\tau$$

as described for the *completely* homogeneous case of Chapter Five. Then select and apply integral transforms on χ, η, γ, and τ, having kernels which are the appropriate eigenfunctions of these operators, either finite or infinite as the ranges of the respective variables in D dictate, taking account of the boundary conditions, even or odd character, periodicity, etc., in this selection. Let T denote this *composite* transform. Then

$$(9.150) \qquad TOU = (a_\chi + a_\eta + a_\gamma + a_\tau)u = f$$

where $a_\chi, a_\eta, a_\gamma, a_\tau$ are eigenvalues of the eigenfunctions in the kernels of the transforms, and u is just T applied directly to U, i.e., the multiple integral transform of U and f is just TF, the multiple transform of F. As we have seen in our examples these eigenvalues generally appear through one or more integrations by parts.

Finally we apply the inversions of the transforms, here denoted as for the composite transform inverse, to obtain,

$$(9.151) \qquad U(\chi, \eta, \gamma, \tau) = T^{-1}\left\{ \frac{TF}{a_\chi + a_\eta + a_\gamma + a_\tau} \right\}$$

where we have used TF inside instead of f to emphasize that F is acted on directly by T.

An important point to note here is that we can also apply this same procedure to *homogeneous equations*. The major difference appears in Eq. (9.150). If the equation is homogeneous $f = 0$ and this then provides a *constraint* on the eigenvalues, one can be expressed in terms of the others and we have the *same* result as would be obtained by simple separation of variables.

PROBLEMS

Set (9–1)

(1) Using finite sine or cosine transforms, and the Laplace transform, solve the heat conduction problem:

$$k\frac{\partial^2 T}{\partial x^2} = C\rho\frac{\partial T}{\partial t} - Q(x, t), \qquad 0 < x < L$$

subject to

$$-k\frac{\partial T}{\partial x} = q = \text{constant} \quad \text{at} \quad x = 0, \qquad t > 0$$

$$T = 0, \quad \text{at} \quad x = L, \qquad t > 0$$

$$T(x, 0) = T_0(x)$$

(2) A solid cylindrical shaft is rotating at a constant angular speed, ω, when the ends $x = -L/2$, $x = L/2$ are suddenly clamped at time $t = 0$. Use the appropriate *transforms* to solve the problem for the angular displacement $\theta(x, t)$ of a cross section at any time $t > 0$.

$$\frac{\partial^2\theta}{\partial t^2} = a^2\frac{\partial^2\theta}{\partial x^2}, \qquad \frac{-L}{2} < x < \frac{L}{2}, \qquad t > 0$$

$$\theta(x, 0) = 0, \qquad \frac{\partial\theta(x, 0)}{\partial t} = \omega$$

$$\theta\left(\frac{-L}{2}, t\right) = \theta\left(\frac{L}{2}, t\right) = 0$$

(3) A circular membrane is elastically loaded so that a force $-k(z - z_0)$ is applied to each element, with z_0 being the plane of equilibrium of the loading. Also a prescribed displacement is given to the circular boundary, $r = R$. Find the displacement, $z(r, \theta, t)$.

$$\rho\frac{\partial^2 z}{\partial t^2} = T\left[\frac{1}{r}\frac{\partial}{\partial r}\left(r\frac{\partial z}{\partial r}\right) + \frac{1}{r^2}\frac{\partial^2 z}{\partial\theta^2}\right] - k(z - z_0)$$

with ρ and T constants and,

$$z(R, \theta, t) = z_0 + F(t)$$

$$z(r, \theta, 0) = z_0 + A_0(r, \theta)$$

$$\frac{\partial z(r, \theta, 0)}{\partial t} = V_0(r, \theta)$$

Assume A_0 and V_0 to both be *even* functions in θ. [Hint: make the substitution

$$y(r, \theta, t) = z - z_0 - F(t)$$

and solve the resulting inhomogeneous equation using appropriate transforms.]

(4) Solve the problem of vibration of a circular membrane with edge clamped and a mass m rigidly attached to the center of the membrane. Approximate this loading as a *point* loading of force $-mg$. The formulation is exactly as in Problem (3) above with $k(z - z_0)$ replaced by $(-mg/a)[\delta(r)/r]$, where a is the actual area of contact of m on the membrane and $\delta(r)$ is the ordinary linear weight, or Dirac, delta function. Also we put $z_0 + F(t) = 0$ for the clamped edge. Assume the motion starts from *equilibrium*, $\partial^2 z/\partial t^2 = 0$ at $t = 0$, with some initial velocity imparted to the single point at which the mass is attached, i.e.,

$$\frac{\partial z}{\partial t} = v_0\frac{\delta(r)}{r}, \qquad \text{at } t = 0$$

where $v_0 = \text{constant}$. [Hint: set up the problem with the term $\partial^2 z/\partial t^2 = 0$ in the differential equation. The solution of this gives the *initial* shape.]

(5) A chemical reaction occurs in a closed cylindrical vessel of radius R and length L. A chemical resulting from the reaction is formed at the rate $Q(r, t)$. The r dependence results from a non-uniform temperature distribution. The concentration of this chemical is described by

$$D\left[\frac{1}{r}\frac{\partial}{\partial r}\left(r\frac{\partial C}{\partial r}\right) + \frac{\partial^2 C}{\partial z^2}\right] = \frac{\partial C}{\partial t} - Q(r, t)$$

with

$$C(r, z, 0) = 0$$

$$\frac{\partial C}{\partial r} = 0, \qquad r = R, \qquad \frac{-L}{2} < z < \frac{L}{2}$$

$$\frac{\partial C}{\partial z} = 0, \qquad z = \pm\frac{L}{2}, \qquad 0 \le r < R$$

Find $C(r, z, t)$ for the general case and for $Q(r, t) = A[1 - (r/R)e^{-at}$.

(6) A simply supported rectangular plate has a concentrated load p at one point, (ξ, η). Find the deflection z of the plate, in general,

$$\nabla^4 z = \frac{\partial^4 z}{\partial x^4} + 2\frac{\partial^4 z}{\partial x^2 \partial y^2} + \frac{\partial^4 z}{\partial y^4} = \frac{P(x, y)}{f}$$

$$0 < x < a, \qquad 0 < y < b$$

where P is the loading stress and $f = Eh^3/[12(1 - v)]$ is the flexural rigidity of a plate of thickness h, Young's modulus, E, and Poisson ratio, v. Here,

$$P = p\delta(x - \xi)\delta(y - \eta)$$

with the δ being delta functions of linear weight, and

$$0 = z(0, y) = z(a, y) = \frac{\partial^2 z(0, y)}{\partial x^2} = \frac{\partial^2 z(a, y)}{\partial x^2}$$

and

$$0 = z(x, 0) = z(x, b) = \frac{\partial^2 z(x, 0)}{\partial y^2} = \frac{\partial^2 z(x, b)}{\partial y^2}$$

(7) Any function $F(x)$ defined on $-L < x < L$ can be represented as the sum of an odd and even part, i.e.,

$$F(x) = F_o(x) + F_e(x)$$

where

$$F_e(x) = \frac{F(x) + F(-x)}{2} = \text{even part}$$

$$F_o(x) = \frac{F(x) - F(-x)}{2} = \text{odd part}$$

Show that if we write

$$\frac{1}{L} \int_{-L}^{L} e^{i\frac{n\pi x'}{L}} F(x') \, dx' = f_n$$

then the real part of f_n,

$$\text{Re}(f_n) = \text{cosine transform of } F_e(x)$$

and the imaginary part of f_n,

$$\text{Im}(f_n) = \text{sine transform of } F_0(x)$$

all *except* for $n = 0$ where an extra factor of *two* appears in the cosine transform, as defined in Eq. (9.9). Show, however, that if we let f_n be called the complex finite Fourier transform on $-L < x < L$, then the appropriate inversion is:

$$F(x) = \frac{1}{2} \sum_{n=-\infty}^{+\infty} f_n e^{-i\frac{n\pi x}{L}}$$

i.e., compare to Problem (4) of Set (3-5) of Chapter Three.

Set (9–2)

(1) Verify Formula (3) of Table 9.1.
(2) Verify Formula (6) of Table 9.1.
(3) Find the infinite Fourier transform of $(\cos ax/x^2)$ and also of $(\sin ax/x^2)$.
(4) Solve the diffusion problem in the half-space in two dimensions,

$$D\left(\frac{\partial^2 C}{\partial x^2} + \frac{\partial^2 C}{\partial y^2}\right) = \frac{\partial C}{\partial t} - Q(x, y), \qquad \begin{matrix} y > 0 \\ -\infty < x < +\infty \end{matrix}$$

with the source function

$$Q(x, y) = Ae^{-\alpha y^2 - \beta x^2}$$

and the conditions

$$\frac{\partial C}{\partial y} = 0 \quad \text{at} \quad y = 0, \qquad -\infty < x < +\infty$$

$$\lim_{\substack{y \to \infty \\ |x| \to \infty}} C \to 0$$

and

$$C(x, y, 0) = 0$$

(5) Show that:

$$\int_{-\infty}^{+\infty} F(x)G(x) \, dx = \frac{1}{2\pi} \int_{-\infty}^{+\infty} f(k)g(-k) \, dk$$

This is known as Parseval's relation (f and g are Fourier Transforms of F and G.)

(6) A meteorite entering the earth's atmosphere is vaporized. Treat this as a *moving point source* from which material is liberated at the constant rate of "q" grams per second. This material diffuses out into the atmosphere. The concentration at any point at any time is

given by $C(\mathbf{r}, t)$, where,

$$DV^2 C(\mathbf{r}, t) = \frac{\partial C(\mathbf{r}, t)}{\partial t} - q\delta(\mathbf{r} - \mathbf{v}t) H(t, 0)$$

with δ being a "δ function." For simplicity let the meteorite travel along the z axis so,

$$\delta(\mathbf{r} - \mathbf{v}t) = \delta(x)\,\delta(y)\,\delta(z - vt)$$

where on the right each "δ" is a regular "Dirac" delta function with unit weight function. For approximate treatment assume an infinite atmosphere and require $C \to 0$ as $\mathbf{r} \to \infty$. Find $C(x, y, z, t)$, when $C(x, y, z, 0)$ is zero, i.e., consider $q = 0$ for $t < 0$, $t = 0$ being the time of "appearance" of the meteorite in the atmosphere.

(7) Since the response function $\overline{R}(t - t')$ of a linear, causal system with constant properties must have a Fourier transform $\mathscr{R}_F(\omega)$ that is analytic in the lower half-plane what does this imply for the circuit problem discussed in Example (3) of this chapter, i.e., what can be said about the polynomial,

$$\frac{1}{\mathscr{R}_F(\omega)} = \sum_{j=0}^{n} a_j(i\omega)^j$$

Set (9–3)

(1) Solve

$$\frac{1}{r}\frac{\partial}{\partial r}\left(r\frac{\partial V}{\partial r}\right) + \frac{\partial^2 V}{\partial z^2} = \rho(r, z)$$

in the infinite domain $r \geq 0$, $-\infty < z < +\infty$.
   In particular consider

$$\rho(r, z) = Ae^{-\alpha z^2 - \beta r}$$

as a special case.

(2) Consider the half-space in which diffusion occurs

$$D\left[\frac{1}{r}\frac{\partial}{\partial r}\left(r\frac{\partial C}{\partial r}\right) + \frac{1}{r^2}\frac{\partial^2 C}{\partial \theta^2} + \frac{\partial^2 C}{\partial z^2}\right] = \frac{\partial C}{\partial t} - Q$$

$$r \geq 0, \qquad z > 0, \qquad 0 \leq \theta < 2\pi$$

and Q is a source function. Suppose $\partial C/\partial z$ is specified as a function of r, θ, and t, say $S(r, \theta, t)$, on the surface $z = 0$. Let $C = 0$ everywhere at $t = 0$. After constructing the general solution for this case take

$$Q = Ae^{-\alpha z - \beta r}$$

and

$$S = Be^{-\gamma r - \varepsilon t}$$

and form the solution.

Set (9–4)

(1) Show that if $S_1(nt)$ and $S_2(nt)$ have z-transforms $s_1(z)$ and $s_2(z)$, respectively, then

$$z^{-1}\{s_1(z)s_2(z)\} = \sum_{k=0}^{\infty} S_1(kT)S_2(t - kT)$$

(2) Assume the response function of a linear instrument, as used in Eq. (9.133) $\bar{R}(nt)$, is represented by the scaled sequence:

$$R_0 = 0$$

$$R_1 = 2$$

$$R_2 = 3$$

$$R_3 = 2$$

$$R_4 = 1, R_k = 0, k \geq 5$$

Then for an instrument response represented by the scaled data, I_n:

$I_0 = 0,$	$I_1 = 1,$	$I_2 = 2,$	$I_3 = 3,$	$I_4 = 2,$	$I_5 = 1,$
$I_6 = -2,$	$I_7 = -3,$	$I_8 = -2,$	$I_9 = -1,$	$I_{10} = -2,$	$I_{11} = -3,$
$I_{12} = -1,$	$I_{13} = 0,$	$I_{14} = +1,$	$I_{15} = 2,$	$I_{16} = 1,$	$I_{17} = 0,$

$$I_k = 0, \quad k > 17$$

find the scaled input signal using Eq. (9.147). (Note synthetic division is the best route.)

REFERENCES

H. S. Carslaw, *Introduction to the Theory of Fourier Series and Integrals*, (Macmillan and Company Ltd., London, 1930).

R. V. Churchill, *Fourier Series and Boundary Value Problems* (McGraw-Hill Book Company, Inc., New York, 1963).

Avner Friedman, *Generalized Functions and Partial Differential Equations* (Prentice-Hall, Inc., Englewood Cliffs, N.J., 1963).

J. Irving and N. Mullineux, *Mathematics in Physics and Engineering* (Academic Press, Inc., New York, 1963).

E. A. Kraut, *Fundamentals of Mathematical Physics* (McGraw-Hill Book Company, Inc., New York, 1967).

A. Kryala, *Theoretical Physics: Applications of Vector, Matrices, Tensors and Quaternions* (W. B. Saunders Company, Philadelphia, 1967).

N. N. Lebedev, I. P. Skalskaya and Y. S. Uflyand, *Problems of Mathematical Physics* (translated by R. A. Silverman (Prentice-Hall, Inc., Englewood Cliffs, N.J., 1965).

J. Mathews and R. L. Walker, *Mathematical Methods of Physics* (W. A. Benjamin, Inc., New York, 1965).

P. M. Morse and H. Feshbach, *Methods of Theoretical Physics* (McGraw-Hill Book Company, Inc., New York, 1953).

I. N. Sneddon, *Fourier Transforms* (McGraw-Hill Book Company, Inc., New York, 1951).

CHAPTER TEN

Inhomogeneous boundary conditions; Green's functions

We have found how to solve boundary value problems in the case of homogeneous equations subject to homogeneous boundary conditions and also *inhomogeneous* equations subject to *homogeneous* boundary conditions. We now consider the solution of boundary value problems involving *inhomogeneous boundary conditions*. That is, problems in which the unknown function sought as the solution is specified or its normal derivative is specified, as a *function of position* on the boundary S of the domain D in which the solution is defined.

The solution of such problems is constructed in terms of *Green's Functions*. As a matter of fact not only does the Green's function technique apply to the problem here considered but it also solves those cases considered earlier of homogeneous boundary conditions.

While the ideas and techniques of solving boundary value problems in terms of Green's functions can be developed in a very sophisticated fashion in terms of a general operator symbolism, we choose here the route best suited for actual computational application.

To develop the method of Green's functions for linear partial differential equations involving the *Laplacian operator* we require *Green's Theorem* and we also require a general definition for a delta function in a three dimensional domain.

GREEN'S THEOREM

Green's theorem relates a surface integral to a volume integral. Consider the vector $[U\nabla V - V\nabla U]$, where U and V are *continuous* scalar functions having continuous first derivatives in a domain D. If we integrate this vector over the closed surface S which bounds D and then apply the *divergence theorem* to the surface integral we obtain,†

$$(10.1) \qquad \oint_S [U\nabla V - V\nabla U] \cdot \mathbf{dS} = \int_D [U\nabla^2 V - V\nabla^2 U] \, d\tau$$

where $\mathbf{dS}$ is the element of area of S with outward normal and $d\tau$ is the differential volume element in D. This integral relationship is Green's theorem.

DELTA FUNCTION

Let $\mathbf{r}$ be the position vector of a point *in the domain D*, and $\mathbf{r}'$ that of another point also in D. Then the function,

$$(10.2) \qquad \delta(\mathbf{r}, \mathbf{r}') = \delta(\mathbf{r}', \mathbf{r}) = \begin{cases} \infty, & \mathbf{r} = \mathbf{r}' \\ 0, & \mathbf{r} \neq \mathbf{r}' \end{cases}, \quad \text{in } D$$

such that

$$(10.3) \qquad \int_D F(\mathbf{r}') \, \delta(\mathbf{r}, \mathbf{r}') \, d\tau' = F(\mathbf{r}), \quad \mathbf{r} \text{ and } \mathbf{r}' \text{ in } D$$

where the *volume* integral is over the whole domain, is called the *delta function of the domain*.

This is a generalization of the delta function defined on a *linear* one-dimensional domain, i.e., $\delta(x - x')$, as we have employed in previous chapters. Later we will see that $\delta(\mathbf{r}, \mathbf{r}')$ can be represented in terms of a product of such linear delta functions with appropriate weighting factors.

GREEN'S FUNCTION FOR A GENERAL TIME-INDEPENDENT EQUATION

Many important physical problems can be reduced to solving an equation of the form,

$$(10.4) \qquad \nabla^2 \psi + V\psi = W, \quad \text{in } D$$

where V and W are known functions of the coordinates, or else are constants. For example the Laplace transform applied to the diffusion equation, with or without a source term, yields this form with V being the negative of the transform parameter, $-p$, and W is a composite of the negative of the initial function and the transform of the source function. Then ψ is the transform of the function we seek. This is also explicitly the form of the *Schrödinger* equation of quantum theory, if $W = 0$.

In any event we seek to find ψ in D subject to some boundary conditions given for ψ on S. Our approach here is to define another function, $G(\mathbf{r}, \mathbf{r}')$, which we call the Green's function of the problem, as the solution of

$$(10.5) \qquad \nabla^2 G + VG = \delta(\mathbf{r}, \mathbf{r}'), \quad \text{in } D$$

with some conditions on G over S to be chosen later.

Here we look upon $G(\mathbf{r}, \mathbf{r}')$ as a function of the unprimed coordinates, $\mathbf{r}$, having the primed coordinates $\mathbf{r}'$ as parameters. However, since $\delta(\mathbf{r}, \mathbf{r}')$ is by definition symmetric in $\mathbf{r}, \mathbf{r}'$ we can interchange these in Eq. (10.5) above to yield,

$$(10.6) \qquad \nabla'^2 G' + V'G' = \delta(\mathbf{r}', \mathbf{r}), \quad \text{in } D'$$

† The generalizations of this relationship to linear operators other than those formed on the Laplacian are found in the references, particularly Morse and Feshback. See *adjoint operator*.

where $V' = V(\mathbf{r}')$, $G' = G(\mathbf{r}', \mathbf{r})$ and now $\mathbf{r}'$ is variable in D' while $\mathbf{r}$ is a parameter. Also write our original equation in the primed coordinates, which is simply a change of notation,

$$(10.7) \qquad\qquad \nabla'^2\psi' + V'\psi' = W', \quad \text{in } D'$$

where primes denote functions of $\mathbf{r}'$ and ∇' operates only on primed variables.

Multiply Eq. (10.7) by $G(\mathbf{r}', \mathbf{r})$ and Eq. (10.6) by $\psi(\mathbf{r}')$ then subtract the first result from the second and integrate over D'; the result is:

$$(10.8) \qquad \int_{D'} [\psi'\nabla'^2 G' - G'\nabla'^2\psi']\, d\tau' = \int_{D'} \psi(\mathbf{r}')\, \delta(\mathbf{r}', \mathbf{r})\, d\tau' - \int_{D'} W'G'\, d\tau'$$

We then apply Green's theorem to the integral on the left and the definition of the delta function to the first integral on the right, rearrange and have:

$$(10.9) \qquad \psi(\mathbf{r}) = \oint_{S'} \left[\psi(\mathbf{r}')\frac{\partial G(\mathbf{r}', \mathbf{r})}{\partial n'} - G(\mathbf{r}', \mathbf{r})\frac{\partial \psi(\mathbf{r}')}{\partial n'} \right] dS' + \int_{D'} W(\mathbf{r}')G(\mathbf{r}', \mathbf{r})\, d\tau'$$

Where we use $\partial/\partial n'$ to denote the *outward* normal derivative on S'.

Now we examine the forms this will have for different *given* boundary conditions on ψ and *chosen* boundary conditions on G.

Case (1)

If ψ is a specified *function of position* on S, then $\partial\psi/\partial n$ is not known on S. For this case we choose G to be zero on $S(G' = 0$ on S' in the primed space); then our Eq. (10.9) reduces to,

$$(10.10) \qquad\qquad \psi(\mathbf{r}) = \oint_{S'} \psi(\mathbf{r}')\frac{\partial G(\mathbf{r}', \mathbf{r})}{\partial n'}\, dS' + \int_{D'} W(\mathbf{r}')G(\mathbf{r}', \mathbf{r})\, d\tau'$$

Here, *if* we have solved the differential equation for G subject to $G = 0$ on S then *both* integrals on the right contain only *known* quantities; then $\psi(\mathbf{r})$ is determined in terms of $G(\mathbf{r}, \mathbf{r}')$, the "source" function $W(\mathbf{r})$ and the *boundary values* of ψ on the boundary S. The advantage gained is that now we need solve a problem subject only to the *simple homogeneous boundary condition*, $G = 0$, instead of the original *inhomogeneous boundary condition* on ψ.

Case (2)

If $\partial\psi/\partial n$ is a specified function of position on S, then ψ is not known on S and we make the obvious *choice* of $\partial G/\partial n = 0$ on S (again a *homogeneous* boundary condition for G) and have,

$$(10.11) \qquad\qquad \psi(\mathbf{r}) = -\oint_{S'} \frac{\partial \psi(\mathbf{r}')}{\partial n'} G(\mathbf{r}', \mathbf{r})\, dS' + \int_{D'} W(\mathbf{r}')G(\mathbf{r}', \mathbf{r})\, d\tau'$$

Hence again, if we have solved the boundary value problem in G, we find the solution of our problem, $\psi(\mathbf{r})$, in terms of G, the source function, W, and the *boundary values* of ψ.

However, this choice of boundary condition for $\partial\psi/\partial n$ on S cannot be used for a *finite* domain in the special case of $V = 0$, i.e., Poisson's equation. The reason is evident if we integrate Eq. (10.5) over the volume of D, apply the divergence theorem to the integral over

$\nabla^2 G$ and use the definition of the delta function. We obtain,

(10.12)
$$\oint_{S'} \frac{\partial G(\mathbf{r}',\mathbf{r})}{\partial n'} dS' + \int_D V(\mathbf{r})G(\mathbf{r}',\mathbf{r})d\tau' = 1$$

thus the choice of $\partial G/\partial n' = 0$ on S', if $V = 0$, would yield $0 = 1$. To avoid this contradiction when $V = 0$ in the differential equation we use the value $\partial G/\partial n' = 1/S'$ on S', where S' is the total surface area of D'. Note that for an infinite or semi-infinite region $S' \to \infty$ and we again have $\partial G/\partial n' = 0$. Also note that $\partial G/\partial n' = 1/S'$ in Eq. (10.9) reduces the integral of the first term to a constant, thus for Neumann conditions, ($\partial \psi/\partial n'$ given on S') in Poisson's equation ψ is determined to within one additive constant.

Case (3)

If the boundary condition for ψ on S is : ($a = $ constant)

(10.13)
$$a\psi(\mathbf{r}) + \frac{\partial \psi(\mathbf{r})}{\partial n} = f(\mathbf{r}), \quad \text{on } S$$

then solving this for $\partial \psi/\partial n$ and substituting into Eq. (10.9) gives, if we *choose* for this case the homogeneous boundary condition for G,

(10.14)
$$aG + \frac{\partial G}{\partial n} = 0, \quad \text{on } S$$

the form,

(10.15)
$$\psi(\mathbf{r}) = -\oint_{S'} f(\mathbf{r}')G(\mathbf{r}', \mathbf{r}) \, dS' + \int_{D'} W(\mathbf{r}')G(\mathbf{r}', \mathbf{r}) \, d\tau'$$

Again $\psi(\mathbf{r})$ is given in terms of G and known quantities by integrals.

HOMOGENEOUS CASES

The advantage of Green's functions is that we replace a problem with *inhomogeneous* boundary conditions by one having *homogeneous* conditions. The latter class of problems can be solved (for G) by methods we have treated in previous chapters.

The Green's function method can also be applied to problems homogeneous in either or both the differential equation and the boundary conditions. The advantage gained is that since G can sometimes be obtained in *closed form* we obtain ψ in *closed form*, i.e., simply by carrying out the integrations in the above equations.

PROPERTIES OF THE GREEN'S FUNCTION

Here we exhibit some of the basic properties of the Green's function $G(\mathbf{r}, \mathbf{r}')$ for the Eq. (10.4) above.

First we prove symmetry, i.e., $G(\mathbf{r}, \mathbf{r}') = G(\mathbf{r}', \mathbf{r})$. We write,

(10.16)
$$\nabla^2 G(\mathbf{r}, \mathbf{r}_1) + V(\mathbf{r})G(\mathbf{r}, \mathbf{r}_1) = \delta(\mathbf{r}, \mathbf{r}_1)$$

and also,

(10.17)
$$\nabla^2 G(\mathbf{r}, \mathbf{r}_2) + V(\mathbf{r})G(\mathbf{r}, \mathbf{r}_2) = \delta(\mathbf{r}, \mathbf{r}_2)$$

Now multiply Eq. (10.17) by $G(\mathbf{r}, \mathbf{r}_1)$, and Eq. (10.16) by $G(\mathbf{r}, \mathbf{r}_2)$, subtract and integrate over the domain using Green's theorem. Thus, since the term in V subtracts out,

$$(10.18) \qquad \oint_S \left\{ G(\mathbf{r}, \mathbf{r}_1) \frac{\partial G(\mathbf{r}, \mathbf{r}_2)}{\partial n} - G(\mathbf{r}, \mathbf{r}_2) \frac{\partial G(\mathbf{r}, \mathbf{r}_1)}{\partial n} dS \right\} = G(\mathbf{r}_2, \mathbf{r}_1) - G(\mathbf{r}_1, \mathbf{r}_2)$$

in view of the property of delta functions. Since $G(\mathbf{r}, \mathbf{r}_1)$ and $G(\mathbf{r}, \mathbf{r}_2)$ satisfy the *same* boundary conditions on S the integral on S vanishes; thus,

$$(10.19) \qquad G(\mathbf{r}_2, \mathbf{r}_1) = G(\mathbf{r}_1, \mathbf{r}_2)$$

or G *is symmetric in "source" and "field" coordinates*.

Now we show the nature of the *singularity* in $G(\mathbf{r}, \mathbf{r}')$ at $\mathbf{r} = \mathbf{r}'$. Since in

$$(10.20) \qquad \nabla^2 G(\mathbf{r}, \mathbf{r}') + V(\mathbf{r})G(\mathbf{r}, \mathbf{r}') = \delta(\mathbf{r}, \mathbf{r}')$$

$\delta(\mathbf{r}, \mathbf{r}')$ is infinite at $\mathbf{r} = \mathbf{r}'$ we see that G must be singular at $\mathbf{r} = \mathbf{r}'$ also, i.e., $V(\mathbf{r})$ is assumed to be regular everywhere. Now in the neighborhood of $\mathbf{r}'$ we can put $V(\mathbf{r}) \approx V(\mathbf{r}')$ and then G will have spherical symmetry near $\mathbf{r}'$. This is valid since $V(\mathbf{r})$ is slowly varying compared to $G(\mathbf{r}, \mathbf{r}')$ near $\mathbf{r}'$. Thus put

$$(10.21) \qquad \mathbf{R} = \mathbf{r} - \mathbf{r}'$$

and have

$$(10.22) \qquad \frac{1}{R^2} \frac{\partial}{\partial R} \left(R^2 \frac{\partial G}{\partial R} \right) + V(\mathbf{r}')G \approx \delta(R)$$

Now if G is singular at $R = 0$ then $\partial G/\partial R$ and $\partial^2 G/\partial R^2$ are more singular, thus we neglect the term in G and have

$$(10.23) \qquad \frac{1}{R^2} \frac{\partial}{\partial R} \left(R^2 \frac{\partial G}{\partial R} \right) \approx \delta(R)$$

Multiply by the volume element $4\pi R^2 \, dR$ and integrate over a sphere of radius R, to obtain,

$$(10.24) \qquad 4\pi R^2 \frac{\partial G}{\partial R} \approx 1$$

by definition of the delta function. Then integrate with limits, R to ∞, to obtain,

$$(10.25) \qquad G \approx -\frac{1}{4\pi R} = \frac{-1}{4\pi|\mathbf{r} - \mathbf{r}'|}$$

near the singularity.

Actually this result is the key to solving the Green's function equation in many cases. Referring to Eq. (10.20) note that for $\mathbf{r} \neq \mathbf{r}'$ the equation is *homogeneous*. This fact is the basis for the image method below. We now *outline* the two main procedures for obtaining $G(\mathbf{r}, \mathbf{r}')$.

OBTAINING THE GREEN'S FUNCTION, EIGENFUNCTION METHOD

To obtain the Green's function one must solve the inhomogeneous equation,

$$(10.26) \qquad \nabla^2 G + VG = \delta(\mathbf{r}, \mathbf{r}') \quad \text{in } D$$

subject to a *homogeneous boundary* condition on S. One way to do this is the method of *eigenfunctions* with which we are already familiar, i.e., let the functions U_i be eigenfunctions of the operator $O = \nabla^2 + V$, thus,

(10.27) $$[\nabla^2 + V(\mathbf{r})]U_i = a_i U_i$$

where the a_i are the eigenvalues, i.e., when $\nabla^2 + V$ is separable a_i is a function of the eigenvalues of the eigenfunctions of the separate operators as noted at the end of Chapter Five, and again at the end of Chapter Nine. Also in this case U_i is a *product* such as say $X_i(x)Y_i(\eta)Z_i(\xi) \ldots$.

Then suppose

(10.28) $$G(\mathbf{r}, \mathbf{r}') = \sum_i A_i U_i(\mathbf{r})$$

when we put this into Eq. (10.26) we have

(10.29) $$\sum_i A_i a_i U_i(\mathbf{r}) = \delta(\mathbf{r}, \mathbf{r}')$$

and if we multiply both sides by $U_j(\mathbf{r})$, then integrate over the volume of D we get, by *orthogonality* of the U_i,†

(10.30) $$N_j A_j a_j = U_j(\mathbf{r}')$$

Thus,

(10.31) $$G(\mathbf{r}, \mathbf{r}') = \sum_i \frac{U_i(\mathbf{r})U_i(\mathbf{r}')}{a_i N_i}$$

For an infinite domain we would be using integrals in lieu of sums and hence integral transforms and inversions in place of orthogonality but these are equivalent to the above procedure as will be seen later. This is only an *outline* intended to convey the *concepts* of the method.

METHOD OF IMAGES

Another method of obtaining G stems from a general property of inhomogeneous equations. The solution of the inhomogeneous equation, Eq. (10.26) can be given as the sum of a *particular integral* of the inhomogeneous equation and the solution of the homogeneous equation. This *sum* must satisfy the boundary condition on S. Thus let $g(\mathbf{r}, \mathbf{r}')$ be a solution of Eq. (10.26) which does not, however, satisfy the conditions given on S, and let $F(\mathbf{r})$ be a solution of

(10.32) $$\nabla^2 F + VF = 0$$

then

(10.33) $$G(\mathbf{r}, \mathbf{r}') = g(\mathbf{r}, \mathbf{r}') + F(\mathbf{r})$$

provided that this satisfies the conditions on S.

Under some conditions this can be accomplished in a very simple way. Consider the special case of the *infinite* domain. Here we will denote the solution of Eq. (10.26) for this case by $g(\mathbf{r}, \mathbf{r}')$, [$G(\mathbf{r}, \mathbf{r}')$ for *infinite* domain.] Now $g(\mathbf{r}, \mathbf{r}')$ must vanish at infinity but will not in general satisfy any given condition on a finite boundary.

Consider then the trial solution,

(10.34) $$G(\mathbf{r}, \mathbf{r}') = g(\mathbf{r}, \mathbf{r}') + \sum_j A_j g(\mathbf{r}, \mathbf{r}_j)$$

† Here N_j is the *norm* over *all* the orthogonal function $X_i(x)Y_i(\zeta)\ldots$ of the product solution, when the operator is separable.

Since $\mathbf{r}'$ is *in the finite domain D* the term $g(\mathbf{r}, \mathbf{r}')$ is a solution of Eq. (10.26). The sum, however, will be a solution of the homogeneous Eq. (10.32) provided that all the $\mathbf{r}_j$ are *outside* of the domain D, i.e., $g(\mathbf{r}, \mathbf{r}_j)$ which satisfies Eq. (10.26) is regular except at $\mathbf{r} = \mathbf{r}_j$. Thus if the A_j and $\mathbf{r}_j$ can be found in terms of $\mathbf{r}'$ such that G satisfies the boundary conditions on S then this is the proper Green's function. In many cases having the proper *symmetry* of domain this is readily accomplished. In this way a Green's function for a *finite domain* is constructed from that for an *infinite* domain. This is the fundamental structure of the method of images.

HELMHOLTZ EQUATION

If one applies the Laplace or Fourier transform to the time variable of the diffusion equation or the wave equation, or if the time dependence of the wave function and the source function in the *inhomogeneous* wave equation is $e^{i\omega t}$, then there results,

$$(10.35) \qquad \nabla^2 \psi + k^2 \psi = W, \qquad k = \text{constant}$$

for a space function $\psi(\mathbf{r})$; this is the inhomogeneous Helmholtz equation.

   This is a special case of the equation already considered. The Green's function equation is,

$$(10.36) \qquad \nabla^2 G + k^2 G = \delta(\mathbf{r}, \mathbf{r}')$$

which we now examine.

HELMHOLTZ EQUATION IN THE INFINITE DOMAIN

For the *infinite domain G* is spherically symmetric about $\mathbf{r} = \mathbf{r}'$. We have already seen that near this singular point G has a particular form. Now for the *infinite domain* let

$$(10.37) \qquad g(R) = -\frac{U(R)}{4\pi R}, \qquad R = |\mathbf{r} - \mathbf{r}'|$$

where $U(R)$ is to be determined. Then, since for spherical symmetry we have,

$$(10.38) \qquad \frac{1}{R^2}\frac{\partial}{\partial R}\left(R^2 \frac{\partial g}{\partial R}\right) + k^2 g = \delta(R)$$

we find for the function $U(R)$,

$$(10.39) \qquad \frac{d^2 U}{dR^2} + k^2 U = 0, \qquad R \geq 0$$

Thus since we must have $U(R) \to 1$ as $R \to 0$, in view of Eq. (10.25), the possibilities for U are:

$$e^{ikR}, \qquad e^{-ikR}, \qquad \cos kR$$

In most cases the form is dictated by other considerations (i.e., require an outgoing wave). Usually we use

$$(10.40) \qquad g(R) = -\frac{e^{ikR}}{4\pi R}$$

in most applications to wave problems.

SEMI-INFINITE DOMAIN FOR THE HELMHOLTZ EQUATION

Consider the semi-infinite domain, $z > 0$, in which we require $G = 0$ on the boundary $z = 0$. We use the method of images, thus,

(10.41)
$$G = -\frac{e^{ikR}}{4\pi R} + \frac{e^{ikR'}}{4\pi R''}, \qquad z > 0$$

where

(10.42)
$$\begin{cases} R = \sqrt{(x - x')^2 + (y - y')^2 + (z - z')^2} \\ R' = \sqrt{(x - x')^2 + (y - y')^2 + (z + z')^2} \end{cases}$$

Note that G does indeed vanish on the boundary $z = 0$. Also note that in accordance with Eq. (10.34) the singularity of the second term is not in the domain $z > 0$.†

The alternate case of $\partial G / \partial n = 0$ on the boundary is satisfied by

(10.43)
$$G = -\frac{e^{ikR}}{4\pi R} - \frac{e^{ikR'}}{4\pi R'}$$

with R and R' as given above, indeed since we have,

(10.44)
$$\frac{\partial R}{\partial z'} = -\frac{\partial R'}{\partial z'}, \qquad z' = 0 \quad \text{or} \quad z = 0$$

we see that

(10.45)
$$\frac{\partial G}{\partial z'} = 0 \quad \text{on} \quad z' = 0$$

as required.

Example (1)

An important problem in the half-space is acoustic or optical diffraction at a hole in a solid sheet. For a plane acoustic wave incident on a plane surface, $z = 0$, having a hole in it, the boundary condition on the amplitude ψ of the velocity potential is $\partial \psi / \partial n = 0$ on solid surface and

$$\frac{\partial \psi}{\partial n} = A = \text{constant}$$

in the hole. In particular we have the boundary value problem for a *circular hole*:

$$\nabla^2 \psi + k^2 \psi = 0, \qquad z > 0$$

(10.46)
$$\left(\frac{\partial \psi}{\partial z} \right)_{z=0} = \begin{cases} 0, & r > a \\ A, & r < a \end{cases}$$

The Green's function satisfies Eq. (10.36) with in this case $\partial G / \partial z = 0$ on $z = 0$. Thus the solution is: [by use of Eq. (10.11)]

(10.47)
$$\psi(r, \theta, z) = \int_0^\infty \int_0^{2\pi} \frac{\partial \psi(r', \theta', 0)}{\partial z} G(r, \theta, z; r', \theta', 0) r' \, d\theta' \, dr'$$

† Note that only *one* term of the sum in our Eq. (10.34) is required and $A_1 = -1$.

where we have used cylindrical coordinates. This is the integral over S'. G is given by Eq. (10.43). Thus we have

$$(10.48) \qquad \psi(r, \theta, z) = \frac{-A}{2\pi} \int_0^a \int_0^{2\pi} \frac{e^{ik\sqrt{r^2+r'^2-2rr'\cos(\theta-\theta')+z^2}}}{\sqrt{r^2+r'^2-2rr'\cos(\theta-\theta')+z^2}} \cdot r'\, d\theta'\, dr'$$

as the solution. (R in cylindrical coordinates is the radical here, but with $z' = 0$.)

For the particular case of very large distances from the hole this is readily evaluated in approximate form. Consider the picture in Fig. 10-1. Let

$$(10.49) \qquad \rho^2 = r^2 + z^2, \qquad \sin \eta = \frac{r}{\rho}$$

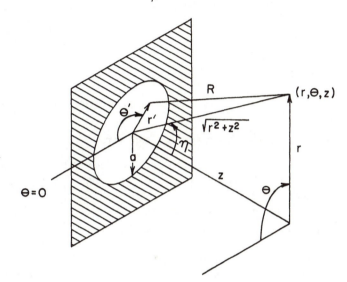

FIGURE 10.1. Geometry for the Problem of Wave Diffraction at a Circular Hole.

then,

$$(10.50) \qquad R = \sqrt{r^2 + r'^2 - 2rr'\cos(\theta - \theta') + z^2}, \qquad z' = 0$$

becomes

$$(10.51) \qquad R = \rho\left[1 + \frac{r'^2}{\rho^2} - 2\frac{r'}{\rho}\sin\eta\cos(\theta - \theta')\right]^{\frac{1}{2}}$$

or, if $r'/\rho \ll 1$, approximately,

$$(10.52) \qquad R \approx \rho - r'\sin\eta\cos(\theta - \theta')$$

Furthermore when we put this into Eq. (10.48) we obtain:

$$(10.53) \qquad \psi \approx \frac{-A}{2\pi} \int_0^a \int_0^{2\pi} \frac{e^{ik\rho - ikr'\sin\eta\cos(\theta-\theta')}r'\, d\theta'\, dr'}{\rho - r'\sin\eta\cos(\theta - \theta')}$$

Again use the fact that r'/ρ is very small and obtain,

$$(10.54) \qquad \psi \approx \frac{-A}{2\pi}\frac{e^{ik\rho}}{\rho} \int_0^a \int_0^{2\pi} e^{-ikr'\sin\eta\cos(\theta-\theta')}r'\, d\theta'\, dr'$$

Now we see that

$$\int_0^{2\pi} e^{ix\cos(\theta-\theta')}\, d\theta' = \int_\theta^{\theta+2\pi} e^{ix\cos(\theta-\theta')}\, d\theta' = \int_0^{2\pi} e^{ix\cos\gamma}\, d\gamma = 2\int_0^\pi e^{ix\cos\gamma}\, d\gamma = 2\pi J_0(x)$$

(10.55) $$= 2\pi J_0(-x)$$

from the formulas of Chapter Four. Thus we have,

(10.56) $$\psi(r,z) \approx -A\frac{e^{ik\rho}}{\rho}\int_0^a J_0(kr'\sin\eta)r'\, dr'$$

and this can be evaluated to yield:

(10.57) $$\psi(r,z) \approx -Aa\frac{e^{ik\rho}J_1(ka\sin\eta)}{\rho k\sin\eta}$$

where the J's are Bessel functions.

EXECUTE PROBLEM SET (10-1)

SYMMETRY AND GREEN'S FUNCTIONS

We now show how symmetry in a boundary value problem modifies the required Green's function and the delta function is also modified, i.e., as we go from three to two-dimensional domains.

We continue with the equation

(10.58) $$\nabla^2\psi + V\psi = W$$

as our example. If for example V, W, and the boundary conditions are independent of the cylindrical angle coordinate then the problem is axially symmetric and is independent of θ. In this case G is a solution of,

(10.59) $$\frac{1}{r}\frac{\partial}{\partial r}\left(r\frac{\partial G}{\partial r}\right) + \frac{1}{r^2}\frac{\partial^2 G}{\partial\theta^2} + \frac{\partial^2 G}{\partial z^2} + V(r,z)G = \delta(\mathbf{r},\mathbf{r}')$$

and in view of the boundary conditions G is a function of $r, r', z, z',$ and θ, θ'.

The solution $\psi(r,z)$ is given still by Eq. (10.9) but now since ψ, $\partial\psi/\partial n$, and W are all independent of θ, the integrations on θ can be taken in to be on G only. Thus for example the integral,

(10.60) $$\mathscr{I} = \int_D W(\mathbf{r}')G'(\mathbf{r}',\mathbf{r})\, d\tau' = \int_0^{2\pi}\int_r\int_z W(r',\theta',z')G(r',\theta',z';r,\theta,z)\cdot r'\, d\theta'\, dr'\, dz'$$

becomes,

(10.61) $$\mathscr{I} = \int_r\int_z W(r',z')\bar{G}'(r',z';r,z)r'\, dr'\, dz'$$

where we now define,

(10.62) $$\bar{G} = \int_0^{2\pi} G(r,\theta,z;r',\theta',z')\, d\theta'$$

Similarly if ψ, $\partial\psi/\partial n$, and W were also independent of z, the integration on z' could be taken in on G only. Then one would have,

$$(10.63) \qquad \int_D W(\mathbf{r}')G(\mathbf{r}', \mathbf{r})\, d\tau' = \int_r W(r')\bar{\bar{G}}(r\,;r')r'\, dr'$$

where we define,

$$(10.64) \qquad \bar{\bar{G}}(r, r') = \int_{-\infty}^{+\infty} \bar{G}(r, z\,; r', z')\, dz'$$

Now examining the meaning of this in the G equation we multiply Eq. (10.59) by $d\theta$ and integrate from zero to 2π and use the definition given in Eq. (10.62), and the symmetry property of G, thus:

$$(10.65) \qquad \frac{1}{r}\frac{\partial}{\partial r}\left(r\frac{\partial\bar{G}}{\partial r}\right) + \frac{1}{r^2}\frac{\partial G}{\partial\theta}\Big|_0^{2\pi} + \frac{\partial^2\bar{G}}{\partial z^2} + V(r, z)G = \delta(\mathbf{r}, \mathbf{r}')$$

since G must have period 2π, to be single valued in θ, and G is symmetric in θ and θ', the integrated terms in $\partial G/\partial\theta$ add to zero. Now the quantity

$$(10.66) \qquad \bar{\delta}(r, r', z, z') = \int_0^{2\pi} \delta(\mathbf{r}, \mathbf{r}')\, d\theta$$

must have in view of Eq. (10.3) the property,

$$(10.67) \qquad \int_D \psi(r, z)\,\bar{\delta}(r, r', z, z')r\, dr\, dz = \psi(r', z')$$

i.e., multiply Eq. (10.66) by $\psi(r, z)\, r\, dr\, dz$ and note that on the right $r\, dr\, dz\, d\theta$ is the volume element, thus integrate over the domain and apply the definition of the delta function given in Eq. (10.3).

In fact if ψ is independent of z as well as θ then another integration over z can be carried out, i.e., define

$$(10.68) \qquad \bar{\bar{\delta}}(r, r') = \int_{-\infty}^{+\infty}\int_0^{2\pi} \delta(\mathbf{r}, \mathbf{r}')\, d\theta\, dz$$

and then, by argument similar to that above, we find.

$$(10.69) \qquad \int_D \psi(r)\,\bar{\bar{\delta}}(r, r')r\, dr = \psi(r')$$

Thus we see that it is *convenient* to define for cylindrical coordinates,

$$(10.70) \qquad \delta(\mathbf{r}, \mathbf{r}') = \delta_r(r, r')\,\delta_\theta(\theta, \theta')\,\delta_z(z, z')$$

such that, for any functions $F(r)$, $G(\theta)$, $H(z)$, in the domain,

$$(10.71) \qquad \begin{cases} F(r') = \displaystyle\int_{R_1}^{R_2} F(r)\,\delta_r(r, r')r\, dr, & R_1 < r < R_2 \\[2ex] G(\theta') = \displaystyle\int_{\theta_1}^{\theta_2} G(\theta)\,\delta_\theta(\theta, \theta')\, d\theta, & \theta_1 < \theta' < \theta_2 \\[2ex] H(z') = \displaystyle\int_{z_1}^{z_2} H(z)\,\delta_z(z, z')\, dz, & z_1 < z' < z_2 \end{cases}$$

where the *domain* is as specified to the right in each equation. Then for example in *axial symmetry* as seen in Eq. (10.65),

(10.72)
$$\frac{1}{r}\frac{\partial}{\partial r}\left(r\frac{\partial \bar{G}}{\partial r}\right) + \frac{\partial^2 \bar{G}}{\partial z^2} + V\bar{G} = \delta_r(r, r')\,\delta_z(z, z')$$

Similar definitions of delta functions apply in other coordinate systems.

Example (2)

As an example we solve the *diffraction problem* already considered by another method (i.e., the eigenfunction method).

The boundary value problem is: (in cylindrical coordinates)

(10.73)
$$\frac{1}{r}\frac{\partial}{\partial r}\left(r\frac{\partial \psi}{\partial r}\right) + \frac{\partial^2 \psi}{\partial z^2} + k^2\psi = 0, \qquad \begin{cases} z > 0 \\ r \geq 0 \end{cases}$$

with

(10.74)
$$\frac{\partial \psi}{\partial z} = \begin{cases} 0, & r > a, \quad z = 0 \\ A, & r < a, \quad z = 0 \end{cases}$$

Thus we have *axial symmetry*. The solution is: (from our general formula of Green's theorem)

(10.75)
$$\psi(r, z) = \int_0^\infty \frac{\partial \psi(r', 0)}{\partial z}\,\bar{G}(r, r', z, 0)r'\, dr'$$

where $\bar{G}$ satisfies,

(10.76)
$$\frac{1}{r}\frac{\partial}{\partial r}\left(r\frac{\partial \bar{G}}{\partial r}\right) + \frac{\partial^2 \bar{G}}{\partial z^2} + k^2\bar{G} = \delta_r(r, r')\,\delta_z(z, z')$$

i.e., Eq. (10.75) is Eq. (10.9) reduced by the special conditions of the problem and definition of $\bar{G}$. That is, this is the surface integral already given. The boundary condition on $\bar{G}$ is

(10.77)
$$\frac{\partial \bar{G}}{\partial z} = 0 \quad \text{at} \quad z = 0$$

Now we consider the eigenfunction problem (this is the analog of Eq. (10.27) for this special case),

(10.78)
$$\frac{1}{r}\frac{\partial}{\partial r}\left(r\frac{\partial U_i}{\partial r}\right) + \frac{\partial^2 U_i}{\partial z^2} + k^2 U_i = a_i U_i$$

We want those solutions of this that are regular at $r = 0$, vanish as $r \to \infty$ and which satisfy the boundary condition

(10.79)
$$\frac{\partial U_i}{\partial z} = 0 \quad \text{at} \quad z = 0$$

These are

(10.80)
$$U_i = J_0(\beta_i r) \cos(\lambda_i z)$$

and the eigenvalues are

(10.81)
$$a_i = k^2 - \lambda_i^2 - \beta_i^2$$

Since no restrictions apply on β and λ we have as a trial solution,

(10.82)
$$\bar{G} = \int_0^\infty \int_0^\infty f(\beta, \lambda) J_0(\beta r) \cos \lambda z \beta \, d\beta \, d\lambda$$

putting this into Eq. (10.76) yields,

(10.83)
$$\int_0^\infty \int_0^\infty f(\beta, \lambda)(k^2 - \lambda^2 - \beta^2) J_0(\beta r) \cos \lambda z \beta \, d\beta \, d\lambda = \delta_r(r, r') \delta_z(z, z')$$

Here we invert the Hankel and Fourier transforms, i.e., multiply by $J_0(\bar{\beta} r) r \, dr$ and integrate from 0 to ∞; also multiply by $\cos \bar{\lambda} z \, dz$ and integrate from 0 to ∞. We obtain by definition of δ_r and δ_z

(10.84)
$$\frac{\pi}{2} f(\bar{\beta}, \bar{\lambda})(k^2 - \bar{\lambda}^2 - \bar{\beta}^2) = J_0(\bar{\beta} r') \cos \bar{\lambda} z'$$

or

(10.85)
$$f(\beta, \lambda) = \frac{2}{\pi} \frac{J_0(\beta r') \cos(\lambda z')}{k^2 - \lambda^2 - \beta^2}$$

and hence,†

(10.86)
$$\bar{G}(r, r', z, z') = \frac{2}{\pi} \int_0^\infty \int_0^\infty \frac{J_0(\beta r) J_0(\beta r') \cos \lambda z \cos \lambda z' \beta \, d\beta \, d\lambda}{k^2 - \lambda^2 - \beta^2}$$

Now apply this to our particular problem, i.e., using this in Eq. (10.75) we have, with the boundary conditions in Eq. (10.77),

(10.87)
$$\psi(r, z) = \frac{2A}{\pi} \int_0^\infty \int_0^\infty \frac{J_0(\beta r) \cos \lambda z}{k^2 - \lambda^2 - \beta^2} \int_0^a J_0(\beta r') r' \, dr' \, \beta \, d\beta \, d\lambda$$

or,

(10.88)
$$\psi(r, z) = \frac{2Aa}{\pi} \cdot \int_0^\infty \int_0^\infty \frac{J_0(\beta r) J_1(\beta a) \cos \lambda z \, d\beta \, d\lambda}{k^2 - \lambda^2 - \beta^2}$$

as the final form of the solution for $\psi(r, z)$. Of course this too can be analyzed, particularly at large distance from the hole as we did before.

These results should be compared to those obtained previously by the method of images. In particular we point out how $\bar{G}$, the axially symmetric Green's function is obtained by images.

We have as the G for the half-space, $z > 0$,

(10.89)
$$G = -\frac{e^{ikR}}{4\pi R} - \frac{e^{ikR'}}{4\pi R'}$$

obtained previously. Thus according to Eq. (10.62) we must have

(10.90)
$$\bar{G} = \frac{1}{4\pi} \int_0^{2\pi} \left(-\frac{e^{ikR}}{4\pi R} - \frac{e^{ikR'}}{4\pi R'} \right) d\theta$$

as the axially symmetric Green's function. Here of course R and R' must be expressed in cylindrical coordinates before the integration can be executed.

† Note that we could obtain this same result by applying the appropriate integral transforms *directly* to Eq. (10.76) as in Chapter Nine.

GREEN'S FUNCTIONS AND "SOURCES" OR "IMPULSES"

One can visualize the Green's functions for various symmetries as follows. The original Green's function, $G(\mathbf{r}, \mathbf{r}')$, is the "Field" due to the "unit point source" in the three dimensional domain. The axially symmetric Green's function, $\bar{G}(r, r', z, z')$ defined by Eq. (10.62) is the "field" due to the "ring source" of radius r' located at $z = z'$. This is obtained by superposing point sources on the ring with strength $d\theta'$ for the source located at r', θ', z'. Similarly the cylindrical Green's function $\bar{\bar{G}}(r, r')$, defined by Eq. (10.64) is the "field" due to the cylindrical sheet source of radius r'. This is obtained by superposing ring sources with the number of rings in the interval dz' being equal to dz'.

While we have used the Helmholtz equation and cylindrical coordinates to illustrate these ideas it must be emphasized that these methods apply in general. For example the spherically symmetric Green's function would be obtained as the superposition of point sources on a sphere of radius r' with the number of point sources in the area element, $r'^2 \sin \theta' \, d\theta' \, d\phi'$, being equal to the area of the element.

In this context the student should recognize that in some of the example problems treated in previous chapters we actually constructed Green's functions. Example (6) of Chapter Five, with $\lambda = 1$ is the *axially symmetric* Green's function for Laplace's equation expressed in spherical coordinates.‡ This same Green's function is constructed again in cylindrical coordinates in Example (3) of Chapter Six, by other methods.

We should also recognize the particular case of a *single* blow with *unit* impulse in Example (4) of Chapter Six, as the Green's function for the general problem of the driven oscillator, Eq. (6.20) in the "pendulum" format there.

Another Green's function was constructed in Example (2) of Chapter Nine; there for the one dimensional wave equation, the final form being Eq. (9.85) if we put the impulse $E_0 = 1$ in this case. However, this will be examined further in a later part of this chapter.

EXECUTE PROBLEM SET (10-2)

DIFFUSION-TYPE EQUATIONS

Here we will consider the Green's function problems for an equation of the type

$$(10.91) \qquad \nabla^2 \psi + V\psi - \frac{1}{a} \frac{\partial \psi}{\partial t} = -W, \qquad \text{in } D, \qquad t > 0$$

We begin by deriving as an example of this form the equation of heat conduction with sources.† This will aid in using *physical* concepts in the treatment of Green's functions.

Consider a domain of conductivity k, specific heat C, and density ρ in which heat is generated at the rate s (heat per volume per time).

The quantity of heat in a volume τ is changing at the rate

$$(10.92) \qquad \frac{d}{dt} \int_\tau C\rho T \, d\tau = \int_\tau C\rho \frac{\partial T}{\partial t} \, d\tau$$

This rate of increase is contributed by a net rate of heat flow into τ and heat generated in τ. Heat flow is given by $-k\nabla T$, and the generated heat by s, thus,

$$(10.93) \qquad \int_\tau C\rho \frac{\partial T}{\partial t} \, d\tau = -\oint_S (-k\nabla T) \cdot \mathbf{dS} + \int_\tau s \, d\tau$$

---

† This was assigned as an exercise in Chapter One.
‡ Except for a constant multiplier.

where the surface S incloses τ. Applying the divergence theorem to the surface integral and noting that this must hold for any volume τ we obtain,

$$(10.94) \qquad C\rho \frac{\partial T}{\partial t} = k\nabla^2 T + s$$

or

$$(10.95) \qquad \nabla^2 T - \frac{1}{a}\frac{\partial T}{\partial t} = -W$$

where $a = k/C\rho$ is the thermal diffusivity and

$$(10.96) \qquad W = \frac{s}{k}$$

Thus see that W has the physical characteristics of a *source*.

Now *define* the Green's function, $G(\mathbf{r}, t, \mathbf{r}', t')$ for the problem in Eq. (10.91) as the solution of,

$$(10.97) \qquad \nabla^2 G + VG - \frac{1}{a}\frac{\partial G}{\partial t} = -\delta(\mathbf{r}, \mathbf{r}')\,\delta(t, t')$$

Here $\delta(t, t')$ is the delta function defined so that

$$(10.98) \qquad F(t') = \int_{-\infty}^{+\infty} F(t)\,\delta(t - t')\,dt$$

Since

$$(10.99) \qquad \int_{-\infty}^{+\infty}\int_{D} \delta(\mathbf{r}, \mathbf{r}')\,\delta(t - t')\,d\tau\,dt = 1$$

we see that the right member of Eq. (10.97) has the physical characteristics of a *quantity* of "something" liberated at the point $\mathbf{r} = \mathbf{r}'$ at the instant $t = t'$ (obviously, in the heat equation the *units* are not correct, i.e., the factor of $1/k$).

We choose the initial condition on $G(\mathbf{r}, t; \mathbf{r}', t')$ to be $G = 0$ at $t = 0$. In effect G is *analogous* to the temperature distribution at time t resulting from a unit quantity of heat released at $\mathbf{r} = \mathbf{r}'$ at $t = t'$. Obviously there should be no effect of the source *before* it came into existence. Thus require,

$$(10.100) \qquad G(\mathbf{r}, t; \mathbf{r}', t') = 0 \quad \text{for} \quad t < t'$$

Now note the effect of interchanging t and t' in the function G. Then the time of observation and the time the source occurs are exchanged. Then Eq. (10.100) would say there is no effect of the source *after* it occurs. To avoid this contradiction we require that,

$$(10.101) \qquad G(\mathbf{r}, t; \mathbf{r}', t') = G(\mathbf{r}', -t'; \mathbf{r}, -t)$$

Then we see that in the primed variables, Eq. (10.97) becomes,

$$(10.102) \qquad \nabla'^2 G' + V'G' + \frac{1}{a}\frac{\partial G'}{\partial t'} = -\delta(\mathbf{r}, \mathbf{r}')\,\delta(t', t)$$

However, the form of the ψ equation in the primed variables is unchanged,

(10.103)
$$\nabla'^2 \psi' + V' \psi' - \frac{1}{a} \frac{\partial \psi'}{\partial t'} = - W'$$

Thus now multiply Eq. (10.102) by $\psi(\mathbf{r}', t')$ and Eq. (10.103) by $G(\mathbf{r}', -t'; \mathbf{r}, -t)$, subtract, and integrate over the space domain. Also integrate over t' from 0 to a value t^+ *just greater than t*. Apply Green's theorem to the volume integral of $G'\nabla'^2 \psi' - \psi' \nabla'^2 G'$ and use the definitions of the delta functions. This yields,

$$\psi(\mathbf{r}, t) = \int_0^{t^+} \int_D W(\mathbf{r}, t) G(\mathbf{r}', -t'; \mathbf{r}, -t) \, d\tau' \, dt'$$

$$- \frac{1}{a} \int_0^{t^+} \int_D \frac{\partial}{\partial t'} \left[\psi(\mathbf{r}', t') G(\mathbf{r}', -t'; \mathbf{r}, -t') \right] d\tau' \, dt'$$

$$+ \int_0^{t^+} \int_S \left[\frac{\partial \psi'}{\partial n'} G' - \psi' \frac{\partial G'}{\partial n'} \right] dS' \, dt'$$

In the second integral above the time integration can be performed to yield

$$\psi(\mathbf{r}', t') G(\mathbf{r}', -t^+; \mathbf{r}, -t) - \psi(\mathbf{r}', 0) G(\mathbf{r}', 0; \mathbf{r}, -t)$$

inside the volume integral. But in view of Eqs. (10.100) and (10.101) the first term is zero. Furthermore, we can use Eq. (10.101) to write our equations as:

$$\psi(\mathbf{r}, t) = \int_0^t \int_D W(\mathbf{r}', t') G(\mathbf{r}, t; \mathbf{r}', t') \, d\tau' \, dt' + \frac{1}{a} \int_D \psi(\mathbf{r}', 0) G(\mathbf{r}, t; \mathbf{r}', 0) \, d\tau'$$

(10.105)
$$+ \int_0^t \int_S \left[\frac{\partial \psi(\mathbf{r}', t')}{\partial n'} G(\mathbf{r}, t; \mathbf{r}', t') - \frac{\partial G(\mathbf{r}, t; \mathbf{r}', t')}{\partial n'} \psi(\mathbf{r}', t') \right] dS' \, dt'$$

Thus if G is known and the boundary conditions on G are suitably chosen we can obtain ψ in terms of G and the *initial* and boundary conditions on ψ just as we did for the time-independent equation earlier.

INFINITE DOMAIN FOR THE DIFFUSION EQUATION

We now find G for the infinite domain in the case of $V = 0$. Thus,

(10.106)
$$\sum_{i=1}^{3} \frac{\partial^2 g}{\partial x_i^2} - \frac{1}{a} \frac{\partial g}{\partial t} = - \delta(\mathbf{r}, \mathbf{r}') \, \delta(t, t')$$

were x_1, x_2, x_3 are rectangular coordinates. Let g be represented by the multiple Fourier integral,

(10.107)
$$g_n = \frac{1}{(2\pi)^n} \int_{-\infty}^{+\infty} \cdots \int_{-\infty}^{+\infty} \eta(\mathbf{k}, t) e^{i\mathbf{k} \cdot \mathbf{r}} \, dk_1 \, dk_2 \ldots dk_n$$

for n dimensions. This substituted into Eq. (10.106) yields

(10.108)
$$\frac{1}{(2\pi)^n} \int_{-\infty}^{+\infty} \cdots \int_{-\infty}^{+\infty} \left(-k^2 \eta - \frac{1}{a} \frac{\partial \eta}{\partial t} \right) e^{i\mathbf{k} \cdot \mathbf{r}} \, dk_1 \, dk_2 \cdots dk_n = - \delta(\mathbf{r}, \mathbf{r}') \delta(t, t')$$

Now multiply both sides by $e^{-i\mathbf{k}\cdot\mathbf{r}}$ and integrate over all of space. This inverts the Fourier integrals on the left.† On the right the definition of the delta function is used and there results:

$$(10.109) \qquad k^2\eta + \frac{1}{a}\frac{\partial\eta}{\partial t} = \delta(t - t')e^{-i\mathbf{k}\cdot\mathbf{r}'}$$

This is integrated with $\eta = 0$ at $t = 0$ to yield,

$$(10.110) \qquad \eta(\mathbf{k}, t) = aH(t, t')e^{-ak^2(t-t')-i\mathbf{k}\cdot\mathbf{r}'}$$

where $H(t, t')$ is the unit step function. Then,

$$(10.111) \qquad g_n = \frac{a}{(2\pi)^n}\int_{-\infty}^{+\infty}\cdots\int_{-\infty}^{+\infty} H(t, t')e^{-ak^2(t-t')+i\mathbf{k}\cdot\mathbf{R}}\, dk_1\, dk_2\ldots dk_n$$

where,

$$(10.112) \qquad \mathbf{R} = \mathbf{r} - \mathbf{r}' \quad \text{and} \quad k^2 = k_1^2 + k_2^2 + \ldots + k_n^2$$

This is integrated as follows. Consider just one of the integrals, i.e.,

$$(10.113) \qquad I_1 = \int_{-\infty}^{+\infty} e^{-ak_1^2(t-t')+ik_1(x_1-x_1')}\, dk_1$$

we write the exponent, by completing the square, as:

$$(10.114) \qquad ak_1^2(t - t') - ik_1(x_1 - x_1') = a(t - t')\left[k_1 - \frac{i(x_1 - x_1')}{2a(t - t')}\right]^2 + \frac{(x_1 - x_1')^2}{4a(t - t')}$$

Then denoting the bracket by ξ_1, we have,‡

$$(10.115) \qquad I_1 = e^{-\frac{(x_1-x_1')^2}{4a(t-t')}}\int_{-\infty}^{+\infty} e^{-a(t-t')\xi_1^2}\, d\xi_1 = \sqrt{\pi}\,\frac{e^{-\frac{(x-x')^2}{4a(t-t')}}}{\sqrt{a(t-t')}}$$

Thus finally, using this in Eq. (10.111) above,

$$(10.116) \qquad g_n = \frac{ae^{-\frac{R_n^2}{4a(t-t')}}H(t, t')}{(2\sqrt{\pi a(t-t')}\,)^n}$$

where

$$(10.117) \qquad R_n^2 = (x_1 - x_1')^2 + \ldots + (x_n - x_n')^2$$

METHOD OF IMAGES FOR THE DIFFUSION EQUATION

The method of images is essentially the same here as for the Helmholtz equation.

Example (3)

Suppose we require the Green's function for the domain $z > 0$ (now using x, y, z) with say $G = 0$ on the boundary $z = 0$. We put a positive source at x', y', z' *in the domain $z > 0$* and a negative source at x', y', $-z'$ which is in the image space, $z < 0$. Thus the desired Green's

---

† See Chapter Nine.
‡ Note that the integral is $[\pi/a(t - t')]^{\frac{1}{2}}$ onto $\mathrm{erf}(\gamma)$ and $\gamma \to \infty$ by appropriate change of variables.

function is:

(10.118) $$G = g_3(x, y, z, t; x', y', z', t') - g_3(x, y, z, t; x', y', -z', t')$$

which indeed does vanish for $z = 0$ as is seen from the form of g_3 in Eq. (10.116) above.
Similarly if we require say, $\partial G/\partial n = 0$ on $z = 0$, or $z' = 0$ we employ,

(10.119) $$G = g_3(x, y, z, t; x', y', z', t) + g_3(x, y, z, t; x', y', -z', t')$$

This can be employed to solve a problem such as:

(10.120) $$\nabla^2 \psi - \frac{1}{a}\frac{\partial \psi}{\partial t} = 0, \qquad z > 0, \qquad t > 0$$

with

(10.121) $$\frac{\partial \psi}{\partial z} = f(x, y, t) \quad \text{on} \quad z = 0$$

and

(10.122) $$\psi(x, y, z, 0) = 0, \qquad t = 0$$

Then we choose $\partial G/\partial n = 0$, on S and thus use G as given by Eq. (10.119) with the above in
Eq. (10.105) to obtain

(10.123) $$\psi(x, y, z, t) = \frac{a}{4(\pi a)^{\frac{3}{2}}} \int_0^t \int_{-\infty}^{+\infty} \int_{-\infty}^{+\infty} \frac{f(x', y', t')e^{-\frac{(x-x')^2 + (y-y')^2 + z^2}{4a(t-t')}}}{(t-t')^{\frac{3}{2}}} \, dx'\, dy'\, dt'$$

as the solution.

SOURCES, SYMMETRY, AND GREEN'S FUNCTIONS

The solution of the Green's function problem in heat conduction for the infinite domain, Eq.
(10.116), denoted by g_n, has the same physical characteristics as an instantaneous point source
of heat at $r = r'$ at time $t = t'$, of strength k where k is conductivity, i.e., the correspondence is
according to Eqs. (10.96) and (10.97),

(10.124) $$\frac{S}{k} \to W \to \delta(\mathbf{r}, \mathbf{r}')\,\delta(t - t')$$

and the integral over volume and time is 1 on the right so,

(10.125) $$\int\int \frac{S}{k}\, d\tau\, dt \to 1$$

i.e., k units of heat liberated at $\mathbf{r} = \mathbf{r}'$ at time $t = t'$. This *physical analogy* allows us to talk about
Green's functions as "fields" due to sources.

With regard to spatial symmetry the same arguments apply to the diffusion-type equation
as to the Helmholtz-type equation. Note that in Eq. (10.65) if $W, \psi, \partial\psi/\partial n$, etc. do not depend
on a coordinate then the integrations over that coordinate can be taken in to apply to G only.
Thus for example,

(10.126) $$\bar{G} = \int_0^{2\pi} G\, d\theta$$

is the appropriate Green's function for a problem of axial symmetry. This would represent an *instantaneous ring source*. (Note, here G is the instantaneous point source in the given domain, the solution of Eq. (10.97) say.)

However, we note that now we also have the time domain to consider. Thus note in Eq. (10.105) if W and ψ and/or $\partial \psi / \partial n$ on *the boundary* are independent of time then the time integrations can be taken in to apply to G only. Thus see the usefulness of *continuous point sources*.

$$(10.127) \qquad \tilde{G}(\mathbf{r}, \mathbf{r}', t) = \int_0^t G(\mathbf{r}, t; \mathbf{r}', t') \, dt'$$

Example (4)

The continuous line source is constructed as follows. This would be appropriate to *two dimensional* problems in which the boundary conditions do not depend on time. We could use for the infinite domain,

$$(10.128) \qquad g_2 = \int_{-\infty}^{+\infty} g_3 \, dx_3' = g_2(x_1, x_2, t; x_1', x_2', t')$$

but the integration is not necessary since we already have g_2. All we need to do is integrate over t, thus

$$(10.129) \qquad \tilde{g}_2(x_1, x_2, t; x_1', x_2') = \int_0^t g_2(x_1, x_2, t; x_1', x_2', t') \, dt'$$

This is, using Eq. (10.116) for the heat conduction equation,

$$(10.130) \qquad \tilde{g}_2 = \frac{1}{4\pi} \int_0^{t'} \frac{e^{-\frac{R_2^2}{4a(t - t')}}}{t - t'} \, dt'$$

where in plane polar coordinates,

$$(10.131) \qquad R_2^2 = r^2 + r'^2 - 2rr' \cos(\theta - \theta')$$

If we make the change of variables,

$$(10.132) \qquad \lambda = \frac{R_2^2}{4a(t - t')}$$

we obtain,

$$(10.133) \qquad \tilde{g}_2 = \frac{1}{4\pi} \int_{\frac{R_2^2}{4a(t - t')}}^{\infty} \frac{e^{-\lambda}}{\lambda} \, d\lambda$$

and this integral is defined as the exponential integral function: (see Chapter Four)

$$(10.134) \qquad -Ei(-\xi) = \int_\xi^\infty \frac{e^{-\lambda} \, d\lambda}{\lambda}$$

For small values of ξ this can be approximated by:

$$(10.135) \qquad -Ei(-\xi) \approx C - \ln \xi, \qquad \xi \ll 1$$

and $C \approx 0.577215665$ is Euler's Constant. A *physical application* of this result is the following.

Example (5)

Consider a thin wire embedded in a medium of thermal conductivity k. Let a current pass through the wire so that heat is generated in the wire at the rate, h, per *length of wire* (cal/sec · cm). Thus we have the equation to solve,

$$\text{(10.136)} \qquad \frac{\partial^2 T}{\partial x^2} + \frac{\partial^2 T}{\partial y^2} = \frac{C\rho}{k} \frac{\partial T}{\partial t} - \frac{h}{k} \delta(x, 0)\, \delta(y, 0)$$

Here we see that with the wire on the z axis,

$$\text{(10.137)} \qquad s = h\, \delta(x, 0)\, \delta(y, 0)$$

is the volume source function, i.e.,

$$\text{(10.138)} \qquad \int_{z_1}^{z_2} \int_{-\infty}^{+\infty} \int_{-\infty}^{+\infty} s\, dx\, dy\, dz = h(z_2 - z_1)$$

is the rate of heat generation in the volume bounded by planes at z_1 and z_2.

Now we have with g_2 as the Green's function, and with $T = 0$ at $t = 0$,

$$\text{(10.139)} \qquad T = \frac{1}{4\pi k} \int_0^t \int_{-\infty}^{+\infty} \int_{-\infty}^{+\infty} h(t') \frac{\delta(x', 0)\, \delta(y', 0)}{t - t'} e^{-\frac{(x - x')^2 + (y - y')^2}{4a(t - t')}}\, dx'\, dy'\, dt'$$

from Eq. (10.105), or

$$\text{(10.140)} \qquad T = \frac{1}{4\pi k} \int_0^t h(t') \frac{e^{-\frac{x^2 + y^2}{4a(t - t')}}}{t - t'}\, dt'$$

This is exactly the same as $\tilde{g}_2/k$ if $h = 1$.

Now note a particular case of practical interest. Let current flow in the wire for a time and then turn it off, i.e.,

$$\text{(10.141)} \qquad h = \begin{cases} h = \text{cons't.}, & 0 < t < t_0 \\ 0, & t > t_0 \end{cases}$$

then

$$\text{(10.142)} \qquad T = \frac{h}{4\pi k} \left\{ -Ei\left(-\frac{r^2}{4at}\right) + Ei\left[-\frac{r^2}{4a(t - t_0)}\right] \right\}$$

for $t > t_0$. Here $r^2 = x^2 + y^2$. Now we can use, for $r^2/4at$ and $r^2/4a(t - t_0)$ small, the logarithmic approximation to obtain,

$$\text{(10.143)} \qquad T \approx \frac{h}{4\pi k} \ln \frac{t}{t - t_0}, \qquad t \gg t_0$$

Thus if we had measurements of T near the wire at various times after the current is turned off we could plot T versus $\ln[t/(t - t_0)]$. The slope of the resulting line would be $h/4\pi k$ from which we could determine the value of k. Such techniques are of practical utility.

One can readily generalize the above result as follows. See that if $T_u(r, t)$ is the solution for $h = 1$ then for a heating history given by

$$
\begin{cases}
h = h_1, & 0 < t < t_1 \\
h = h_2, & t_1 < t < t_2 \\
\quad \vdots \\
h = h_n, & t_{n-1} < t < t_n \\
\quad \vdots
\end{cases}
$$

(10.144)

the solution is:

(10.145)
$$
T = \frac{1}{4\pi k} \sum_{n=1}^{N} (h_n - h_{n-1}) T_u(r, t - t_{n-1})
$$

for $t_{N-1} < t < t_N$ and $t_0 = 0$, $h_0 = 0$. Similar treatment applies to a variety of problems.

Example (6)

Another example of interest is that of a semi-infinite medium with the temperature or the heat flux (temperature gradient) given on the boundary. We consider the heat flux given here. In particular we solve,

(10.146)
$$
\nabla^2 T - \frac{C\rho}{k} \frac{\partial T}{\partial t} = 0 \qquad \begin{matrix} z > 0 \\ t > 0 \end{matrix}
$$

with a circle of radius ε on $z = 0$ heated at constant rate, i.e.,

(10.147)
$$
\frac{\partial T}{\partial z} = \begin{cases} 0, & r > \varepsilon \\ A, & r < \varepsilon, \end{cases} \qquad t > 0
$$

By images,

(10.148)
$$
G = g_3(x, y, z, t; x', y', z', t') + g_3(x, y, z, t; x', y', -z', t')
$$

is appropriate, i.e., $\partial G / \partial z' = 0$ on $z' = 0$. Since we have axial symmetry we integrate g_3 over θ to construct on instantaneous ring source, i.e.,

(10.149)
$$
g_A = \int_0^{2\pi} g_3 \, d\theta
$$

Thus

(10.150)
$$
g_A = \int_\theta^{\theta + 2\pi} \frac{a e^{-\frac{r^2 + r'^2 - 2rr' \cos(\theta - \theta') + (z - z')^2}{4a(t - t')}}}{[2\sqrt{\pi a(t - t')}]^3} \, d\theta'
$$

i.e., we choose $\theta' = 0$ at θ. Here we are using the natural cylindrical coordinates.
Now note that

(10.151)
$$
\int_\theta^{\theta + 2\pi} e^{\frac{2rr' \cos(\theta - \theta')}{4a(t - t')}} \, d\theta' = 2\pi \frac{1}{\pi} \int_0^\pi e^{\frac{rr'}{2a(t - t')} \cos \gamma} \, d\gamma
$$

$$
= 2\pi I_0 \left[\frac{rr'}{2a(t - t')} \right]
$$

where I_0 is the modified Bessel function of first kind of order zero.† Thus

(10.152)
$$g_A = \frac{\pi a I_0\left[\dfrac{rr'}{2a(t-t')}\right]}{4[\sqrt{\pi a(t-t')}]^3} e^{-\frac{r^2 + r'^2 + (z-z')^2}{4a(t-t')}}$$

which has the character of an instantaneous ring source. Thus we use the G defined by

(10.153)
$$\bar{G} = g_A(r, r', z - z', t - t') + g_A(r, r', z + z', t - t')$$

and then according to Eq. (10.105) and our other results,

(10.154)
$$T = \int_0^t \int_0^\infty A(r', t')\bar{G}(r. r', z, t - t')r' \, dr' \, dt'$$

or, if $A(r', t')$ is a *constant* in r' and t', for $r < \varepsilon$,

(10.155)
$$T = A \int_0^t \frac{\pi a \, e^{-\frac{r^2 + z^2}{4a(t-t')}}}{4[\sqrt{\pi a(t-t')}]^3} \int_0^\varepsilon e^{-\frac{r'^2}{4a(t-t')}} I_0\left[\frac{rr'}{2a(t-t')}\right] r' \, dr' \, dt'$$

This can be put in other forms by performing the integrations in a different order.

EXECUTE PROBLEM SET (10-3)

WAVE-TYPE EQUATIONS

For equations of the wave-equation type,

(10.156)
$$\nabla^2 \psi - \frac{1}{c^2}\frac{\partial^2 \psi}{\partial t^2} + V\psi = -W, \quad \text{in } D,$$

we define $G(\mathbf{r}, t; \mathbf{r}', t)$ as the solution of

(10.157)
$$\nabla^2 G - \frac{1}{c^2}\frac{\partial^2 G}{\partial t^2} + VG = -\delta(\mathbf{r}, \mathbf{r}') \, \delta(t, t')$$

Here as in the diffusion-type equations we must require that a source not produce an effect before it comes into existence, i.e.,

(10.158)
$$G(\mathbf{r}, t; \mathbf{r}', t') = 0, \qquad t' > t$$

as a consequence of this we impose the "casuality requirement"

(10.159)
$$G(\mathbf{r}, t; \mathbf{r}', t') = G(\mathbf{r}', -t'; \mathbf{r}, -t)$$

Just as for the diffusion equation we write our equations in the primed variables as

(10.160)
$$\nabla'^2 \psi' - \frac{1}{c^2}\frac{\partial^2 \psi'}{\partial t'^2} + V'\psi' = -W'$$

and

(10.161)
$$\nabla'^2 G' - \frac{1}{c^2}\frac{\partial^2 G'}{\partial t'^2} + V'G' = -\delta(\mathbf{r}', \mathbf{r}) \, \delta(t', t)$$

† See integral representations of $I_0(x) = J_0(ix)$ in Chapter Four.

Now we multiply Eq. (10.160) by G', multiply Eq. (10.161) by ψ' and subtract the last result from the first. Then integrate over the volume of the domain and over time (t') from 0 to t^+, *where t^+ is just greater than t*. Use Green's theorem and the definition of the delta functions. Also note that

$$(10.162) \qquad \frac{\partial}{\partial t'}\left(\psi'\frac{\partial G'}{\partial t'} - G'\frac{\partial \psi'}{\partial t'}\right) = \psi'\frac{\partial^2 G'}{\partial t'^2} - G'\frac{\partial^2 \psi'}{\partial t'^2}$$

so the integral involving the terms on the right here can be integrated, but since at the upper limit, $t = t'$, we must have in view of Eqs. (10.158) and (10.159),

$$(10.163) \qquad G' = 0 \quad \text{and} \quad \frac{\partial G'}{\partial t'} = 0, \quad \text{at} \quad t' = t^+ > t$$

Thus we obtain finally, using Eq. (10.159) inside all integrals,

$$(10.164) \qquad \psi(\mathbf{r}, t) = \int_0^{t^+}\int_D W(\mathbf{r}', t')G(\mathbf{r}, t; \mathbf{r}', t')\, d\tau'\, dt'$$

$$+ \frac{1}{c^2}\int_D\left\{\psi(\mathbf{r}', 0)\frac{\partial G}{\partial t'}(\mathbf{r}, t; \mathbf{r}', 0) - \frac{\partial\psi}{\partial t'}(\mathbf{r}', 0)G(\mathbf{r}, t; \mathbf{r}', 0)\right\}d\tau'$$

$$+ \int_0^{t^+}\oint_S\left\{G(\mathbf{r}, t; \mathbf{r}', t')\frac{\partial\psi}{\partial n'}(\mathbf{r}', t') - \frac{\partial G}{\partial n'}(\mathbf{r}, t; \mathbf{r}', t')\psi(\mathbf{r}', t')\right\}dS'\, dt'$$

This is similar in structure to the result obtained for the diffusion-type equation.

INFINITE DOMAIN FOR THE SCALAR WAVE EQUATION

For the scalar wave equation, $V = 0$ in Eq. (10.156) and we have the Green's equation in three dimensions.

$$(10.165) \qquad \nabla^2 g - \frac{1}{c^2}\frac{\partial^2 g}{\partial t^2} = -\delta(\mathbf{r}, \mathbf{r}')\,\delta(t, t')$$

Furthermore we have spherical symmetry so we write this as:

$$(10.166) \qquad \frac{1}{R^2}\frac{\partial}{\partial R}\left(R^2\frac{\partial g}{\partial R^2}\right) - \frac{1}{c^2}\frac{\partial^2 g}{\partial\lambda^2} = -\delta(R)\,\delta(\lambda)$$

where $R = |\mathbf{r} - \mathbf{r}'|$, $\lambda = t - t'$. Now for the moment assume that in the neighborhood of $R = 0$ the first term on the left dominates, thus

$$(10.167) \qquad \frac{1}{R^2}\frac{\partial}{\partial R}\left(R^2\frac{\partial g}{\partial R}\right) \approx -\delta(R)\,\delta(\lambda)$$

Multiply by the volume element $4\pi R^2\, dR$ and integrate from 0 to R to obtain,

$$(10.168) \qquad 4\pi R^2\frac{\partial g}{\partial R} \approx -\delta(\lambda)$$

then divide by $4\pi R^2$ and integrate on R from R to ∞. This gives:

$$(10.169) \qquad g \approx \frac{\delta(\lambda)}{4\pi R}, \quad \text{near} \quad R = 0$$

Now let

(10.170)
$$U = Rg$$

and obtain upon substitution in Eq. (10.166),

(10.171)
$$\frac{\partial^2 U}{\partial R^2} - \frac{1}{c^2} \frac{\partial^2 U}{\partial \lambda^2} = -R\,\delta(R)\,\delta(\lambda)$$

For $R > 0$, $\lambda > 0$ this has the general solution (i.e., for the *homogeneous* equation).

(10.172)
$$U = f_1\left(\frac{R}{c} - \lambda\right) + f_2\left(\frac{R}{c} + \lambda\right)$$

where f_1 and f_2 are arbitrary functions of their arguments. The first corresponds to a distur-
bance moving away from the origin while the second is moving toward the origin. Thus for
casuality we must have

(10.173)
$$U = f_1\left(\frac{R}{c} - \lambda\right)$$

Then note that, as $R \to 0$, our previous result, Eq. (10.169), shows that we must have,

(10.174)
$$U \to \frac{\delta(\lambda)}{4\pi}, \quad \text{as} \quad R \to 0$$

thus we have finally,

(10.175)
$$g = \frac{\delta(\lambda - R/c)}{4\pi R}$$

That is, we see that f_1 must be

(10.176)
$$f_1 = \frac{1}{4\pi}\delta\left(\lambda - \frac{R}{c}\right)$$

Now we can use

(10.177)
$$g_3 = \frac{\delta\left[(t - t') - \frac{|\mathbf{r} - \mathbf{r}'|}{c}\right]}{4\pi|\mathbf{r} - \mathbf{r}'|}$$

for the three dimensional *infinite* domain. By similar procedures or integrations over one or
two coordinates, the g_2 and g_1 for the infinite two and one dimensional domains can be
constructed.

All other techniques, such as methods of images, eigenfunctions, etc., apply here as for the
diffusion equation and Helmholtz equation. We leave the development of these to the student.

EXECUTE PROBLEM SET (10-4)

METHODS EMPLOYING MULTIPLE IMAGES

The method of images introduced earlier and employed in some of our examples becomes most
powerful as a tool in problems involving multiple plane boundaries and equations with
constant coefficients. Thus for the Laplace, Helmholtz, Diffusion, and Wave equations we can
follow the general method to be described here.

Bearing in mind that $g(\mathbf{r}, t; \mathbf{r}', t')$ for *any* of these equations is analogous to an instantaneous source in the infinite domain which comes into existence at $t = t'$ (or exists for all time in the Laplace or Helmholtz case) we see that if the point $\mathbf{r}'$ is located at some normal distance l_1 from an infinite plane surface it will generate at each point of this surface a certain effect at time t. An *identical* source located at an equal normal distance, l_1, on the opposite side of this plane, but on the normal line drawn from the point $\mathbf{r}'$, will produce the same *magnitude* of effect at each point of this plane at time t, but the *normal derivatives* of the fields due to these two sources will have opposite *signs*. Thus, the *net* normal derivative will vanish on the plane. This fact has been used in our examples.

Also note that if the "image" is identical to the "source" but with *opposite sign*, then the field of the source on the plane is identical to that of the image, but, of *opposite sign* and the *net field* vanishes on this plane.

Now consider some other plane at *right* angles to this plane as shown in Fig. 10-2. Here we see the source as S and its image I located as just described. If we wish to assure that either the field or its normal derivative, vanish on Plane Number 2 then we must by the same argument place an image of S' at the distance l_2 on the normal line to Plane Number 2 on the opposite side of Plane Number 2. The *sign* of the image will depend on whether the field or its normal derivative is to vanish on this plane. However, we must *also* locate an image, I', of the first image I as indicated and choose its sign so that the net effect of I and I' on the plane Number 2 gives either field, or normal derivative equal to zero as required.

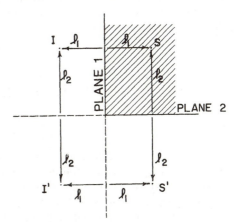

FIGURE 10.2. Source and Images for Two Perpendicular Plane Boundaries.

It is evident that in this particular case I' will be the *appropriate* image of S' across Plane Number 1 so that the desired condition on Plane Number 1 is maintained.

With this arrangement of "four sources" the shaded quadrant in the figure is a *semi-infinite* domain in which the appropriate Greens function is the sum of *four* terms of the same type, $g(\mathbf{r}, t; \mathbf{r}', t')$, with coefficient, ± 1, and value of $\mathbf{r}'$ chosen as just described.

This process can be extended to multiple plane boundaries very readily. For example, the region between two parallel planes with $\partial G/\partial n = 0$ on the plane at $x = 0$, and also that at $x = L$, is represented as in Fig. 10.3. Here it should be evident *by inspection* that $\partial G/\partial n = 0$ on *all* of the vertical planes indicated in this figure. Also, we see G for the strip as given by,

$$(10.178) \qquad G = \sum_{n=-\infty}^{+\infty} \{g(\mathbf{r}, t; \mathbf{r}' + \mathbf{1}_x(2nL + l), t') + g(\mathbf{r}, t; \mathbf{r}' + \mathbf{1}_x(2nL - l), t')\}$$

Quite obviously further generalization is possible and is evident in the exercises for the student.

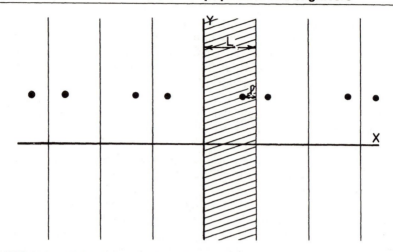

FIGURE 10.3. **Multiple Images for a Source Between Parallel Plane Boundaries.**

CONFORMAL MAPPING AND GREEN'S FUNCTION

Green's functions for Poisson's equation can in many cases be constructed very easily in *closed form* by a combination of conformal mapping of the domain and the method of images. Thus, suppose we have

$$(10.179) \qquad \nabla^2 \psi = W$$

in some domain D of the x, y plane, where ψ must satisfy some condⁱᵗⁱⁱ:ⁱⁱⁱ:ⁱⁱⁱⁱⁱ conditions on the boundary curve C. Here, the Green's function must satisfy

$$(10.180) \qquad \frac{\partial^2 G}{\partial x^2} + \frac{\partial^2 G}{\partial y^2} = \delta(x - x') \, \delta(y - y')$$

with appropriate conditions for G on C. Here we have written $\delta(\mathbf{r}, \mathbf{r}')$ as the product, $\delta(x - x') \, \delta(y - y')$ of two "Dirac", or linear delta functions. Thus, G is *analytic* everywhere in D except at the single point $\mathbf{r}'$. Thus, we see G as the solution of Laplace's equation in D, with the given conditions on C, due to a "point source," S, at $\mathbf{r}' = (x', y')$. We have already seen in our chapter on conformal mapping that a point source maps as a point source of the *same strength* under a conformal mapping, $w = f(z)$, $z = x + iy$, with $f(z)$ analytic. Thus, we map the domain D containing the source S into the upper half-plane in the $w = u + iv$ plane as shown in Fig. 10.4. Then simply by adding the image I below the u axis, which is the mapping of the contour C, we have G given as:

$$(10.181) \qquad G = -\frac{1}{2\pi} \ln[(u - u')^2 + (v - l)^2]^{\frac{1}{2}} \pm \frac{1}{2\pi} \ln[(u - u')^2 + (v + l)^2]^{\frac{1}{2}}$$

Here, the first term is the solution of

$$\frac{\partial^2 G}{\partial u^2} + \frac{\partial^2 G}{\partial v^2} = 0$$

with a *point* source at u', l, where u' and $v' = l$ are the coordinates of the source S under the

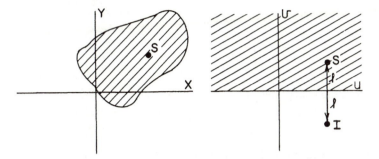

Figure 10.4. Source and Its Image in a Mapped Domain.

mapping, i.e.,

(10.182) $$u' + il = f(x' + iy')$$

where x', y' are the coordinates of S in the x–y plane, and $f(z)$ is the mapping function. The choice of sign on the second term in Eq. (10.183) depends on whether $G = 0$ or $\partial G/\partial n = 0$ on C is required. Finally, we simply express u, v, in terms of x, y for the form of G in the original domain.

EXECUTE PROBLEM SET (10-5)

PROBLEMS FOR CHAPTER X

Set (10-1)

(1) In the discussion of the various forms that Eq. (10.9) must have for various possible boundary conditions for ψ and $\partial\psi/\partial n$ on S we described *three* cases in which *one* "kind" of condition applied over *all* of S. Show what forms Eq. (10.9) must take, and what conditions must be imposed on G and/or $\partial G/\partial n$ when S is composed of *parts* S_1, S_2 etc. on which different kinds of conditions are specified for ψ and/or $\partial\psi/\partial n$, i.e., ψ a given function on part S_1 and $\partial\psi/\partial n$ a given function on S_2 for example.

(2) Consider the diffraction problem as in Example (1) of this chapter, however suppose the radiation incident on the circular hole is from a *point source* on the normal line through the center of the hole, i.e., the boundary condition is:

$$\left(\frac{\partial\psi}{\partial z}\right)_{z=0} = \begin{cases} 0, & r > a \\[2mm] \dfrac{-Ale^{ik\sqrt{l^2+r^2}}}{4\pi\sqrt{l^2+r^2}} & r < a \end{cases}$$

where l is the distance from the source to the center of the hole. Make the approximation $a/l \ll 1$ and obtain the diffraction field far from the hole, $z/a \gg 1$, i.e., make the same order of approximation as in our example.

Set (10-2)

(1) Construct the Green's function for Laplace's equation in the half-space, $z > 0$, then find the potential due to a charged disk in the plane $z = 0$ with radius R, i.e., the boundary

condition

$$\left(\frac{\partial V}{\partial z}\right)_{z=0} = \begin{cases} 0, & r > R \\ -\dfrac{\sigma}{\varepsilon}, & r < R \end{cases}$$

where $\sigma(r, \theta)$ is the charge density. Consider the special case of

$$\sigma = \sigma_0\left(1 - \frac{r^2}{R^2}\right), \qquad \sigma_0 = \text{constant}$$

and compare your result to that obtained in Problem (5) of Set (6–2) in Chapter Six.

(2) Formulate the Green's function for the Helmholtz equation for the region *external* to a sphere with the boundary condition,

$$\frac{\partial \psi}{\partial n} = 0, \quad \text{on} \quad r = R$$

where R is the radius of the sphere, then use this to solve the problem of the scattering of a spherical wave originating from a point source a distance r' from the center of the sphere. Compare this with your solution in Problem (7) of Set (5-3) in Chapter Five.

Set (10-3)

(1) Construct the Green's function for,

$$\nabla^2\psi - \frac{1}{a}\frac{\partial \psi}{\partial t} = -W$$

in the domain having the form of a segment of a cylinder $0 < r < R$, $0 < \theta < \pi/4$, $-L/2 < z < +L/2$ where ψ is some specified function, or functions, on all the *plane* surfaces but $\partial \psi/\partial r$ is a given function of position and time on the curved surface $r = R$. Let the initial value of ψ be zero. Do this problem *two* ways: First, by reducing it to the Helmholtz equation by Laplace transform and, secondly, by direct construction of the time-dependent Green's function. For the special case of ($\psi = 0$) on all plane surfaces and

$$\frac{\partial \psi}{\partial r} = \frac{A}{L}\left(z - \frac{L}{2}\right)t \quad \text{on} \quad r = R$$

write out the solution for $W = 0$.

(2) If we suppose the earth to be a homogeneous sphere of constant thermal properties and to have a spherically symmetric distribution of heat sources $W(r, t)$, and also an initial temperature distribution which is spherically symmetric, find the temperature $T(r, t)$ at any time t if the condition at the surface is $aT + b(\partial T/\partial r) = 0$ at $r = R$. Solve by constructing the Green's function for the region *inside* the sphere for the time-dependent equation.

Set (10-4)

(1) Construct the Green's function g_2 for the *scalar wave equation* in the case of *two dimensions*, for the *infinite domain* by integrating Eq. (10.177) over the z' coordinate from $-\infty$ to $+\infty$, i.e.,

$$g_2(\mathbf{r}_2,t;\mathbf{r}'_2,t') = \int_{-\infty}^{+\infty} \frac{\delta[(t - t') - R/c]\,dz'}{4\pi R}$$

where $\mathbf{r}_2$ and $\mathbf{r}_2'$ are the position vectors in the *plane* and,

$$R = |\mathbf{r} - \mathbf{r}'| = \sqrt{|\mathbf{r}_2 - \mathbf{r}_2'|^2 + (z - z')^2}$$

[Hint: use the substitutions,

$$dz' = \frac{R}{|z - z'|} dR$$

and

$$|z - z'| = \sqrt{R^2 - |\mathbf{r}_2 - \mathbf{r}_2'|^2}$$

to obtain:

$$g_2(\mathbf{r}_2, t; \mathbf{r}_2', t') = \begin{cases} 0, & c(t - t') < |\mathbf{r}_2 - \mathbf{r}_2'| \\[2ex] \dfrac{c}{2\pi\sqrt{c^2(t - t')^2 - |\mathbf{r}_2 - \mathbf{r}_2'|^2}}, & c(t - t') > |\mathbf{r}_2 - \mathbf{r}_2'| \end{cases}$$

Note: the difference in the "physical" nature of g_3 and g_2 for the scalar wave equation, i.e., g_3 is a spherical *shell* of radius $R = c(t - t')$ expanding at speed c, being *infinite* on the shell and zero *everywhere* else. On the other hand, g_2 is an expanding *cylinder* of radius $R_2 = c(t - t')$ which expands at speed c also, being zero *outside* the cylinder, but *not* zero *inside* the cylinder.

(2) Using the results of Problem (1) just above construct the Green's function for the *scalar wave equation*, in the *infinite domain*, for *one dimension*, i.e.,

$$g_1(x, t; x', t') = \int_{-\infty}^{+\infty} g_2(x, y, t; x', y', t')\, dy'$$

where we express $\mathbf{r}_2$ and $\mathbf{r}_2'$ in terms of their x and y components. Demonstrate that the result is:

$$g_1(x, t; x', t') = \begin{cases} 0, & c(t - t') < |x - x'| \\[2ex] \dfrac{c}{2}, & c(t - t') > |x - x'| \end{cases}$$

Hint: use the substitution $\xi = y' - y$ in g_2 and note that the quantity inside the radical in g_2 must always be positive.

Note: the significant difference in g_1 here as compared to g_3 and g_2 above. Here g_1 is a step function of height $c/2$, being of this height inside the interval

$$x' - c(t - t') < x < x' + c(t - t').$$

This interval expands in *both* directions with speed c.

(3) Consider the problem of forming the two-dimensional Green's function for the scalar wave equation in the *bounded circular domain* of radius a. Let the condition on G be $G = 0$ on the outer boundary of the circular region. [Hint: solve the problem in the Fourier transform domain on t, then invert.]

(4) Use the solution of the above problem to construct the solution of the problem of the *driven membrane*.

$$\frac{1}{r}\frac{\partial}{\partial r}\left(r\frac{\partial z}{\partial r}\right) + \frac{1}{r^2}\frac{\partial^2 z}{\partial \theta^2} = \frac{1}{c^2}\frac{\partial^2 z}{\partial t^2} - F(r, \theta, t)$$

for $0 < r < a, 0 < \theta < 2\pi, t > 0$, and

$$z = \partial z/\partial t = 0 \quad \text{at} \quad t = 0$$

for all r, θ, and $z = 0$ at $r = a$, all t. Examine the special case of

$$F = \begin{cases} Ae^{i\omega t}, & 0 < r < b, \quad (b < a), \quad t > 0 \\ 0, & b < r < a, \quad t > 0 \end{cases}$$

and $F = 0$ for $t < 0$.

Set (10-5)

(1) Construct the Green's function for the diffusion, or heat flow, equation in the rectangular domain $0 < x < a, 0 < y < b, 0 < z < c$, by the method of images, for $G = 0$ on all boundaries.

(2) Show that for Laplace's or Poisson's, equation for the region *exterior* to a sphere, subject to the function being given on the sphere, and hence $G = 0$ on the sphere, G can be formed as $g_3(\mathbf{r}, \mathbf{r}')$ with $\mathbf{r}'$ being a point outside the sphere, plus an "image" inside the sphere at a point on the line from the $\mathbf{r}'$ point passing through the center of the sphere. This "image" must however have a modified "strength", i.e., be multiplied by some factor.

(3) Use the method of conformal mapping and images to construct the Green's function for Poisson's equation in the *plane* region bounded by a circle for the function given on the circle, i.e., $G = 0$ on the circle.

REFERENCES

P. Dennery and A. Krzywicki, *Mathematics for Physicists* (Harper and Row, Publishers, Inc., New York, 1967).

J. W. Dettman, *Mathematical Methods in Physics and Engineering* (McGraw-Hill Book Company, Inc., New York, 1958).

G. Goertzel and N. Tralli, *Some Mathematical Methods of Physics* (McGraw-Hill Book Company, Inc., New York, 1960).

E. A. Kraut, *Fundamentals of Mathematical Physics* (McGraw-Hill Book Company, Inc., New York, 1967).

A. Kyrala, *Theoretical Physics: Application of Vectors, Matrices, Tensors and Quaternions* (W. B. Saunders Co., Philadelphia, 1967).

N. N. Lebedev, I. P. Skalskaya, and Y. S. Uflyand, *Problems of Mathematical Physics* (translated by R. A. Silverman) (Prentice-Hall, Inc., Englewood Cliffs, N.J., 1965).

J. Mathews and R. L. Walker, *Mathematical Methods of Physics* (W. A. Benjamin, Inc., New York, 1953).

P. M. Morse and H. Feshbach, *Methods of Theoretical Physics* (McGraw-Hill Book Company, Inc., New York, 1953).

CHAPTER ELEVEN

Introduction to integral equations

An integral equation is an equation in which the unknown function appears within an integral. We have already seen several examples of such integral equations arise in the form of *convolution integrals* in the discussion of linear systems and in the discussion of population growth.

TYPES OF INTEGRAL EQUATIONS

The general linear integral equation involving a *single* unknown appears as

$$(11.1) \qquad h(x)F(x) = g(x) + \lambda \int_a^b K(x, y)F(y)\, dy$$

with $h(x)$, $g(x)$ being known functions of x, λ is a constant parameter, which in some cases plays the role of an eigenvalue, and $K(x, y)$ is the "kernel" of the integral equation.

If $h(x) = 0$ this becomes a *Fredholm equation of the first kind*, while if $h(x) = 1$ it has the form of a *Fredholm equation of the second kind*. In either case, if $g(x) = 0$, the equation is homogeneous. If $K(x, y)$ is zero for $y > x$ then the upper limit on the integral becomes x instead of b and the equation is called a *Volterra equation*.

These same *types* of equations exist when the unknown function is a function, $\psi(x, y, z)$, of *several* variables, instead of being a function, $F(x)$, of the single variable. In particular we will consider the general form,

$$(11.2) \qquad \psi(x, y, z) = f(x, y, z) + \lambda \int \int_\tau \int \psi(x', y', z')G(x, y, z; x', y', z')\, d\tau'$$

which we will see is closely linked to a boundary value problem.

DEGENERATE KERNELS

A kernel of the form, (n finite)

(11.3)
$$K(x, y) = \sum_{i=1}^{n} U_i(x)V_i(y)$$

is said to be degenerate. In this case the integral equation can be solved by elementary methods. This is best seen by an example.

Example (1)

The integral equation is given as:

(11.4)
$$F(x) = \sin x + \lambda \int_0^\pi \sin(x + y)F(y)\,dy$$

Here the kernel is degenerate since,

(11.5)
$$K(x, y) = \sin(x + y) = \sin x \cos y + \cos x \sin y$$

If we let,

(11.6)
$$A = \int_0^\pi \sin y F(y)\,dy$$

and

(11.7)
$$B = \int_0^\pi \cos y\, F(y)\,dy$$

then we have our integral equation in the form,

(11.8)
$$F(x) = \sin x + \lambda(A \cos x + B \sin x)$$

Putting this into our expressions for A and B then gives:

(11.9)
$$A = (1 + \lambda B)\frac{\pi}{2}$$

and

(11.10)
$$B = \lambda A \frac{\pi}{2}$$

or,

(11.11)
$$A = \frac{\dfrac{\pi}{2}}{1 - \dfrac{\pi^2}{4}\lambda^2}$$

and

(11.12)
$$B = \frac{\lambda\dfrac{\pi^2}{4}}{1 - \dfrac{\pi^2}{4}\lambda^2}$$

Thus we put these now back into our Eq. (11.8) for $F(x)$ and have:

(11.13)
$$F(x) = \frac{1 + \lambda^2 \frac{\pi^2}{4}}{1 - \lambda^2 \frac{\pi^2}{4}} \sin x + \frac{\lambda \frac{\pi}{2}}{1 - \lambda^2 \frac{\pi^2}{4}} \cos x$$

as our final solution for $F(x)$.

Observe that this solution becomes unbounded, "blows up", for $\lambda = \pm 2/\pi$. These are the "eigenvalues" of the *homogeneous* equation, that is, solutions of

(11.14)
$$F(x) = \lambda \int_0^\pi \sin(x + y)F(y)\,dy$$

which can be obtained in a manner just as above for the inhomogeneous equation and exist only for $\lambda = \pm 2/\pi$. The *general nature* of this feature of *homogeneous* equations will become evident in the discussion to follow.

FREDHOLM'S THEOREMS

On the basis that any reasonably well behaved kernel can be represented as an infinite series of degenerate kernels of the type cited in Eq. (11.3) Fredholm established that for *real* kernels, *either* the *inhomogeneous* equation,

(11.15)
$$F(x) = g(x) + \lambda \int_a^b K(x, y)F(y)\,dy$$

has a unique solution for any function $g(x)$ (with λ *not* an eigenvalue), *or* the *homogeneous* equation,

(11.16)
$$F(x) = \lambda \int_a^b K(x, y)F(y)\,dy$$

has at least one nontrivial solution for some value of λ [λ is an eigenvalue of the kernel $K(x, y)$]. In the latter case the solution, $F(x)$, is the eigenfunction of $K(x, y)$ corresponding to the eigenvalue λ.

If λ is *not* an eigenvalue, the first alternative above, then λ is also *not* an eigenvalue of the transposed equation [i.e., x and y interchanged in $K(x, y)$].

If λ is an eigenvalue then the *inhomogeneous* equation, Eq. (11.15), has a solution if and only if,

(11.17)
$$\int_a^b g(x)F(x)\,dx = 0$$

for every solution, $F(x)$, of the *transposed homogeneous equation*, i.e., Eq. (11.16) with $K(x, y) \rightarrow K(y, x)$.

We will not investigate the proofs of Fredholm's theorems but these are available in the references.

NEUMANN SERIES SOLUTION

A most obvious approach to try for constructing the general solution of the inhomogeneous equation above is an "iterative" procedure. Thus, as a *first* approximation,

$$(11.18) \qquad F_{(1)}(x) \approx g(x)$$

then the *second* approximation is,

$$(11.19) \qquad F_{(2)} = g(x) + \lambda \int_a^b K(x, y)g(y)\, dy$$

and the process is continued,

$$(11.20) \qquad F(x) = g(x) + \lambda \int_a^b K(x, y)g(y)\, dy + \lambda^2 \int_a^b \int_a^b K(x, y)K(y, y')g(y')\, dy'\, dy + \ldots$$

However, this will converge only if λ is sufficiently small and provided $K(x, y)$ is bounded. A more general series approach called the *Fredholm series*[†] can be formulated which converges under much broader conditions but will not be considered here. Instead we will show the general connection of integral equations to boundary value problems and then examine the Schmidt–Hilbert theory.

INTEGRAL EQUATIONS AND BOUNDARY VALUE PROBLEMS

An intimate relation exists between certain types of integral equations and boundary value problems. This is best illustrated in the following general example. Consider again our general second-order linear equation involving the Laplacian operator,

$$(11.21) \qquad \nabla^2 \psi + V\psi = F, \quad \text{in } D$$

with

$$(11.22) \qquad \alpha\psi + \beta\frac{\partial\psi}{\partial n} = 0, \quad \text{on } S$$

This is the same as Eq. (10.4) of the previous chapter but with a general boundary condition. Here α and β are constants.

Here we will define a Green's *function* but in a *different* manner than used in the previous chapter. Consequently, we will use $K(\mathbf{r}, \mathbf{r}')$ for this function and define $K(\mathbf{r}, \mathbf{r}')$ such that,

$$(11.23) \qquad \nabla^2 K = \delta(\mathbf{r}, \mathbf{r}'), \quad \text{in } D$$

with

$$(11.24) \qquad \alpha K + \beta\frac{\partial K}{\partial n} = 0, \quad \text{on } S$$

Note that the significant difference here is that the term VK *does not* appear in Eq. (11.23).

Now if we multiply Eq. (11.21) by K, Eq. (11.23) by ψ, subtract the former from the latter, and integrate over the domain D we obtain[‡]

$$(11.25) \qquad \int_D (\psi\nabla^2 K - K\nabla^2\psi)\, d\tau - \int_D KV\psi\, d\tau = \int_D \psi\, \delta(\mathbf{r}, \mathbf{r}')\, d\tau - \int_D KF\, d\tau$$

† See References.
‡ Our *volume* element is $d\tau$ in the unprimed space as usual.

Then we apply Green's theorem to the terms involving the Laplacian and use the definition of the delta function to write:

$$(11.26) \quad \psi(\mathbf{r}') = \int_D K(\mathbf{r}, \mathbf{r}')F(\mathbf{r}) \, d\tau - \int_D V(\mathbf{r})K(\mathbf{r}, \mathbf{r}')\psi(\mathbf{r}) \, d\tau + \int_S \left[\psi \frac{\partial K}{\partial n} - K \frac{\partial \psi}{\partial n} \right] dS$$

Here the surface integral terms vanish by virtue of the *common boundary conditions* on K and ψ. Thus, we have then the *integral equation*,

$$(11.27) \qquad \psi(\mathbf{r}') = \int_D K(\mathbf{r}, \mathbf{r}')F(\mathbf{r}) \, d\tau - \int_D V(\mathbf{r})K(\mathbf{r}, \mathbf{r}')\psi(\mathbf{r}) \, d\tau$$

In particular if we let,

$$(11.28) \qquad f(\mathbf{r}') = \int_D K(\mathbf{r}, \mathbf{r}')F(\mathbf{r}) \, d\tau$$

and assume

$$(11.29) \qquad V(\mathbf{r}) = -\lambda = \text{constant}$$

we have the form,

$$(11.30) \qquad \psi(\mathbf{r}') = f(\mathbf{r}') + \lambda \int_D K(\mathbf{r}, \mathbf{r}')\psi(\mathbf{r}) \, d\tau$$

which is of the general form of the inhomogeneous Fredholm equation for a function $\psi(x, y, z)$ of three *independent variables*. Of course, we can consider in the *same* way *one* or *two* independent variables just by proper definition of the delta function as pointed out in the last chapter.

This shows that the general integral equation of Fredholm type is intimately related to a boundary value problem. Consequently, *we may solve the integral equation simply by constructing the solution of the corresponding boundary value problem.*

We also see that for the homogeneous case,

$$(11.31) \qquad \psi(\mathbf{r}') = \lambda \int_D K(\mathbf{r}, \mathbf{r}')\psi(\mathbf{r}) \, d\tau$$

λ is indeed an eigenvalue in exactly the same sense discussed in connection with differential operators. In particular, we have,

$$(11.32) \qquad \nabla^2 \psi = \lambda \psi, \quad \text{in } D$$

as the form of our Eq. (11.21). Thus, λ is an eigenvalue of the operator, ∇^2, in our usual terminology. Here, we *also* call λ an eigenvalue of the kernel $K(\mathbf{r}, \mathbf{r}')$, which, as we see now, is a particular Green's function built on the Laplacian operator, ∇^2. In *one dimension* the differential equation reduces to an equivalent equation of the *Sturm–Liouville* type; we examine this special case as follows.

FREDHOLM INTEGRAL EQUATION FOR THE STURM–LIOUVILLE DIFFERENTIAL EQUATION

In the one-dimensional case our Eq. (11.21) becomes the inhomogeneous Sturm–Liouville equation,†

$$(11.33) \qquad \frac{1}{\omega(x)} \frac{d}{dx}\left(h(x)\frac{dy}{dx} \right) - \frac{g(x)}{\omega(x)}y - \lambda y = f(x) \qquad a < x < b$$

† Note that this is the same equation discussed in Chapter Three written now in a slightly different form.

with now a variety of possible end conditions to be considered. Here, we define a Green's function as the solution of the equation:

$$(11.34) \qquad \frac{1}{\omega}\frac{d}{dx}\left(h\frac{dG}{dx}\right) - \frac{g(x)}{\omega}G = \delta(x - x')$$

with $\delta(x - x')$ being the linear delta function for a weight function $\omega(x)$ on $a < x < b$‡ As above, we multiply Eq. (11.33) by $G(x, x')$ and Eq. (11.34) by $y(x)$, but here we also multiply by the weight function $\omega(x)$. We obtain, after subtracting and integrating on x from a to b,

$$(11.35) \quad y(x') = \int_a^b f(x)\omega(x)G(x, x')\,dx + \lambda \int_a^b G(x, x')\omega(x)y(x)\,dx + \left[h(x)y\frac{dG}{dx} - h(x)G\frac{dy}{dx}\right]_a^b$$

Then, if as above, $y(x)$ and $G(x, x)$ satisfy the same *linear boundary conditions*, the integrated terms vanish, and we have:

$$(11.36) \qquad y(x') = \int_a^b f(x)\omega(x)G(x, x')\,dx + \lambda \int_a^b G(x, x')\omega(x)y(x)\,dx$$

Then we multiply through by $\sqrt{\omega(x')}$ and define

$$(11.37) \qquad \psi = \sqrt{\omega}\,y$$

and

$$(11.38) \qquad K(x, x') = \sqrt{\omega(x')}G(x, x')\sqrt{\omega(x)}$$

so that Eq. (11.36) now appears as,

$$(11.39) \qquad \psi(x') = U(x') + \lambda \int_a^b K(x, x')\psi(x)\,dx$$

which is in the standard form of the inhomogeneous equation of the second kind with kernel $K(x, x')$ defined by Eq. (11.38), and

$$(11.40) \qquad U(x') = \int_a^b K(x, x')f(x)\sqrt{\omega(x)}\,dx$$

In the usual formulation of the Sturm–Liouville equation there are five different types of boundary conditions on $y(x)$ at a and b which lead to a *complete set* of orthonormal functions on $a < x < b$ for the *homogeneous* problem, these are:

$$y(a) = y(b) = 0 \qquad (1)$$
$$y'(a) = y'(b) = 0 \qquad (2)$$
$$y'(a) + \alpha y(a) = y'(b) + \alpha y(b) = 0 \qquad (3)$$
$$y' \text{ and } y \text{ finite at } a \text{ and } b, \; h(a) = h(b) = 0 \qquad (4)$$
$$y(a) = y(b) \text{ and } h(a)y'(a) = h(b)y'(b) \qquad (5)$$

Any one of these may also apply to G in our formulation of the Green's function here. [Here $y'(a) = dy(a)/dx$ etc.]

‡ Here, $G(x, x') = G(x', x)$ is required to be continuous at $x = x'$ although its first and higher derivatives are discontinuous.

Obviously we have here the *homogeneous Fredholm equation* in one independent variable,

$$(11.41) \qquad \psi(x') = \lambda \int_a^b K(x, x')\psi(x)\,dx$$

corresponding to the homogeneous Sturm–Liouville equation, and λ is an eigenvalue under any of the above set of possible boundary conditions on $G(x, x')$ which is used to define $K(x, x')$.

Example (2)

Formulate an integral equation corresponding to the Sturm–Liouville problem:

$$(11.42) \qquad \frac{d^2y}{dx^2} - \lambda y = 0, \qquad 0 < x < 1$$

with

$$(11.43) \qquad y(0) = y(1) = 0$$

According to our Eq. (11.34) we have,

$$(11.44) \qquad \frac{d^2G}{dx^2} = \delta(x - x')$$

and the weight function for this delta function is unity. Thus, for $x < x'$ the right member is zero, we integrate to get

$$(11.45) \qquad G(x, x') = ax + b, \qquad 0 \le x < x'$$

and for $x > x'$

$$(11.46) \qquad G(x, x') = cx + d, \qquad x' < x \le 1$$

The boundary condition $G = 0$ at $x = 0$ and $G = 0$ at $x = 1$ gives,

$$(11.47) \qquad \begin{cases} G(x, x') = ax, & 0 \le x < x' \\ G(x, x') = c(x - 1), & x' < x \le 1 \end{cases}$$

but direct integration of Eq. (11.44) gives, with ε an infinitesimal,

$$(11.48) \qquad \int_{x'-\varepsilon}^{x'+\varepsilon} \frac{d^2G}{dx^2} = \frac{dG}{dx}(x' + \varepsilon, x') - \frac{dG}{dx}(x' - \varepsilon, x') = 1$$

Thus, using our results of Eq. (11.47) we have,

$$(11.49) \qquad c - a = 1$$

Also $G(x, x')$ is continuous at $x = x'$ so

$$(11.50) \qquad ax' = c(x' - 1)$$

Thus we have

$$(11.51) \qquad G(x, x') = \begin{cases} x(x' - 1), & 0 \le x < x' \\ x'(x - 1), & x' < x \le 1 \end{cases}$$

Hence, the integral equation is:

(11.52) $$y(x') = \lambda\left\{\int_0^{x'} x(x'-1)y(x)\,dx + \int_{x'}^1 x'(x-1)y(x)\,dx\right\}$$

In this case $K(x, x')$, or $G(x, x')$ is separated in two distinct representations for $x < x'$ and $x > x'$.

EXECUTE PROBLEM SET (11-1)

HILBERT-SCHMIDT THEORY

In our discussion of the relation of integral *equations of Fredholm type* to boundary value problems above we found the kernel $K(\mathbf{r}, \mathbf{r}')$, or $K(x, x')$ for the Sturm–Liouville case, to be built on a Green's function and therefore, is *symmetric,*

(11.53) $$K(x, x') = K(x', x)$$

If we generalize our discussion to include *complex* functions, $\psi(\mathbf{r})$, or $\psi(x)$, then we must consider complex kernels. In this case we call a kernel with the property,

(11.54) $$K(x, x') = K^*(x', x)$$

a *Hermitian kernel*, i.e., it is equal to its transposed complex conjugate.† Certainly, if a kernel is real and symmetric it is also Hermitian. Here, we consider just the *real symmetric kernel.*

In the homogeneous equation, Eq. (11.41) above, λ is an eigenvalue of $K(x, x')$ to be determined simultaneously with the function $\psi(x)$, while in the inhomogeneous equation, Eq. (11.39), λ is a known constant.

The Hilbert–Schmidt theory is based on consideration of the eigenfunctions and eigenvalues of the homogeneous equation. To get the general idea of this approach consider a *degenerate kernel.*

(11.55) $$K(x, x') = \sum_{i=1}^N U_i(x)U_i(x')$$

The corresponding homogeneous integral equation will be,

(11.56) $$\psi(x') = \lambda\int_a^b K(x, x')\psi(x)\,dx$$

or,

(11.57) $$\psi(x') = \lambda\sum_{i=1}^N U_i(x')\int_a^b U_i(x)\psi(x)\,dx$$

where we define now,

(11.58) $$C_i = \int_a^b U_i(x)\psi(x)\,dx$$

and hence have:

(11.59) $$\psi(x') = \lambda\sum_{i=1}^N C_i U_i(x')$$

† Recall this terminology in regard to matrix operators in Chapter Two.

Here we see that any eigenfunction of $K(x, x')$, $\psi(x')$, can be expressed as a *linear* combination of the N functions $U_i(x)$ used to form $K(x, x')$. It should then be obvious that in *this case* there can be no more than N *linearly independent eigenfunctions*.

If we substitute Eq. (11.59) back into Eq. (11.58) we obtain the *set of linear equations* for the C_i, $i = 1, 2, \ldots N$,

$$(11.60) \qquad C_i^m = \lambda_m \sum_{=1}^{N} a_{ij} C_j \, m$$

where,

$$(11.61) \qquad a_{ij} = \int_a^b U_i(x) U_j(x) \, dx$$

or, in a *matrix notation*,

$$(11.62) \qquad \left[\frac{1}{\lambda}\right] C = AC$$

Thus the eigenvalues, λ, of the homogeneous integral equation are the reciprocals of the eigenvalues of the matrix A with elements defined by Eq. (11.61). We know that the eigenvectors of a real symmetric matrix are orthogonal† so we have,

$$(11.63) \qquad \sum_{l=1}^{N} C_l^{(i)} C_l^{(j)} = 0, \qquad i \neq j$$

where $C_l^{(i)}$ and $C_l^{(j)}$, $l = 1, 2, \ldots N$ correspond to eigenvectors Number (i) and (j), respectively, with the eigenvalues λ_i and λ_j.

We now show that as a consequence of the orthogonality of the eigenvectors above we also have orthogonality of the functions $\psi^{(i)}(x)$ and $\psi^{(j)}(x)$. Thus we write,

$$(11.64) \qquad \int_a^b \psi^{(i)}(x) \psi^{(j)}(x) \, dx = \lambda_i \lambda_j \sum_{l=1}^{N} \sum_{n=1}^{N} C_l^{(i)} C_n^{(j)} \int_a^b U_l(x) U_n(x) \, dx$$

using Eq. (11.59), adding the appropriate superscript (i) or (j) on ψ and the C_n with either λ_i and λ_j. Here we then write a_{ln} for the integral on the right, as defined previously and have,

$$(11.65) \qquad \int_a^b \psi^{(i)}(x) \psi^{(j)}(x) \, dx = \lambda_i \sum_{l=1}^{N} C_l^{(i)} \left(\lambda_j \sum_{n=1}^{N} a_{ln} C_n^{(j)}\right)$$

with rearrangement. The quantity in parenthesis here is just $C_l^{(j)}$ according to Eq. (11.60) and therefore we have

$$(11.66) \qquad \int_a^b \psi^{(i)}(x) \psi^{(j)}(x) \, dx = \lambda_i \sum_{l=1}^{N} C_l^{(i)} C_l^{(j)}$$

which according to Eq. (11.63) is zero for $i \neq j$. Thus the orthogonality of the eigenfunctions is established. If we also wish to impose a normalization condition so that the above integral is unity for $i = j$ we must require,

$$(11.67) \qquad \sum_{l=1}^{N} C_l^{(i)} C_l^{(j)} = \frac{\delta_{ij}}{\lambda_i} \begin{cases} 0, & i \neq j \\ \dfrac{1}{\lambda_i}, & i = j \end{cases}$$

† See Problem (2) of Set (2-3) in Chapter Two.

and then

(11.68)
$$\int_a^b \psi^{(i)}(x)\psi^{(j)}(x)\,dx = \delta_{ij}$$

Here we must emphasize that although the $\psi^{(i)}(x)$ form an orthonormal set, it is *not a complete set*.

With these results we can then show that,[†]

(11.69)
$$U_j(x) = \sum_i C_j^{(i)} \psi^{(i)}(x)$$

and express the kernel $K(x, x')$ in terms of its eigenfunctions,[‡]

(11.70)
$$K(x, x') = \sum_i \frac{\psi^{(i)}(x)\psi^{(i)}(x')}{\lambda_i}$$

This result, although constructed here only for a degenerate symmetric kernel holds in the more general form,

(11.71)
$$K(x, x') = \sum_i \frac{\psi^{(i)}(x)\psi^{*(i)}(x')}{\lambda_i}$$

for nondegenerate *Hermitian* kernels. Also, one can show that the eigenvalues of a Hermitian kernel are *all real*. The student should consult the references, particularly Courant and Hilbert, for details and proofs.

We continue to restrict our consideration to *real* symmetric kernels and extend this approach to the inhomogeneous equation. From our Eq. (11.71) it follows that any function $\phi(x)$ representable by,

(11.72)
$$\phi(x') = \int_a^b K(x, x')W(x)\,dx$$

can be expanded in the eigenfunctions of $K(x, x')$, i.e.,

(11.73)
$$\phi(x') = \sum_i b_i \psi^{(i)}(x')$$

where

(11.74)
$$b_i = \frac{1}{\lambda_i}\int_a^b \psi^{(i)}(x)W(x)\,dx$$

and $W(x)$ is a "source function." We say $\phi(x)$ is "source wise" representable in terms of $K(x, x')$. Here, we should emphasize that in general the $\psi^{(i)}(x)$ *do not form a complete set* of ortho-normal functions and hence, not just any reasonably well behaved function can be represented in a series of the $\psi^{(i)}(x)$; only those of the type given here.

Now to solve the equation

(11.75)
$$F(x') = g(x') + \lambda \int_a^b K(x, x')F(x)\,dx$$

where $g(x')$ is "source wise" representable in terms of $K(x, x')$[¶], we expand,

(11.76)
$$F(x') - g(x') = \sum_i \bar{b}_i \psi^{(i)}(x')$$

---

† i.e. using the fact that equation 11.67 shows the matrix with elements $\sqrt{\lambda_i}\,C_j^{(i)}$ to be orthogonal.
‡ Compare this form with that of $G(\mathbf{r}, \mathbf{r}')$ in Eq. (10.31) of Chapter Ten.
¶ The origin of the terminology "source wise" is evident in that the $g(x)$ term arises from the corresponding boundary value problem as a *source* term.

where the $\psi^{(i)}(x)$ are eigenfunctions of $K(x, x')$. Here, according to Eq. (11.74) we have,

(11.77)
$$\bar{b}_i = \int_a^b \psi^{(i)}(x)[F(x) - g(x)] \, dx$$

This can also be written as

(11.78)
$$\bar{b}_i = \frac{\lambda}{\lambda_i} \int_a^b \psi^{(i)}(x)F(x) \, dx$$

by substituting Eq. (11.71) for $K(x, x')$ in Eq. (11.75). Thus, define here

(11.79)
$$\alpha_i = \int_a^b \psi^{(i)}(x)F(x) \, dx, \qquad \beta_i = \int_a^b \psi^{(i)}(x)g(x) \, dx$$

and we have from the above two equations,

(11.80)
$$\bar{b}_i = \alpha_i - \beta_i = \frac{\lambda}{\lambda_i}\alpha_i$$

Then,

(11.81)
$$\alpha_i = \frac{\lambda_i}{\lambda_i - \lambda}\beta_i$$

and

(11.82)
$$\bar{b}_i = \frac{\lambda}{\lambda_i - \lambda}\beta_i$$

so we have, putting these back into Eq. (11.76).

(11.83)
$$F(x) = g(x) + \lambda \sum_i \beta_i \frac{\psi^{(i)}(x)}{\lambda_i - \lambda}$$

or,

(11.84)
$$F(x) = g(x) + \lambda \sum_i \frac{\psi^{(i)}(x)}{\lambda_i - \lambda} \int_a^b \psi^{(i)}(x')g(x') \, dx'$$

Thus, we have obtained the solution of the *inhomogeneous* Fredholm equation in terms of the eigenfunctions, $\psi^{(i)}(x)$, of the corresponding *homogeneous* equation (i.e., having the same kernel). It is quite obvious that *the method fails when the fixed constant λ is an eigenvalue of the homogeneous equation.* For further details the student should consult the references.

Generally, whenever possible we transform a given Fredholm equation into a *differential* equation and solve the differential equation, but in some cases the *formalism* given above is of value in the analysis of the theoretical *structure* of a physical problem.

THE RELATION OF THE FREDHOLM INTEGRAL EQUATION TO CERTAIN INTEGRAL TRANSFORMS

While the formal analysis of the Hilbert–Schmidt theory provides the solution of the inhomogeneous Fredholm equation there are special forms of this equation which can be solved by other procedures. Some of these are discussed in a following section.

Here we consider a special class of kernels in an integral transform pair: i.e.,

(11.85) $$f(x') = \int_a^b K(x, x')F(x)\,dx, \quad \text{and} \quad F(x) = \int_a^b K'(x, x')f(x')\,dx'$$

Then if in the Fredholm equation, Eq. (11.75) above, we look upon the integral as a transform with kernel $K(x, x')$, having an inverse with the *same limits of integration* and a kernel $K'(x, x')$ such that,

(11.86) $$\alpha \int_a^b F(x)K(x, x')\,dx = \int_a^b F(x)K'(x, x')\,dx$$

where α is a real constant, then the solution of the inhomogeneous Fredholm equation will have the form,

(11.87) $$F(x') = \frac{1}{\alpha - \lambda^2}\left[\alpha g(x') + \lambda \int_a^b g(x)K(x, x')\,dx\right]$$

for $\alpha \neq \lambda^2$, when $\alpha > 0$. The student can verify that this applies to the infinite Fourier sine and cosine transforms, the infinite Hankel transforms and the infinite Hilbert transforms.†

EXECUTE PROBLEM SET (11-2)

SPECIAL FORMS

Volterra integral equations can frequently be reduced to differential equations by differentiation.

Example (3)

The Volterra equation,

(11.88) $$F(x) = x + \int_0^x xyF(y)\,dy = x + xg(x)$$

where,

(11.89) $$g(x) = \int_0^x yF(y)\,dy$$

becomes upon differentiation with respect to x,

(11.90) $$\frac{dg}{dx} = xF(x) = x[x + xg]$$

upon using Eq. (11.88) to substitute for $F(x)$. Integration then yields,

(11.91) $$g(x) = -1 + Ce^{\frac{x^3}{3}}$$

and substitution back into Eq. (11.88) yields $C = 1$.

† See the definition in the discussion of dispersion relations, page 223.

Convolution-type kernels, $K(x - x')$, when appearing in *Fredholm equations* can be solved by use of Fourier transforms. Thus, for equations of the form,

$$(11.92) \qquad F(x) = g(x) + \lambda \int_{-\infty}^{+\infty} K(x - x')F(x')\,dx'$$

we can apply the *Fourier transform* to yield

$$(11.93) \qquad \bar{F} = \bar{g} + \lambda \bar{K} \bar{F}$$

where bars denote the *Fourier transforms*. Then, we rearrange to solve explicitly for $\bar{F}$ and invert the resulting form by the inversion integral (if we can find the right trick to perform the integration).

In the particular case of the *convolution-type kernel* with the *Volterra* equation,

$$(11.94) \qquad F(x) = g(x) + \lambda \int_{0}^{x} K(x - x')F(x')\,dx'$$

we have exactly the same procedure using the *Laplace transform*. Thus, an equation of exactly the same form as Eq. (11.94) results with the bar in this case denoting the *Laplace transform*. Again, the problem is then to successfully execute the inversion integral on a form such as,

$$(11.95) \qquad \bar{F} = \frac{\bar{g}}{1 - \lambda \bar{K}}$$

This can be a very difficult problem.

Another special form involving the *convolution kernel* is,

$$(11.96) \qquad F(x) = g(x) + \lambda \int_{0}^{\infty} K(x - x')F(x')\,dx'$$

The special features here are the limits 0 to ∞ on the integral and the form, $K(x - x')$, of the kernel. In this case the *Fourier transform* method, as modified in the *Wiener–Hopf* technique,† can be applied. Thus, we define

$$(11.97) \qquad F(x) = F_{+}(x) + F_{-}(x)$$

where,

$$(11.98) \qquad \begin{cases} F_{+}(x) = 0, & x > 0 \\ F_{-}(x) = 0, & x < 0 \end{cases}$$

and we can write,

$$(11.99) \qquad F_{+}(x) + F_{-}(x) = g(x) + \lambda \int_{-\infty}^{+\infty} K(x - x')F_{-}(x')\,dx'$$

Then taking *Fourier transforms* (again using bars as our notation for the transform of a function)

$$(11.100) \qquad \bar{F}_{+} + \bar{F}_{-} = \bar{g} + \lambda \bar{K} \bar{F}_{-}$$

In order to proceed to find the proper expressions for $\bar{F}_{+}$ and $\bar{F}_{-}$ we must make some assumptions about the analyticity of $\bar{K}$ and the behavior of $\bar{F}$ and $\bar{K}$ at infinity. This method of analysis becomes quite long and tedious and is rarely employed. The student can find some discussion of the method in the references.

† See Appendix and References.

Dispersion integrals as described in Chapters Eight and Nine give rise to integral equations of the *Fredholm type* of the *first kind*, i.e.,

$$(11.101) \qquad F(z) = \frac{1}{2\pi} \int \frac{G(x')\, dx'}{x' - z}$$

and the integration is along some part of the real axis. Here, $G(x')$ is the *unknown function*. Note that we have stated the *form* of the *dispersion integral* in a more general form than stated in Chapter Eight. Here, however, as in our earlier discussion $F(z)$ is analytic everywhere *except* for this part of the real axis over which the integration is executed. [This integral form is often called the *Hilbert transform* of $G(x)$.]

For one special case of this form an interesting solution procedure exists. If the range of integration is $-a < x' < a$, and $F(x)$ is known on this interval, $-a < x < a$, [actually $F(x + i\varepsilon)$], then set:

$$(11.102) \qquad x = -a \cos \phi, \qquad y = -a \cos \theta$$

and *assume*

$$(11.103) \qquad \sin \phi F(-a \cos \phi) = - \sum_{n=1}^{\infty} \frac{b_n}{2} \sin n\phi$$

is a convergent series representation of F. Then, if we make use of the formula,†

$$(11.104) \qquad \text{Principal part of } \int_0^{\pi} \frac{\cos n\theta\, d\theta}{\cos \theta - \cos \phi} = \frac{\pi \sin n\phi}{\sin \phi}$$

we find that,

$$(11.105) \qquad G(-a \cos \phi) = \frac{1}{\sqrt{1 - \cos^2 \phi}} \left[\tfrac{1}{2} b_0 + \sum_1^{\infty} b_n \cos n\phi \right]$$

and b_0 is to be determined so that G is properly behaved at $x = \pm a$. Of course, several variations of this technique are possible.

EXECUTE PROBLEM SET (11-3)

PROBLEMS

Set (11-1)

(1) Solve the integral equation,

$$\text{(a)} \quad f(x) = x^2 + \int_0^1 xy f(y)\, dy$$

(2) Find the eigenfunctions and eigenvalues of the equation,

$$y(x) = \lambda \int_0^1 e^{x - \xi} y(\xi)\, d\xi$$

(3) Solve the integral equation,

$$y(x) = e^x + \lambda \int_0^1 e^{x - \xi} y(\xi)\, d\xi$$

---

† This can be established by the technique of integration around the unit circle discussed in Chapter Eight with a modification for a simple pole *on* the unit circle.

(4) Convert the differential equation,

$$\frac{d^2 y}{dx^2} + \frac{1}{x}\frac{dy}{dx} + \lambda y = 0$$

with $y(1) = dy\,(0)/dx = 0$ into a Fredholm integral equation.

(5) Use the Neumann series method to find an approximate solution to the equation,

$$\phi(x) = \sinh x + e^{-x}\int_0^x e^t \phi(t)\, dt$$

Carry this to the third-order approximation, then show that the result is almost exact for small x.

(6) Construct a Neumann series solution for the Fredholm equation in Problem (3) above and compare the result to the exact solution.

(7) Construct a Fredholm equation for the differential equation,

$$\frac{d^2 y}{dx^2} + (ae^{-x^2} + b)y = 0$$

with $y(0) = y(\infty) = 0$.

(8) In Chapter Ten we solved the problem of diffraction of a plane wave by a circular hole in an infinite plane barrier. Set up an *integral* equation whose solution leads to the field in the diffraction region.

Set (11-2)

(1) Verify Eqs. (11.69) and (11.70) of the text.

(2) Show that if $\psi(x)$ is a nontrivial solution of the equation

$$\psi(x) = \lambda_1 \int_a^b K(x, \xi)\psi(\xi)\, d\xi$$

and $\bar{\psi}(x)$ is a nontrivial solution of the *associated* equation,

$$\bar{\psi}(x) = \lambda_2 \int_a^b K(\xi, x)\bar{\psi}(\xi)\, d\xi$$

with $\lambda_1 \neq \lambda_2$ then

$$\int_a^b \psi(x)\bar{\psi}(x)\, dx = 0$$

(3) If the kernel in the homogeneous Fredholm equation is nonsymmetric the equation may have no eigenfunctions. Show that

$$\psi(x) = \lambda \int_0^\pi \sin x \cos \xi \psi(\xi)\, d\xi$$

has no nontrivial solutions.

(4) Using the connection, pointed out in the text, of certain integral transforms to integral equations show that,

$$\psi(x) = \sin x + a \int_0^\infty J_0(\xi x)\psi(\xi)\xi\, d\xi$$

has the solution,

$$\psi(x) = \frac{1}{1 - a^2}\left[\sin x + a \int_0^\infty J_0(\xi x)(\sin \xi)\xi \, d\xi\right]$$

Set (11-3)

(1) Show that the solution of the Volterra equation,

$$\psi(x) = 1 + \int_0^x (\xi - x)\psi(\xi) \, d\xi$$

satisfies,

$$\frac{d^2\psi}{dx^2} + \psi = 0$$

with $\psi(0) = 1$, $d\psi(0)/dx = 0$ and hence $\psi(x) = \cos x$. However, for illustration also solve this integral equation by the Neumann series method.

(2) Solve Abel's integral equation,

$$f(x) = \int_0^x \frac{y(\xi)}{\sqrt{x - \xi}} \, d\xi$$

Hint: multiply both sides by $(t - x)^{-\frac{1}{2}}$ and integrate both sides from 0 to t, on x, noting that

$$\int_\xi^t \frac{dx}{\sqrt{(x - \xi)(t - x)}} = \int_0^1 \frac{dV}{\sqrt{V(1 - V)}} = \pi$$

(3) Solve the integral equation

$$y(x) = e^{-|x|} + b \int_{-\infty}^{+\infty} e^{-|x - \xi|} y(\xi) \, d\xi$$

REFERENCES

R. Courant and D. Hilbert, *Methods of Mathematical Physics* (Interscience Publishers, Inc., New York, 1953).

J. W. Dettman, *Mathematical Methods in Physics and Engineering* (McGraw-Hill Book Company, Inc., New York, 1958).

J. Irving and N. Mullineux, *Mathematics in Physics and Engineering* (Academic Press, Inc., New York, 1959).

A. Kyrala, *Theoretical Physics: Applications of Vectors, Matrices, Tensors and Quaternions* (W. B. Saunders Co., Philadelphia, 1967).

N. N. Lebedev, I. P. Skalskaya, and Y. S. Uflyand, *Problems of Mathematical Physics* (translated by R. A. Silverman), (Prentice-Hall, Inc., Englewood Cliffs, N.J., 1965).

J. Mathews and R. L. Walker, *Mathematical Methods of Physics* (W. A. Benjamin, Inc., New York, 1965).

P. M. Morse and H. Feshbach, *Methods of Theoretical Physics* (McGraw-Hill Book Company, Inc., New York, 1953).

CHAPTER TWELVE

Variation and perturbation methods; introduction to nonlinear differential equations

In this chapter we develop a few of the basic techniques in the calculus of variations having particular applications in physical problems and illustrate these applications with examples. We also present some of the basic ideas pertinent to the analysis of nonlinear differential equations and develop perturbation methods' for both linear and nonlinear differential equations.

EXTREMUM OF A FUNCTION WITH CONSTRAINTS, LAGRANGIAN UNDETERMINED MULTIPLIERS

A function, $F(x_i)$, of a set of coordinates, $x_i, i = 1, 2, \ldots n$, is said to have an extremum at a point, x_i', if at the point x_i' infinitesimal changes, δx_i, in the values of the coordinates introduce no change in the value of the function, i.e.,

(12.1)
$$F(x_i + \delta x_i) = F(x_i)$$

Thus upon expanding the left member in a Taylor series about the point, x_i', we have to *first-order* in the δx_i,

(12.2)
$$\delta F = \sum_{i=1}^{n} \frac{\partial F(x_i')}{\partial x_i} \delta x_i = 0$$

as the criterion for a first-order extremum at the point x_i'.

If no constraints are specified on the x_i, that is, these are *independent* variables then the δx_i here can be chosen arbitrarily. Thus we may choose all δx_i but any one, say δx_k, to be zero. Then Eq. (12.2) appears as

(12.3)
$$\frac{\partial F}{\partial x_k} \delta x_k = 0, \qquad \delta x_k \neq 0$$

and hence we find

(12.4)
$$\frac{\partial F(x_i)}{\partial x_k} = 0, \qquad k = 1, 2, \ldots n$$

since we could at will take $k = 1, 2, \ldots n$. This set of n equations can then be solved simultaneously to determine the values of the x_i'. However, if *constraints* do exist then the δx_i cannot *all* be chosen arbitrarily and the above procedure does not apply. We classify the constraints on the δx_i in two classes, *integrable* and *nonintegrable*. This classification can be made clear as follows.

Suppose there exists another function of the x_i, say $G_1(x_i)$ which must always have a *constant value*. Then this equation,

(12.5)
$$G_1(x_i) = G_1 = \text{constant}$$

could be solved for one of the x_i, say x_n, in terms of the remaining x_i and substituted into the function $F(x_i)$, thus reducing the number of independent variables in $F(x_i)$ from n to $n - 1$. Then with $F[x_1, x_2, \ldots x_{n-1}, x_n(x_1, x_2, \ldots x_{n-1})]$ as our function the original procedure could be applied. Similarly if in addition to $G_1(x_i)$ there were also other functions, $G_2, \ldots G_K$, which have constant values, we could in principle eliminate K variables in $F(x_i)$ and apply the original procedure.

However, there is another procedure which is much more general and in most cases easier to apply. Note that if $G_j(x_i)$ is to remain constant then for small changes in the x_i we must have,

(12.6)
$$\sum_{i=1}^{n} \frac{\partial G_j(x_i')}{\partial x_i} \delta x_i = 0 \qquad j = 1, 2, 3, \ldots$$

not only at x_i' but at any point actually. This is a *linear relation* in the δx_i which must hold, *in addition to Eq.* (12.2), if $F(x_i)$ is to have an extremum at x_i' and the constraint is also to be fulfilled.

We call *any linear relation* in the δx_i a constraint on the variation; i.e.,

(12.7)
$$\delta C_j = \sum_{i=1}^{n} a_{ji} \delta x_i = 0$$

is a general constraint; here the a_{ji} are functions of the x_i. If δC_j is the *exact differential* of some function, as in Eq. (12.6) above, or if δC_j multiplied by some function $U(x_i)$ is an *exact differential* then we say the constraint is *integrable*. However, if δC_j is not exact, and no function $U(x_i)$ exists such that $U \cdot \delta C_j$ is exact, then the constraint is *nonintegrable*. We call $U(x_i)$ an integrating factor.

Here we recall the definition of an *exact differential form*;

(12.8)
$$\delta C = \sum_{i=1}^{n} B_i \, dx_i$$

is exact if, and only if,

(12.9)
$$\frac{\partial B_i}{\partial x_j} - \frac{\partial B_j}{\partial x_i} = 0, \quad \text{all } i \text{ and } j$$

If δC here were not exact this would not be satisfied. Examining the question of whether $U\delta C$ is exact we have the same criteria with UB_i replacing the B_i in these equations. This can then be shown† to yield the set of equations,

$$(12.10) \qquad 0 = B_i\left(\frac{\partial B_j}{\partial x_k} - \frac{\partial B_k}{\partial x_j}\right) + B_j\left(\frac{\partial B_k}{\partial x_i} - \frac{\partial B_i}{\partial x_k}\right) + B_k\left(\frac{\partial B_i}{\partial x_j} - \frac{\partial B_j}{\partial x_i}\right)$$

for all cyclic permutations of i, j, k, as the criteria for some $U(x_i)$ to exist such that $U\,\delta C$ is exact. Furthermore when one such function U exists it can be shown that an infinity of such functions exist. Note that *every* constraint in *two* variables is integrable.

With this distinction in mind we return to continue our discussion of the extremum with constraints. Here we see that there may be cases in which the constraints, in the form of equations such as Eq. (12.7) are *nonintegrable* and then the algebraic substitution to eliminate the dependent x_i in $F(x_i)$ *cannot* be executed. Thus we use the following method.

We have the set of equations

$$\begin{cases} \delta F = \sum_{i=1}^{n} \frac{\partial F}{\partial x_i}\delta x_i = 0 \\[2mm] \delta C_j = \sum_{i=1}^{n} a_{ji}\,\delta x_i = 0, \qquad j = 1, 2, \ldots K < n \end{cases}$$

which must hold simultaneously. Multiply each δC_j by a parameter λ_j, yet unspecified, and add the result to δF. There results

$$(12.11) \qquad \sum_{i=1}^{n}\left(\frac{\partial F}{\partial x_i} + \sum_{j=1}^{K}\lambda_j a_{ji}\right)\delta x_i = 0$$

Here choose the λ_j such that the coefficients of $\delta x_n, \delta x_{n-1}, \ldots \delta x_{n-K+1}$ all vanish, i.e.,

$$(12.12) \qquad \frac{\partial F}{\partial x_i} + \sum_{j=1}^{K}\lambda_j a_{ji} = 0, \qquad i = n - K + 1, \ldots n$$

Then there remains the sum,

$$(12.13) \qquad \sum_{i=1}^{n-K}\left(\frac{\partial F}{\partial x_i} + \sum_{j=1}^{K}\lambda_j a_{ji}\right)\delta x_i = 0$$

But here now we have precisely $n - K$ *independent* δx_i that may be chosen independently. Hence, we can now apply our original method of setting all but one δx_i equal to zero. We then obtain

$$(12.14) \qquad \frac{\partial F}{\partial x_i} + \sum_{j=1}^{K}\lambda_j a_{ji} = 0, \qquad i = 1, 2, \ldots n - K$$

which are of the same form as the Eq. (12.12) above. Thus we have n equations of this form to determine the n x_i' of the extremum. However, we now have also as unknown quantities the λ_j which are K in number that must be determined. Therefore, to determine $n + K$ unknowns we need additional K equations. These are provided by the constraints themselves, i.e., we can write those as

$$(12.15) \qquad \sum_{i=2}^{n} a_{ji}\frac{dx_i}{dx_1} + a_{ji} = 0, \qquad j = 1, 2, \ldots K$$

† See A. R. Forsyth, *A Treatise on Differential Equations*, Macmillan and Co. Ltd., London 1948, page 325.

which are K differential equations the x_i must also satisfy. Here x_1 is arbitrarily selected as the "independent" variable of the differential equation. When these are simply integrable we have simply

(12.16) $$G_j(x_i) = G_j = \text{constant}, \qquad j = 1, 2, \dots K$$

as the K additional equations for determining the λ_j concurrently with the x_i.

   This is Lagrange's method of undetermined multipliers and the λ_j are the "undetermined" multipliers. Whether the extremum points thus determined correspond to maxima, minima, or saddle points of $F(x_i)$ must be examined in terms of relations in the second derivatives. This will not be considered here.†

Example (1)

Stationary points of a bead on a wire. A bead of mass m slides freely on a wire, the wire is bent into a space curve determined by the pair of simultaneous equations

(12.17) $$\begin{cases} G_1(x, y, z) = G_1 = \text{const.} \\ G_2(x, y, z) = G_2 = \text{const.} \end{cases}$$

i.e., the intersection of two surfaces is a space curve.

   The stationary criterion is that the potential energy, mgz be a minimum,

(12.18) $$\delta F = \delta(mgz) = 0$$

where g is the acceleration of gravity and z is the vertical distance to the bead on the wire. Thus following the format outlined above we have

(12.19) $$\begin{cases} mg + \lambda_1 \dfrac{\partial G_1}{\partial z} + \lambda_2 \dfrac{\partial G_2}{\partial z} = 0 \\[2mm] \lambda_1 \dfrac{\partial G_1}{\partial y} + \lambda_2 \dfrac{\partial G_2}{\partial y} = 0 \\[2mm] \lambda_1 \dfrac{\partial G_1}{\partial x} + \lambda_2 \dfrac{\partial G_2}{\partial x} = 0 \end{cases}$$

along with Eq. (12.17) to solve *simultaneously* for x, y, z and λ_1 and λ_2.

   For a particularly simple illustration let

(12.20) $$\begin{cases} G_1 = \dfrac{x^2 + y^2 + z^2}{2} = \dfrac{R^2}{2} \\[3mm] G_2 = x + y + \dfrac{z^2}{2} = a \end{cases}$$

thus

(12.21) $$\begin{cases} mg + z\lambda_1 + z\lambda_2 = 0 \\ y\lambda_1 + \lambda_2 = 0 \\ x\lambda_1 + \lambda_2 = 0 \end{cases}$$

† See Appendix.

It is then simply a matter of doing a little algebra to find the points, x, y, z at which the bead is stationary, concurrently with the values of λ_1 and λ_2. It is left for the student to show that here, if $R^2 > 2(a - 1)$, stationary points exist and are four in number, two being at maximum values of z on the curve and two at minimum points of z on this curve.

EXTREMUM OF AN INTEGRAL; EULER–LAGRANGE EQUATION

We are familiar with the *line integral* of a vector function along some space curve as encountered in vector calculus in the form†

$$(12.22) \qquad I_C = \int_A^B {}_C\, \mathbf{V}(x_1, x_2, x_3) \cdot \mathbf{dr}$$

where $\mathbf{V}(x_1, x_2, x_3)$ is the vector function, and $\mathbf{dr}$ is the vector displacement along the space curve C connecting the two points A and B, i.e., as depicted in Fig. 12.1.

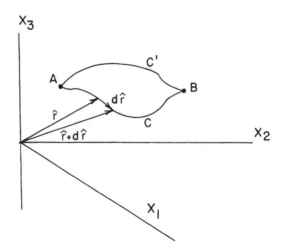

FIGURE 12.1. Two Paths of Integration From Point A to Point B.

Also recall from vector calculus that, according to Stokes' theorem, the integral along the curve C, will have the *same* value along a neighboring curve C' connecting A to B if the vector $\mathbf{V}$ is the *gradient* of some scalar, say $G(x_1, x_2, x_3)$. The criterion that this should be true is just that the *curl* of $\mathbf{V}$ must vanish, i.e., if,

$$(12.23) \qquad \frac{\partial V_i}{\partial x_j} - \frac{\partial V_j}{\partial x_i} = 0, \qquad \text{all } i, j$$

then

$$(12.24) \qquad I_C = I_{C'}$$

Comparing Eq. (12.23) to Eq. (12.9) we see that this is just the condition that the integrand here be an *exact differential*, that is

$$(12.25) \qquad dG = \sum_{i=1}^{3} V_i\, dx_i = \mathbf{V} \cdot \mathbf{dr}$$

is exact.

† See Chapter One.

Here we consider only those integrals for which Eq. (12.23) *do not* hold. Thus we now write the integrand of I_C in scalar form as,

$$(12.26) \qquad \mathbf{V} \cdot \mathbf{dr} = V(x_i) \cos \theta(x_i) \, ds$$

where $V(x_i)$ is the magnitude of V, ds is the magnitude of $\mathbf{dr}$, and $\cos \theta(x_i)$ is the cosine of the angle, θ, between $\mathbf{V}$ and $\mathbf{dr}$, which varies along the curve.

Note that here we can write the element of arc length ds as,

$$(12.27) \qquad ds = \sqrt{1 + \left(\frac{dx_2}{dx_1}\right)^2 + \left(\frac{dx_3}{dx_1}\right)^2} \, dx_1$$

so that the integrand appears as

$$(12.28) \qquad F\left(x_1, x_2, x_3 ; \frac{dx_2}{dx_1}, \frac{dx_3}{dx_1}\right) dx_1 = V \cos \theta \sqrt{1 + v^2} \, dx_1$$

where F is a function of the indicated arguments and we have written the radical of Eq. (12.27) as $\sqrt{1 + v^2}$. Now renaming our variables as, $x_3 \to x_2$, $x_2 \to x_1$, $x_1 \to t$, and at the same time noting that we could have carried all of this through in *more than three coordinates* we see that we are here concerned with an integral of the form,

$$(12.29) \qquad I = \int_{A}^{B} {}_C \, F(x_i, \dot{x}_i, t) \, dt$$

along a curve C, where the integrand is *not* an exact differential. $F(x_i, \dot{x}_i, t)$ is a function of a set of coordinates, $x_i, i = 1, 2, \ldots n$ and their derivatives, $\dot{x}_i = dx_i/dt$, with respect to the parameter t, which is the explicit variable of integration.

We have taken the above line of development to introduce integrals of the form given in Eq. (12.29) because of the connections exhibited to familiar ideas of line integrals, but such integrals arise in other contexts in which the *form* of the function F is other than that constructed above. However, the arguments, x_i, $\dot{x}_i$, and t of F are the same and $F \, dt$ *is not an exact differential*.

Since the integral I_C of Eq. (12.29) now has different values on different *paths* connecting the given points A and B, we seek a characterization of the paths on which I_C has an *extreme value*. We will see that the criterion for an *extremum of an integral* is exactly analogous to that for an extremum of a function.

A path C in an n-dimensional space can be specified by $n - 1$ functions of the x_i, i.e., the intersection of $n - 1$ surfaces. These functions contain constant parameters, say α_j, which are independent. That is, we can introduce a change in α_1 say without altering the value of any other α_j. Thus the curve C can be looked upon as being a function, $C(\alpha_j)$ of this set of parameters, and a change of the α_j to new values, α'_j, defines another curve, C'.

Thus since in the integration we integrate over the coordinates we see that the integral I_C is a function,

$$(12.30) \qquad I_C = I_C(\alpha_j, x_{iA}, x_{iB})$$

of only the parameters α_j specifying the path and the coordinates of the end point, x_{iA}, x_{iB}, $i = 1, 2, \ldots n$.

For *fixed end points* then we say that I_C has an extremum on the path C if infinitesimal changes in the α_j do not change the value of I_C, i.e.,

$$(12.31) \qquad I_C(\alpha_j + \delta\alpha_j, x_{iA}, x_{iB}) = I_C(\alpha_j, x_{iA}, x_{iB})$$

then the path C is an extremum path for the integral. Since the $\delta\alpha_j$ are *infinitesimal* we expand I_C in a Taylor series in the $\delta\alpha_j$ and have to *first order*,

$$(12.32) \qquad \sum_j \frac{\partial I_C}{\partial \alpha_j} \delta\alpha_j = 0$$

in exact analogy to Eq. (12.2). If, as stated above the α_j are independent, meaning that no other constraints are placed on the choice of path, then by the same arguments we employed for the extremum of a function, we get

$$(12.33) \qquad \frac{\partial I_C}{\partial \alpha_j} = 0, \quad \text{all } j$$

as the criteria for the extremum.

Now we look at the effect on the *integrand* of these changes in the α_j defining the path. Since the integration parameter t is *not* involved we see that the effect of the change $\alpha_j \rightarrow \alpha_j + \delta\alpha_j$ is to move each point $x_i(t)$ on C to a neighbor point $x_i' = x_i(t) + \delta x_i$ on the neighbor path corresponding to the same value of t. Similarly $\dot{x}_i(t) \rightarrow \dot{x}_i(t) + \delta\dot{x}_i$ and since

$$(12.34) \qquad \delta x_i = x_i'(t) - x_i(t)$$

and

$$(12.35) \qquad \delta\dot{x}_i = \dot{x}_i'(t) - \dot{x}_i(t)$$

we see that,

$$(12.36) \qquad \frac{d}{dt}\delta x_i = \delta\frac{dx_i}{dt}$$

The type of *first-order* variations considered here are called *weak variations*. For a discussion of *strong* variations the reader should consult the references at the end of the chapter.

Now we have as the criterion for the extremum,

$$(12.37) \qquad \delta I_C = \int_A^B F(x_i + \delta x_i, \dot{x}_i + \delta\dot{x}_i, t)\,dt - \int_A^B F(x_i, \dot{x}_i, t)\,dt = 0$$

as the equivalent of Eq. (12.31). We then expand F in the first integral in a Taylor series in the δx_i *and* the $\delta\dot{x}_i$ and keep only the first-order terms. The result is

$$(12.38) \qquad \delta I_C = \int_A^B \sum_{i=1}^n \left(\frac{\partial F}{\partial x_i}\delta x_i + \frac{\partial F}{\partial \dot{x}_i}\delta\dot{x}_i\right)dt = 0$$

Here the integrals involving the $\delta\dot{x}_i$ are *integrated by parts*,

$$(12.39) \qquad \int_A^B \frac{\partial F}{\partial x_i}\delta\dot{x}_i\,dt = \frac{\partial F}{\partial \dot{x}_i}\delta x_i\Big|_A^B - \int_A^B \frac{d}{dt}\left(\frac{\partial F}{\partial \dot{x}_i}\right)\delta x_i\,dt$$

But the end points of the paths C and C' are the *same* so $\delta x_i = 0$ at A and B, and therefore Eq. (12.38) appears as,

$$(12.40) \qquad \delta I_C = \int_A^B \sum_{i=1}^n \left(\frac{\partial F}{\partial x_i} - \frac{d}{dt}\frac{\partial F}{\partial \dot{x}_i}\right)\delta x_i\,dt = 0$$

Here we must remember that the x_i are independent variables and the changes in the path parameters, $\delta\alpha_j$, introduce *independent* changes, δx_i, in the x_i, i.e., we may view δx_i as,

(12.41)
$$\delta x_i = \sum_j a_{ij}(\alpha_t)\, \delta\alpha_k$$

where the set of functions $a_{ij}(\alpha_k)$ for δx_i and for another, δx_l, are *functionally* independent forms.

Thus we are free to choose the $\delta x_i(t)$ as any functions and we may choose all but say δx_k to be zero. Then only this term appears in the sum in Eq. (12.40) and since the integral must still vanish for *any* $\delta x_k(t)$ we see that we must have,

(12.42)
$$\frac{\partial F}{\partial x_k} - \frac{d}{dt}\frac{\partial F}{\partial \dot{x}_k} = 0, \qquad k = 1, 2, \dots n$$

Here, since we can repeat the same argument for $k = 1, 2, \dots n$ we see that we can write this result, as indicated, for all values of k. These are the *Euler–Lagrange Equations* for the integral I_C. Their solutions, $x_i(t)$, $i = 1, 2, \dots n$ *provide the parametric equation of a path for which I_C is an extremum.*†

EXTREMUM OF A MULTIPLE INTEGRAL

There are a great many ways in which the problem of finding the extremum for a multiple integral arises in physics and engineering.

One defines an extremum for an integral such as

(12.43)
$$I = \iiint\limits_{D} F\left(U, \frac{\partial U}{\partial x}, \frac{\partial U}{\partial y}, \frac{\partial U}{\partial z}, x, y, z\right) dx, dy, dz$$

where $U = U(x, y, z)$ is a function of the coordinates, in precisely the same way as for a single integral. Here D indicates the *fixed domain* of integration.

Generalizing the concept of a *path* of integration we view $U(x, y, z)$ as the *parametric description* of the "path" of integration, then a variation of the "path" is equivalent to changing *parameters* in U so that,

(12.44)
$$\begin{cases} U \to U + \delta U \\ U_x \to U_x + \delta U_x \\ U_y \to U_y + \delta U_y \\ U_z \to U_z + \delta U_z \end{cases}$$

transforms the integral to the neighboring path. Here we use $U_x = \partial U/\partial x$, etc. Thus the condition for an extremum appears as:

$$\delta I = \iiint\limits_{D} F(U + \delta U, U_x + \delta U_x, U_y + \delta U_y, U_z + \delta U_z, x, y, z)\, dx\, dy\, dz$$

(12.45)
$$- \iiint\limits_{D} F(U, U_x, U_y, U_z, x, y, z)\, dx\, dy\, dz = 0$$

† The question of whether the extremum is a maximum or a minimum requires study of the second-order variation. The student should consult the references on this question.

Hence expanding the first integrand in a Taylor series in δU, δU_x, δU_y, and δU_z gives:

$$(12.46) \qquad \delta I = \iiint_D \left\{ \frac{\partial F}{\partial U} \delta U + \frac{\partial F}{\partial U_x} \delta U_x + \frac{\partial F}{\partial U_y} \delta U_y + \frac{\partial F}{\partial U_z} \delta U_z \right\} dx\, dy\, dz = 0$$

to terms of first order.

Here we carry out a transformation of the terms in δU_x etc., in a manner equivalent to the integration by parts used for the single integral. This transformation is based on the *divergence theorem* of vector calculus. Let us define the vector

$$(12.47) \qquad \mathbf{W} = \mathbf{1}_x \frac{\partial F}{\partial U_x} + \mathbf{1}_y \frac{\partial F}{\partial U_y} + \mathbf{1}_z \frac{\partial F}{\partial U_z}$$

and use the fact that δU_x, δU_y, δU_z are the components of

$$(12.48) \qquad \nabla(\delta U) = \mathbf{1}_x\, \delta U_x + \mathbf{1}_y\, \delta U_y + \mathbf{1}_z\, \delta U_z$$

with $\mathbf{1}_x$, $\mathbf{1}_y$, $\mathbf{1}_z$ being our usual unit vectors. Here,

$$(12.49) \qquad \delta \frac{\partial U}{\partial x} = \frac{\partial}{\partial x} \delta U$$

just as we had for weak variations of a single integral.

Now note that the terms in δU_x, δU_y, δU_z in Eq. (12.46) appear as:

$$(12.50) \qquad \mathscr{I} = \iiint_D \mathbf{W} \cdot \nabla(\delta U)\, dx\, dy\, dz$$

But a familiar vector identity is†

$$(12.51) \qquad \nabla \cdot (\mathbf{W}\, \delta U) = (\nabla \cdot \mathbf{W})\, \delta U + \mathbf{W} \cdot \nabla(\delta U)$$

and hence this is equivalent to,

$$(12.52) \qquad \mathscr{I} = \iiint_D \nabla \cdot (\mathbf{W}\, \delta U)\, dx\, dy\, dz - \iiint_D (\nabla \cdot \mathbf{W})\, \delta U\, dx\, dy\, dz$$

Then we apply the divergence theorem to the first integral to yield,

$$(12.53) \qquad \mathscr{I} = \int_S (\delta U)\mathbf{W} \cdot d\mathbf{S} - \iiint_D \nabla \cdot \mathbf{W}\, \delta U\, dx\, dy\, dz$$

where S is the bounding surface of the domain D and $d\mathbf{S}$ is the vector element of area. Since all "paths" must have the same end points, i.e., the boundary conditions on U are *not* varied, we have $\delta U = 0$ on S and the surface integral is zero.

Thus, we have the condition for an extremum of the multiple integral as,

$$(12.54) \qquad \delta I = \iiint_D \left\{ \frac{\partial F}{\partial U} - \frac{\partial}{\partial x} \frac{\partial F}{\partial U_x} - \frac{\partial}{\partial y} \frac{\partial F}{\partial U_y} - \frac{\partial}{\partial z} \frac{\partial F}{\partial U_z} \right\} \delta U\, dx\, dy\, dz = 0$$

† See Chapter One.

Since this should hold for an *arbitrary* variation of U, $\delta U(x, y, z)$, we see that we must have,

$$(12.55) \qquad \frac{\partial F}{\partial U} - \frac{\partial}{\partial x}\frac{\partial F}{\partial U_x} - \frac{\partial}{\partial y}\frac{\partial F}{\partial U_y} - \frac{\partial}{\partial z}\frac{\partial F}{\partial U_z} = 0$$

determining the function $U(x, y, z)$ which yields an extreme value for the multiple integral I. This partial differential equation is also called an Euler–Lagrange equation.

Example (2)

As a simple example of the variational formulation of a partial differential equation by the process outlined above we consider the integral,

$$(12.56) \qquad I = \iiint\limits_{D} \left[\left(\frac{\partial U}{\partial x}\right)^2 + \left(\frac{\partial U}{\partial y}\right)^2 + \left(\frac{\partial U}{\partial z}\right)^2 \right] dx\, dy\, dz$$

Here the function $F(U, U_x, U_y, U_z, x, y, z)$ of the integrand is just $|\nabla U|^2$. Using this form we see,

$$(12.57) \qquad \frac{\partial F}{\partial U_x} = 2U_x, \quad \text{etc}$$

so that Eq. (12.55) appears as,

$$(12.58) \qquad \frac{\partial^2 U}{\partial x^2} + \frac{\partial^2 U}{\partial y^2} + \frac{\partial^2 U}{\partial z^2} = 0, \quad \text{in } D$$

which is just Laplace's equation.

This means that functions U which satisfy Laplaces equation in the domain D will yield extreme values for the integral I of Eq. (12.56). Here we should note that the integrand could be multiplied by *any* constant without altering these results. We also point out that the same results can be formulated in any orthogonal coordinate system.

MULTIPLE INTEGRALS IN SEVERAL FUNCTIONS

Just as we considered the many functions, $x_i(t)$, in the integral formulation leading to the Euler–Lagrange equations from the single integral extremum, we can also consider several functions $U_i(x, y, z)$, $i = 1, 2, \ldots n$, in the multiple integral variational problem. That is we consider,

$$(12.59) \qquad I = \iiint\limits_{D} F(U_i, U_{ix}, U_{iy}, U_{iz}, x, y, z)\, dx\, dy\, dz$$

where there may be $i = 1, 2, \ldots n$ different functions U_i and their partial derivatives, $U_{ix} = \partial U_i/\partial x$, etc., in the integrand.

If we carry through the same sort of analysis as before for the multiple integral we find the result,

$$(12.60) \qquad \delta I = \iiint\limits_{D} \sum_{i=1}^{n} \left\{ \frac{\partial F}{\partial U_i} - \frac{\partial}{\partial x}\frac{\partial F}{\partial U_{ix}} - \frac{\partial}{\partial y}\frac{\partial F}{\partial U_{iy}} - \frac{\partial}{\partial z}\frac{\partial F}{\partial U_{iz}} \right\} \delta U_i\, dx\, dy\, dz = 0$$

and since the $\delta U_i(x, y, z)$ may be chosen *independently* it follows that

$$(12.61) \qquad \frac{\partial F}{\partial U_i} - \frac{\partial}{\partial x}\frac{\partial F}{\partial U_{ix}} - \frac{\partial}{\partial y}\frac{\partial F}{\partial U_{iy}} - \frac{\partial}{\partial z}\frac{\partial F}{\partial U_{iz}} = 0, \qquad i = 1, 2, \ldots n$$

This is a system of partial differential equations determining a set of functions, $U_i(x, y, z)$, $i = 1, 2, \ldots n$ which will yield an extreme value for the integral in Eq. (12.59).

We should emphasize that whereas in all our considerations of multiple integrals we have indicated integration over the three coordinates, x, y, z this is by no means restrictive. We could include an integral over another variable, say t, or any larger or smaller number of variables; in general *all* variables on which the functions U_i may depend. Furthermore, we emphasize again that these extremum problems can be written in other coordinate systems.

VARIATION OF INTEGRALS SUBJECT TO INTEGRAL CONSTRAINTS,† ISOPERIMETRIC PROBLEMS

Here we consider some integral I over a domain D for which an extremum is to be determined, but in which the functions characterizing the "path" are subject to a constraint having the form of another integral over D. We clarify the nature of the problem by stating an example which also explains the name given to this class of problems, i.e., *isoperimetric*.

Example (3)

Consider a string of length L attached to the x axis at two points, x_1 and x_2, such that

(12.62) $$|x_2 - x_1| < L$$

and we ask: if the string is layed out on the x–y plane, what must be the shape of the curve, $y(x)$, so formed such that the area enclosed between the string and the x axis is a maximum. Here the *perimeter*, $L + |x_2 - x_1|$, is the *same* for *every* curve so formed, regardless of the enclosed area; thus the name, isoperimetric.

For the area we obviously have,

(12.63) $$I = \int_{x_1}^{x_2} y(x)\, dx$$

which is to be made a maximum subject to the constraint,

(12.64) $$J = |x_2 - x_1| + \int_{x_1}^{x_2} \sqrt{1 + \left(\frac{dy}{dx}\right)^2}\, dx = \text{constant}$$

Note that $|x_2 - x_1|$ is a constant, so that *only* the integral need be considered. We will now present the general method of attack on such problems.

General method

We have an integral I over a *fixed domain* which may be viewed as a function of a set of parameters which we write as, $\alpha, \beta, \gamma, \ldots$, thus

(12.65) $$I = I(\alpha, \beta, \gamma \ldots)$$

This is to be made an extremum subject to a constraint given in the form of another integral, J, over the *same* domain and containing the *same* functions. This may also be looked upon as a

---

† There are a variety of other forms of constraint which can be considered, however, we leave these to be covered in courses in Lagrangion mechanics, or formal courses in the calculus of variations.

function of the parameters $\alpha, \beta, \gamma, \ldots$, so that

(12.66)
$$J = J(\alpha, \beta, \gamma \ldots) = \text{constant}$$

is the *constraint*.

If we forget for the moment that I and J are constructed as integrals over coordinates then we see the conditions,

(12.67)
$$\delta I = \frac{\partial I}{\partial \alpha} \delta\alpha + \frac{\partial I}{\partial \beta} \delta\beta + \ldots = 0$$

for the extremum, and,

(12.68)
$$\delta J = \frac{\partial J}{\partial \alpha} \delta\alpha + \frac{\partial J}{\partial \beta} \delta\beta + \ldots = 0$$

for the constraint, as *exactly* equivalent to the problem of the extremum of a function $I(\alpha, \beta, \gamma, \ldots)$ subject to a constraint $J(\alpha, \beta, \gamma, \ldots) = \text{constant}$, just as was treated at the beginning of this chapter. Hence, we can apply the method of *undetermined multipliers* just as we did there, i.e., multiply δJ by λ and subtract from δI to obtain,

(12.69)
$$\delta I - \lambda \, \delta J = 0$$

which when written out appears as,

(12.70)
$$\left(\frac{\partial I}{\partial \alpha} - \lambda \frac{\partial J}{\partial \alpha} \right) \delta\alpha + \left(\frac{\partial I}{\partial \beta} - \lambda \frac{\partial J}{\partial \beta} \right) \delta\beta = 0$$

where we have chosen the *independent* parameters, γ, etc., already so $\delta\gamma = 0$ etc. Now choose the undetermined multiplier such that the first term vanishes, and for $\delta\beta \neq 0$ see that the coefficient of $\delta\beta$ *must* vanish. Hence, we get,

(12.71)
$$\lambda = \frac{\dfrac{\partial I}{\partial \alpha}}{\dfrac{\partial J}{\partial \alpha}} = \frac{\dfrac{\partial I}{\partial \beta}}{\dfrac{\partial J}{\partial \beta}} = \lambda(\alpha, \beta)$$

from the resulting two equations. This defines the relation that must exist between the variational parameters, α and β.

Solution to Example (3)

To solve Example (3) we multiply J, Eq. (12.64), by an undetermined multiplier λ, subtract this from I and form†

(12.72)
$$\delta(I - \lambda J) \to \delta \int_{x_1}^{x_2} \left\{ y - \lambda\sqrt{1 + y'^2} \right\} dx = 0$$

(note that $\delta\lambda = 0$). Then the Euler–Lagrange equation for this integral with

(12.73)
$$F(y, y', x) = y - \lambda\sqrt{1 + y'^2}$$

† Noting that $\delta|x_2 - x_1| \equiv 0$.

is,

(12.74)
$$1 + \lambda \frac{d}{dx} \frac{y'}{\sqrt{1 + y'^2}} = 0$$

where y' denotes dy/dx. Since λ is a constant (with regard to x) this can be integrated once to yield

(12.75)
$$\frac{-y'}{\sqrt{1 + y'^2}} = \frac{x}{\lambda} + C$$

Here, since the location of the two tie points on the axis is arbitrary let them be chosen such that the point where $y' = 0$ is at the origin $x = 0$; then $C = 0$.
Now rearrange to read as,

(12.76)
$$y' = \pm \frac{1}{\lambda} \frac{x}{\sqrt{1 - \frac{x^2}{\lambda^2}}}$$

and this integrates to,

(12.77)
$$\frac{y}{\lambda} = \pm \sqrt{1 - \frac{x^2}{\lambda^2}} + \frac{C'}{\lambda}$$

or

(12.78)
$$(y - C')^2 + x^2 = \lambda^2$$

which shows that the string must be in the form of a circle with center O, C', and radius λ. When we put this form for y back into Eq. (12.64) we get λ in terms of x_1, x_2, and L.

EXECUTE PROBLEM SET (12-1)

VARIATIONAL FORMULATION OF THE EIGENVALUE PROBLEM

The general eigenvalue problem can be viewed as finding those numbers, λ_n, for which the operator O, operating on a function U_n produces the constant λ_n times U_n, where the functions U_n are defined in some domain D and satisfy some boundary conditions on the boundary S of D, i.e.,

(12.79)
$$OU_n = \lambda_n U_n$$

Here we show how this problem can be expressed in the context of the calculus of variations.
Consider an integral,

(12.80)
$$I = \iiint_D F(U, U_x, U_y, U_z, x, y, z) \, dx \, dy \, dz$$

which is to be made an extremum subject to the condition that,

(12.81)
$$J = \iiint_D G(U, x, y, z) \, dx \, dy \, dz$$

is to be a *constant*. By previous treatment of such isoperimetric problems we see that by multi-plying J by λ, subtracting the result from I and constructing the Euler–Lagrange equation we get:

$$(12.82) \qquad \frac{\partial F}{\partial U} - \frac{\partial}{\partial x}\frac{\partial F}{\partial U_x} - \frac{\partial}{\partial y}\frac{\partial F}{\partial U_y} - \frac{\partial}{\partial z}\frac{\partial F}{\partial U_z} = \lambda\frac{\partial G}{\partial U}$$

In order that this should have the form of Eq. (12.79) with λ as λ_n we *take G to be a function of U^2 and x, y, z only.*

Here we should emphasize that while we have written most of our equations in rectangular coordinates *any* set of orthogonal coordinates would be equally suitable and we could use any number of dimensions. Thus, here for the eigenvalue problem we write

$$(12.83) \qquad I = \iiint_D \bar{F}(U, U_\chi, U_\eta, U_\xi, \chi, \eta, \xi)\, d\tau$$

where χ, η, ξ are the coordinates and $d\tau$ is the volume element in these coordinates. (Note that the functional form of F changed to a different form, $\bar{F}$, also.) We also have, in view of our dis-cussion above,

$$(12.84) \qquad J = \iiint_D \bar{G}(U^2, \chi, \eta, \xi)\, d\tau$$

for the constraint J.

Example (4)

As an example of the above formulation we exhibit the Helmholtz equation, which is an eigen-value problem,

$$(12.85) \qquad \nabla^2 U_n = -k_n^2 U_n, \quad \text{in } D$$

as a variational problem.
Let

$$(12.86) \qquad I = \iiint_D |\nabla U|^2\, d\tau$$

and,

$$(12.87) \qquad J = \iiint_D U^2\, d\tau$$

Then with the undetermined multiplier $k_n^2 = \lambda$ we form $I - \lambda J$ and write out the Euler–Lagrange equation. The result is Eq. (12.85) if we just attach the subscript, n, to U. Here we should emphasize again that this formulation applies in any number of dimensions.

VARIATIONAL ESTIMATES OF EIGENVALUES; THE RAYLEIGH–RITZ METHOD

In the section just above we have seen the eigenvalue problem exhibited as the extremization of an integral, I, subject to an integral constraint, J. Thus, it is just an isoperimetric problem. This means that those functions U_n which satisfy Eq. (12.79) above will make I an extremum and satisfy J equals a constant.

Previously in our general discussion of such problems we found that the existence of the constraint imposed the relation, Eq. (12.71), on the derivatives $\partial I/\partial \alpha$, $\partial J/\partial \alpha$, etc., with respect to the two variational parameters, α and β. Here we use this to show that we must also have an *extremum* for the undetermined multiplier λ.

Define

$$(12.88) \qquad\qquad \rho = \frac{I}{J}$$

and note that,

$$(12.89) \qquad\qquad \frac{\partial \rho}{\partial \alpha} = \frac{\dfrac{\partial I}{\partial \alpha} - \dfrac{I}{J}\dfrac{\partial J}{\partial \alpha}}{J}$$

or

$$(12.90) \qquad\qquad \frac{\partial \rho}{\partial \alpha} = \frac{\dfrac{\partial I}{\partial \alpha} - \rho \dfrac{\partial J}{\partial \alpha}}{J}$$

upon using Eq. (12.88) for I/J in Eq. (12.89). In exactly the same way we have,

$$(12.91) \qquad\qquad \frac{\partial \rho}{\partial \beta} = \frac{\dfrac{\partial I}{\partial \beta} - \rho \dfrac{\partial J}{\partial \beta}}{J}$$

Now use Eq. (12.71) to express $\partial J/\partial \alpha$ in terms of $\partial I/\partial \alpha$ and $\partial J/\partial \beta$ in terms of $\partial I/\partial \beta$, thus:

$$(12.92) \qquad \frac{\partial \rho}{\partial \alpha} = \frac{\left(1 - \dfrac{\rho}{\lambda}\right)\dfrac{\partial I}{\partial \alpha}}{J} \quad \text{and} \quad \frac{\partial \rho}{\partial \beta} = \frac{\left(1 - \dfrac{\rho}{\lambda}\right)\dfrac{\partial I}{\partial \beta}}{J}$$

Here we see that if we set $\rho = \lambda$ then,

$$(12.93) \qquad\qquad \frac{\partial \rho}{\partial \alpha} = \frac{\partial \rho}{\partial \beta} = 0$$

Thus, we find that,

$$(12.94) \qquad\qquad \lambda = \frac{I}{J} = \text{extremum}$$

without constraints and is another statement of the eigenvalue problem. This means that the λ_n, the eigenvalues, actually represent relative extremums for the ratio I/J. This can be used to advantage in *estimating* values of λ_n.

We simply choose some trial estimate for the *form* of the function, say U_n, which should correspond to the eigenvalue λ_n. If the trial form were *exactly* the proper form of U_n, then, when inserted into I and J in the above equation, the resulting value would be exactly $\lambda = \lambda_n$. However, our estimated function, denoted by $\tilde{U}_n(x, y, z, \alpha, \beta, \gamma \ldots)$, when inserted into I and J

in this equation will yield

(12.95) $$|\lambda - \lambda_n| > 0$$

for *any* values of the parameters $\alpha, \beta, \gamma, \ldots$ in the trial function.

Our procedure then, due to Rayleigh and Ritz, is to vary α, β, γ etc., as follows. After putting $\tilde{U}_n(x, y, z, \alpha, \beta, \gamma \ldots)$ into Eq. (12.94) we have,

(12.96) $$\lambda = \lambda(\alpha, \beta, \gamma, \ldots)$$

and this must be an *extremum*. Thus we require,

(12.97) $$\frac{\partial \lambda}{\partial \alpha} = 0, \quad \frac{\partial \lambda}{\partial \beta} = 0, \quad \text{etc.}$$

and hence obtain a set of equations for determining the *best* estimates of α, β, γ, etc., so that $\lambda \approx \lambda_n$ when these values are used.

The success or failure of this method depends upon how well we can estimate the proper *form* for U_n. We must in general choose approximating functions which satisfy the boundary conditions on the boundary S of the domain. However, the U_n will *not* satisfy the differential equation.

Example (5)

Estimate the frequency of the lowest vibrational mode of a thin circular membrane clamped at the edge.

The eigenvalue equation is,

(12.98) $$\frac{1}{r} \frac{\partial}{\partial r}\left(r \frac{\partial U_n}{\partial r}\right) + \frac{1}{r^2} \frac{\partial^2 U_n}{\partial \theta^2} = -\frac{\omega_n^2}{c^2} U_n$$

for $0 < r < R, 0 < \theta < 2\pi$, and $U_n = 0$ at $r = R$. This is just the two-dimensional Helmholtz equation in cylindrical coordinates and hence we have,

(12.99) $$\frac{\omega_n^2}{c^2} = \frac{\int_0^{2\pi} \int_0^R \left[\left(\frac{\partial U_n}{\partial r}\right)^2 + \frac{1}{r^2}\left(\frac{\partial U_n}{\partial \theta}\right)^2\right] r \, dr \, d\theta}{\int_0^{2\pi} \int_0^R U_n^2 r \, dr \, d\theta}$$

as the equivalent of Eq. (12.94)

Now the lowest mode of vibration must be that with the fewest nodes, i.e., the membrane moves essentially as a whole. Therefore we take as our estimate of U_0, the U_n of the lowest mode,

(12.100) $$\tilde{U}_0 = \{A(R^2 - r^2) + B(R^2 - r^2)^2\}$$

with A and B being constants. Inserting this into Eq. (12.99) above we obtain,

(12.101)
$$\left(\frac{\omega_0 R}{c}\right)^2 \left\{A^2 \int_0^1 (1 - \xi^2)^2 \xi \, d\xi + 2AB \int_0^1 (1 - \xi^2)^3 \xi \, d\xi \right.$$
$$\left. + B^2 \int_0^1 (1 - \xi^2)^4 \xi \, d\xi \right\} = A^2 \int_0^1 \left[\frac{d(1 - \xi^2)}{d\xi}\right]^2 \xi \, d\xi$$
$$+ 2AB \int_0^1 \frac{d(1 - \xi^2)}{d\xi} \frac{d(1 - \xi^2)^2}{d\xi} \xi \, d\xi + B^2 \int_0^1 \left[\frac{d(1 - \xi^2)^2}{d\xi}\right]^2 \xi \, d\xi$$

after a change of variables and rearrangement. We then evaluate these integrals and take the derivatives with respect to A and B of the resulting equation and set each equal to zero; the result is the pair of simultaneous equations for A and B:

(12.102)
$$
\begin{cases}
A\left[\dfrac{1}{6}\left(\dfrac{\omega_0 R}{c}\right)^2 - 1\right] + B\left[\dfrac{1}{8}\left(\dfrac{\omega_0 R}{c}\right)^2 - \dfrac{2}{3}\right] = 0 \\[3mm]
A\left[\dfrac{1}{8}\left(\dfrac{\omega_0 R}{c}\right)^2 - \dfrac{2}{3}\right] + B\left[\dfrac{1}{10}\left(\dfrac{\omega_0 R}{c}\right)^2 - \dfrac{2}{3}\right] = 0
\end{cases}
$$

These have solutions only if the determinant of coefficients vanishes. Thus

(12.103)
$$
\begin{vmatrix}
\dfrac{1}{6}\left(\dfrac{\omega_0 R}{c}\right)^2 - 1 & \dfrac{1}{8}\left(\dfrac{\omega_0 R}{c}\right)^2 - \dfrac{2}{3} \\[3mm]
\dfrac{1}{8}\left(\dfrac{\omega_0 R}{c}\right)^2 - \dfrac{2}{3} & \dfrac{1}{10}\left(\dfrac{\omega_0 R}{c}\right)^2 - \dfrac{2}{3}
\end{vmatrix} = 0
$$

which is a quadratic equation for $(\omega_0 R/c)^2$. This has two solutions. The *smaller* root is,

(12.104)
$$
\left(\frac{\omega_0 R}{c}\right)^2 = 5.7837
$$

which is *very* close to the exact value of 5.7832.

EXECUTE PROBLEM SET (12-2)

VARIABLE END POINTS; TRANSVERSALS

In all our considerations of the conditions for an extremum of an integral thus far we have always assumed that the extremum path, C, and the varied, or neighbor, path, C', had the same end points, or "boundary conditions." Here, we relax this condition and treat the problem of the extremum of a single integral having *variable end points*.

Here, as in the introduction of other topics, it is an advantage to see a physical problem formulated calling for the extremum of an integral with *variable end points* before we examine the mathematical details of the treatment of such problems.

Example (6)

We consider the problem of determining the shape of a liquid surface, in contact with its vapor and air, taking account of *gravity* and the thermodynamic (Helmholtz) free energy of the various interfaces involved. We choose a problem of *axial symmetry* because we restrict our considerations here to *single integrals*.

We have as our example a liquid in a vertical glass vessel, having axial symmetry as in Fig. 12.2. The *equilibrium condition* for the shape of the air–liquid interface is that the total free energy of the *whole* system be a minimum subject to the *constraint* of constant volume.

Let:

(12.105)
$$
z = \xi_l(r)
$$

be the equation giving the depth to the liquid surface at r, and

(12.106)
$$
z = \xi_s(r)
$$

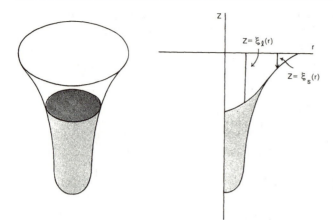

FIGURE 12.2. Geometry of a Liquid in a Vertical Vessel Having Axial Symmetry.

be that for the wall of the containing vessel. Let γ_{as}, γ_{al}, γ_{sl}, be the respective free energies per unit area for air–solid, air–liquid, and solid–liquid interfaces. Let ρg be the weight per unit volume of liquid and neglect that of the air. Then the total free energy of the system is,

$$I = \int_0^{r_\omega} \left\{ \gamma_{sl}\sqrt{1 + \xi_s'^2} + \gamma_{al}\sqrt{1 + \xi_l'^2} + \tfrac{1}{2}\rho g(\xi_l^2 - \xi_s^2) \right\} 2\pi r \, dr + \int_{r_\omega}^{r_m} \gamma_{as}\sqrt{1 + \xi_s'^2} \, 2\pi r \, dr$$

(12.107)

where primes denote d/dr, and the liquid volume is,

(12.108)
$$J = \int_0^{r_\omega} (\xi_l - \xi_s) 2\pi r \, dr$$

Thus, we wish to *extremize* I with J as a *constraint* by varying the *form* of the function, $z = \xi_l(r)$. However, here the *end point* $(r = r_\omega, z = z_\omega)$ at which the interface contacts the solid wall *is not fixed*. Note though that the end point is *not* free to move *completely* arbitrarily, it must move only along the curve defined by $z = \xi_s(r)$, i.e., the wall of the vessel.

The fact that I, as given above, is the sum of two integrals does not materially alter the method here. Also the fact that we have a *constraint* in our example should not confuse the student because this simply calls for the use of a Lagrangian undetermined multiplier. Thus, let us now look at such problems in a general way.

GENERAL METHOD FOR VARIABLE END POINTS

Given an integral, (with $y' = dy/dx$),

(12.109)
$$I = \int_{x_1}^{x_2} F[y(x), y'(x), x] \, dx$$

which is to be an extremum subject to the auxiliary conditions that the end points, x_1, x_2, must lie on the curves,

(12.110)
$$\begin{cases} y = y_1(x), & \text{for } x_1 \\ y = y_2(x), & \text{for } x_2 \end{cases}$$

which are called the *transversals*.

We carry out our variations of I in much the same way we always do; we form $F(y + \delta y, y' + \delta y', x)$ as before, but now also with $x_2 \to x_2 + \delta x_2$ and $x_1 \to x_1 + \delta x_1$, so that the varied integral is,

$$(12.111) \qquad I' = \int_{x_1 + \delta x_1}^{x_2 + \delta x_2} F(y + \delta y, y' + \delta y', x)\, dx$$

Then we require,

$$(12.112) \qquad \delta I = I' - I = 0$$

to *first order* in all the quantities, δy, $\delta y'$, δx_1, and δx_2. Using a Taylor expansion as usual we get,

$$(12.113) \qquad \delta I = F_2\, \delta x_2 - F_1\, \delta x_1 + \int_{x_1}^{x_2} \left(\frac{\partial F}{\partial y}\, \delta y + \frac{\partial F}{\partial y'}\, \delta y' \right) dx$$

where the subscripts on F, F_2, and F_1, indicate that the integrand, $F[y(x), y(x), x]$ is to be evaluated at x_2 and x_1, respectively.

Performing the usual integration by parts on the term in $\delta y'$ in the integral we get:

$$(12.114) \quad \delta I = F_2\, \delta x_2 + \left(\frac{\partial F}{\partial y'}\, \delta y \right)_2 - F_1\, \delta x_1 - \left(\frac{\partial F}{\partial y'}\, \delta y \right)_1 + \int_{x_1}^{x_2} \left(\frac{\partial F}{\partial y} - \frac{d}{dx}\frac{\partial F}{\partial y'} \right) \delta y\, dx$$

where now we have additional terms to be evaluated at the end points.

Recall that the *end points* can only move along the transversals. Thus, using Eq. (12.110) we see that at x_2 *the only possible change in y is to first order*,

$$(12.115) \qquad (\delta y)_{x_2} = \left(\frac{dy_2(x_2)}{dx} - \frac{dy\,(x_2)}{dx} \right) \delta x_2$$

and similarly at x_1. Thus we have,

$$(12.116) \quad \delta I = \left(F + \frac{\partial F}{\partial y'}(y_2' - y') \right)_{x_2} \delta x_2 - \left(F + \frac{\partial F}{\partial y'}(y_1' - y') \right)_{x_1} \delta x_1 + \int_{x_1}^{x_2} \left(\frac{\partial F}{\partial y} - \frac{d}{dx}\frac{\partial F}{\partial y'} \right) \delta y\, dx$$

as our final form for δI. Now the variations δx_1 and δx_2 are here independent; also the *form* of the variation $\delta y(x)$ is independent of these. Hence, since δI must vanish for any δx_1, and δx_2 and any $\delta y(x)$ we see that we must have:

$$(12.117) \qquad \begin{cases} \left(F + \dfrac{\partial F}{\partial y'}(y_2' - y') \right)_{x_2} = 0 \\[2mm] \left(F + \dfrac{\partial F}{\partial y'}(y_1' - y') \right)_{x_1} = 0 \end{cases}$$

and our usual Euler–Lagrange equation

$$(12.118) \qquad \frac{\partial F}{\partial y} - \frac{d}{dx}\frac{\partial F}{\partial y'} = 0$$

In Eqs. (12.117) and (12.118) we must remember that it is the function $y(x)$ which appears, along with its derivative, $y'(x)$, in F. The transversals, $y_1(x)$ and $y_2(x)$, enter only as explicitly indicated in Eq. (12.117).

Here the Euler–Lagrange equation, Eq. (12.118), provides a differential equation which $y(x)$ must satisfy to extremize I. Equations (12.117) provide *special boundary conditions* on $y(x)$, which are also required to extremize the integral I.

EXECUTE PROBLEM SET (12-3)

NONLINEAR ORDINARY DIFFERENTIAL EQUATIONS

A great number of problems in physics and engineering lead to nonlinear differential equations. Recall that for a *linear* differential equation we have the property that if U_1 and U_2 are solutions of the equation then $aU_1 + bU_2$ is also a solution of the equation, where a and b are *any* constants. Any differential equation *not* having this property is *nonlinear*.

Our Example (3) of this chapter led to a nonlinear differential equation, Eq. (12.74). We were able to effect the integration of that equation but usually nonlinear differential equations are not so tractable. In the following sections we present a few methods which assist us in analyzing nonlinear ordinary differential equations, and in some cases obtaining approximate solutions.

SECOND-ORDER EQUATIONS AND PHASE-PLANE DIAGRAMS

The general second-order ordinary differential equation has the form of a function (primes denote derivatives with respect to x)

$$(12.119) \qquad F(y'', y', y, x) = 0$$

where in general this is *not* a linear combination of y, y' and y''. It is convenient, particularly for nonlinear equations, to rewrite this as a *pair* of first order equations by the definition,

$$(12.120) \qquad y' = U(x)$$

so that if we solve Eq. (12.119) for $y'' = dU/dx$ *explicitly* we have the form

$$(12.121) \qquad U' = f(U, y, x)$$

where f is a function of the indicated arguments. In many cases we can effect the integration (once) of the later equation.

Example (7)

The problem of a mass attached to *nonlinear spring* leads to the equation, known as the Duffing equation,

$$(12.122) \qquad m\frac{d^2y}{dt^2} = -ay + by^3$$

where for b positive we call it a "soft-spring problem," while for b negative we call it a "hard-spring" problem. Define

$$(12.123) \qquad U = \frac{dy}{dt}$$

and have

(12.124)
$$\frac{dU}{dt} = -\frac{a}{m}y + \frac{b}{m}y^3$$

Then multiply both sides by U, and we *can* integrate once to give,

(12.125)
$$\frac{1}{2}U^2 = -\frac{a}{2m}y^2 + \frac{b}{4m}y^4 + C$$

where C is a constant of integration fixed by $U(0)$, $y(0)$.

The above example illustrates the general characteristic of Eq. (12.121) which exists when the *independent variable is absent* in the differential equation. That is, it is then often possible to execute *one* integration and hence obtain U as a function of y.

Occasionally, if the independent variable does appear, a change of variables, both dependent and independent, or a different definition of U as a function of y', i.e., other than that in Eq. (12.120), will achieve the same result, so that a quantity can be expressed as a function of y, i.e., *one* integration is possible.

We call the U, y plane the phase plane and the curves, $U = U(y)$, the *trajectories* of the solution in the phase plane. These trajectories tell us a lot about the behavior of the solutions, even when we can not effect the second and final integration to obtain $y(x)$. Furthermore, these trajectories, or phase-plane diagrams are a guide to the construction of *approximate* solutions.

Continuation of Example (7)

For example consider the phase-plane diagram for the *soft-spring problem* given in Example (7) above. We have $U = U(y)$ as in Eq. (12.125) where we now let $\omega^2 = a/m$, $\varepsilon = b/2m$, and $c = 2C$, so that we have

(12.126)
$$U = \pm \sqrt{-\omega^2 y^2 + \varepsilon y^4 + c}$$

We then obtain a *family* of curves by assigning different values for c, one curve for each value of c. Such a family of curves is shown in Fig. 12.3. The *arrows* on the curves are constructed to indicate the direction of the "motion" of the system point as the solution, corresponding to a given value of c, evolves.

Keep in mind that the value of c is fixed by the initial values, $U(0)$ and $y(0)$, at "time" $t = 0$. Also observe that from the pair of Eqs. (12.123) and (12.124), for dy/dt and dU/dt, we can determine at *any point*, (y, U), in the phase plane, whether y is increasing or decreasing, and similarly for U. Thus, we find which way the arrows should point on any trajectory.

Note that the *heavily lined trajectories* passing through the two points $(-\omega^2/2\varepsilon, 0)$ and $(+\omega^2/2\varepsilon, 0)$ *separate regions* in which the arrows on all trajectories are parallel. These particular trajectories are the *separatrices* and separate the regions in which the solutions have a different general character.

Also note the *closed orbits* formed by the trajectories near the origin inside the bounding separatrices. Such closed orbits correspond to *periodic solutions* of the differential equation. From this then we see the criterion for periodic solutions as,

(12.127)
$$y^2 < \frac{\omega^2}{2\varepsilon}, \quad \text{for} \quad U = 0$$

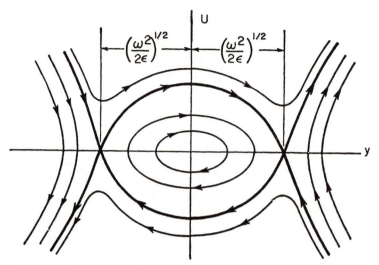

FIGURE 12.3.   Phase Plane Diagram for the Soft Spring Problem.
From "Nonlinear Differential Equations" by R. A. Struble, copyright 1961 by
McGraw-Hill Book Co. Used by permission of McGraw-Hill Book Co.

or, in terms of the initial conditions, [i.e., evaluate c with $U = U_0$, $y = y_0$ in Eq. (12.126)],

$$(12.128) \qquad U_0^2 - \frac{\omega^4}{4\epsilon} + \omega^2 y_0^2 - \epsilon y_0^4 < 0$$

This example illustrates the important characteristic of *separatrices*: solution trajectories never *cross* these curves. The points of *intersection of separatrices* are the *critical points* of the equation and represent points of *instability*. Were a critical point selected as an initial point there could be no criterion for defining the trajectory to follow.

It is sometimes convenient to define the *direction field* for a nonlinear differential equation in a manner similar to that outlined above, i.e., the *direction of the trajectories* in the phase plane can be specified even when the first integration, as used above to obtain $U = U(y)$, is not possible. Note that we have the *forms* in our example above

$$(12.129) \qquad \frac{dy}{dt} = U$$

and

$$(12.130) \qquad \frac{dU}{dt} = f(U, y)$$

so that we can form,

$$(12.131) \qquad \frac{dU}{dy} = \frac{f(U, y)}{U}$$

even though this may not be an exact differential form.

Here, if on the right in these three equations we assign values (y, U) for *many* points in the phase plane, we can construct a small *arrow* at each such point to indicate the *direction* of the trajectory at that point, i.e., $\tan \theta = dU/dy$ is the tangent of the inclination of the arrow to the y axis and the *sense* is determined by the *signs* of y' and U'.

DAMPED SYSTEMS; LIMIT CYCLES

The nonlinear differential equation considered in Example (7) above did not contain the first derivative, y', i.e., in $F(y'', y', y, x)$ as exemplified in Eq. (12.122). The presence of first derivatives introduces *damping* and in certain cases gives rise to *limit cycles*. Again we illustrate these concepts with examples.

Example (8)

The simple pendulum with Newtonian damping is described by the equation,

(12.132)
$$ml\frac{d^2\theta}{dt^2} + \alpha l^2 \frac{d\theta}{dt}\left|\frac{d\theta}{dt}\right| + \frac{mg}{l}\sin\theta = 0$$

where m is the mass of the pendulum bob, l is the length of the pendulum, g is the acceleration of gravity, α a damping constant, and θ is the angular displacement of the pendulum. By suitable change of symbols this appears as

(12.133)
$$\frac{d^2\theta}{dt^2} + k\frac{d\theta}{dt}\left|\frac{d\theta}{dt}\right| + \omega^2 \sin\theta = 0$$

The damping term,

$$k\left|\frac{d\theta}{dt}\right|\frac{d\theta}{dt}$$

has magnitude $(d\theta/dt)^2$ but always the proper sign to *oppose* the motion, acceleration, of the pendulum; for this to be so we have $k > 0$.

The phase-plane diagram of Eq. (12.131) is formed by first defining,

(12.134)
$$U = \frac{d\theta}{dt}$$

then the direction field is given by,

(12.135)
$$\frac{dU}{d\theta} = \frac{-\omega^2 \sin\theta + kU|U|}{U}$$

or,

(12.136)
$$\frac{dU^2}{d\theta} \pm 2kU^2 = -2\omega^2 \sin\theta$$

where the plus sign applies for $U < 0$ and the minus for $U > 0$. This equation is a *linear inhomogeneous* equation in U^2 and can be readily integrated. Then the phase-plane diagram of the trajectories can be drawn as in Fig. 12.4.

The important characteristics of this diagram are the *limit points*. No matter at what point in the phase plane the system point starts it must approach one of the points $\theta = 2n\pi$, $n = 0$, $\pm 1, \pm 2, \ldots$ as a limit for $t \to \infty$. Physically this is as one would expect for a damped pendulum, it must come to rest hanging vertically *downward*. We also note the *critical points* of instability at $\theta = \pm(2n + 1)\pi$, $n = 0, 1, 2, \ldots$, where the *separatrices cross*. These points correspond to

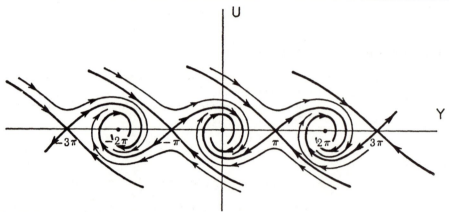

FIGURE 12.4.   Phase Plane Diagram for the Simple Pendulum with Newtonian Damping.
From "Nonlinear Differential Equations" by R. A. Struble, copyright 1961 by McGraw-Hill Book Co.
Used by permission of McGraw-Hill Book Co.

the pendulum being at rest vertically upward, i.e., on the "top of a swing" with zero speed.
Obviously these critical points are points of *physical instability*.

Observe the trajectories, starting from *any* point, spiral in asymptotically to a limit point,
which point simply indicating the number of *complete* revolutions made by the pendulum. In
other types of damped systems the limit that is approached is not a point but a *closed curve* we
call a *limit cycle*. This is shown in the following example.

Example (9)

The *Van der Pol equation* is a classic example of an equation having a *limit cycle*. This equation
has the general form,

(12.137)
$$\frac{d^2y}{dt^2} + \alpha(y^2 - 1)\frac{dy}{dt} + y = 0$$

where α is a positive constant.

This equation is one for which we find it convenient to define U somewhat differently than
we have in previous examples. This possibility was pointed out in our first discussion of the
phase plane. Here we use the definition,

(12.138)
$$U = \frac{dy}{dt} - \alpha\left(y - \frac{y^3}{3}\right)$$

then

(12.139)
$$\frac{dU}{dt} = -y$$

Thus U is still *linear* in dy/dt, but differs considerably from just dy/dt as in previous problems.
The direction field in this U, y space is now given by,

(12.140)
$$\frac{dU}{dy} = -\frac{y}{U + \alpha\left(y - \frac{y^3}{3}\right)}$$

For a detailed description of the construction and analysis of the trajectories from these equations the student is referred to the references. Here we simply show Fig. 12.5.

Observe that every trajectory in this, our regular dy/dt, y space, spirals in, or *out* to the limiting closed curve as indicated. Thus a system described by an equation of this type *always* approaches a periodic motion asymptotically. This curve is called the *limit cycle*.

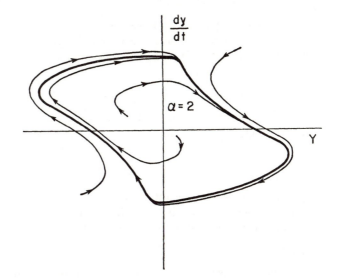

FIGURE 12.5.   Phase Plane Diagram for the Van der Pol Equation Showing the Limit Cycle.
From "Nonlinear Differential Equations" by R. A. Struble, copyright 1961 by McGraw-Hill Book Co. Used by permission of McGraw-Hill Book Co.

PERTURBATION METHODS FOR ORDINARY DIFFERENTIAL EQUATIONS

Here we will develop the general concepts of perturbation methods for ordinary nonlinear differential equations of second order by considering the general equation,

$$(12.141) \qquad F(y'', y', y, x, \varepsilon) = 0$$

to contain some small parameter, ε, such that the equation is *linear* in the limit as $\varepsilon \to 0$. As we will see it is this limiting linearity which makes possible the perturbation process.

We will only be interested in *small* values of ε so we could expand Eq. (12.141) in a Taylor series about $\varepsilon = 0$ as,

$$(12.142) \qquad F(y'', y', y, x, 0) + \varepsilon \frac{\partial F}{\partial \varepsilon}(y'', y', y, 0) + \frac{\varepsilon^2}{2!}\frac{\partial^2 F}{\partial \varepsilon^2}(y'', y', y, 0) + \ldots = 0$$

and in order that the equation be *second order* as $\varepsilon \to 0$ it is *necessary* that $F(y'', y', y, x, 0)$ contain y'', and as stated above, we also require

$$(12.143) \qquad F(y'', y', y, x, 0) = 0$$

to be a *linear differential equation.*

The *essence* of the perturbation method is just this: we assume a solution to Eq. (12.142) in the form of a power series in ε, having functions of x as coefficients, thus,

$$(12.144) \qquad y(x) = y_0(x) + \varepsilon y_1(x) + \varepsilon^2 y_2(x) + \ldots$$

This is then substituted into the differential equation, Eq. (12.142). In order that this should then equal zero as indicated, for *any value of* ε, it is necessary that the *coefficient* of each power of ε vanish separately. This gives rise to a set of differential equations which can usually be solved in a *sequential* manner.

The case of periodic, or limit cycle solutions, requires a slight variation of the above procedure which will be demonstrated in an example shortly. Also the manner in which the *initial conditions* of a problem are to be treated in the perturbation is best made clear in an example.

From what we have said above it is obvious that the terminology, *perturbation method*, stems from the fact that the presence of the small parameter ε in the equation creates a *perturbation* of the solution from its form corresponding to $\varepsilon = 0$.

Example (10)

We consider the problem of the soft spring treated already in Example (7). The differential equation is here written in the form,

(12.145)
$$\frac{d^2 y}{dt^2} + \omega^2 y - 2\varepsilon y^3 = 0$$

and we here consider only the *nonperiodic* motion. As already noted this corresponds to certain constraints on the initial conditions in relation to the magnitude of ω^2/ε.

Assume a solution for this equation of the form,

(12.146)
$$y(t) = y_0(t) + \varepsilon y_1(t) + \varepsilon^2 y_2(t) + \ldots$$

Here we note that, since this must satisfy the initial conditions $y(0)$, $y'(0)$ at $t = 0$ for all values of ε, we must require that $y_0(0)$, $y_0'(0)$ take on these values at $t = 0$, while all other $y_j(0)$, $y_j'(0)$ must *vanish* at $t = 0$.

Now substitute into Eq. (12.145) and with some rearrangement obtain,

(12.147)
$$(y_0'' + \omega^2 y_0) + \varepsilon(y_1'' + \omega^2 y_1 - 2y_0^3) + \varepsilon^2(y_2'' + \omega^2 y_2 - 6y_0 y_1)$$
$$+ \varepsilon^3[y_3'' + \omega^2 y_3 - 6(y_0 y_2 + y_0 y_1^2)] \ldots = 0$$

Hence we must then have,

(12.148)
$$\begin{cases} y_0'' + \omega^2 y_0 = 0 \\ y_1'' + \omega^2 y_1 = 2y_0^3 \\ y_2'' + \omega^2 y_2 = 6y_0 y_1 \\ \vdots \end{cases}$$

where $y_0(0)$ and $y_0'(0)$ are specified numbers, the initial values of y and y', while we require,

(12.149)
$$y_j(0) = y_j'(0) = 0, \quad j > 0$$

as described above. Bear in mind that since we are considering only *nonperiodic motion* and also *small* ε, we must impose the condition stated in Eq. (12.128) with the inequality *reversed*, i.e.,

(12.150)
$$[y_0'(0)]^2 - \frac{\omega^4}{4\varepsilon} + \omega^2[y_0(0)]^2 - \varepsilon[y_0(0)]^4 > 0$$

in the present notation.

With those restrictions we see that the *system* of Eq. (12.148), is a *linear* inhomogeneous system which can be integrated *sequentially*. To be specific we pick the boundary conditions,

(12.151)
$$\begin{cases} y_0'(0) = 0 \\ y_0(0) = A, \qquad A^2 > \dfrac{\omega^2}{2\varepsilon} \end{cases}$$

which is consistent with Eq. (12.150) for nonperiodic motion.

The integral of the first equation in the set of Eq. (12.148) is then,

(12.152)
$$y_0 = A \cos \omega t$$

which inserted into the second equation yields,

(12.153)
$$y_1'' + \omega^2 y_1 = A^3 \cos^3 \omega t$$

The solution of this equation will be the general solution of the homogeneous equation,

(12.154)
$$y_{1H} = a_1 \sin \omega t + b_1 \cos \omega t$$

plus a particular integral. To obtain this, first recall De Moivre's theorem to write,

(12.155)
$$(\cos 3\omega t + i \sin 3\omega t) = (\cos \omega t + i \sin \omega t)^3$$

and obtain,

(12.156)
$$\cos^3 \omega t = \frac{\cos 3\omega t + 3 \cos \omega t}{4}$$

Then we construct a *particular integral* by assuming the linear form,

(12.157)
$$y_{1P} = \frac{\alpha}{\omega^2} \cos 3\omega t + \frac{\beta}{\omega^2} \cos \omega t$$

for y_1, substituting into Eq. (12.153), with $\cos^3 \omega t$ replaced by its equivalent, Eq. (12.156), and equating coefficients of $\cos 3\omega t$ and $\cos \omega t$ on the two sides of the resulting equations. This yields,

(12.158)
$$y_{1P} = \frac{-A^3}{16\omega^2} \cos 3\omega t$$

and hence

(12.159)
$$y_1 = a_1 \sin \omega t + b_1 \cos \omega t - \frac{A^3}{16\omega^2} \cos 3\omega t$$

Now we determine a_1 and b_1 by the conditions of Eq. (12.149). Thus we have the two equations,

(12.160)
$$\begin{cases} 0 = b_1 - \dfrac{A^3}{16\omega^2} \\ 0 = a_1 \end{cases}$$

and hence finally,

(12.161)
$$y_1(t) = \frac{A^3}{16\omega^2}(\cos \omega t - \cos 3\omega t)$$

Now this solution can be inserted, along with the solution for y_0 on the right in the third equation in the set of Eqs. (12.148). The successive integration of the sequence can be continued to whatever degree is desired. In this way we construct an approximate solution to the non-linear equation. Obviously, one effective tool for integrating the successive linear inhomogeneous equations is the Laplace transform. This could have been used above.

Example (11)

Periodic motion—for free oscillation of the soft spring. We have already seen in our phase-plane diagram that periodic motions of the system do exist. However, the factor, ω, appearing in the differential equation, now used in the form Eq. (12.145), is *not* the actual angular frequency except in the limiting case of $\varepsilon = 0$.

Now if a periodic solution for $y(t)$ does exist then

(12.162) $$y(t + nT) = y(t), \qquad n = 0, 1, 2, \ldots$$

defines the periodic character of $y(t)$ and specifies the *period* as T. In general we expect T to depend on *all* parameters in the differential equation and also possibly on the initial conditions of the problem. We note that as $\varepsilon \to 0$, $T \to 2\pi/\omega$ from the form of Eq. (12.145).

Define the new angular frequency

(12.163) $$\bar{\omega} = \bar{\omega}(\varepsilon) = \omega + f_1\varepsilon + f_2\varepsilon^2 + \ldots$$

as the natural frequency of the system when $\varepsilon \neq 0$. Here the f_j are constants. Then introduce the new variable,

(12.164) $$\tau = \bar{\omega}t$$

so that $y(\tau)$ has period 2π even for $\varepsilon \neq 0$. Substituting this change from t to τ in Eq. (12.145) gives,

(12.165) $$\bar{\omega}^2 \frac{d^2y}{d\tau^2} + \omega^2 y - 2\varepsilon y^3 = 0$$

We now require a solution in the series form in ε as in Eq. (12.146) as before, but we also insert the expansion of $\bar{\omega}^2$, Eq. (12.163), into our differential equation.

Note that now since we are going to *insist* that $y(\tau)$ *have period* 2π we have only *one* other condition that can be fixed. We choose this as the initial value of $y(\tau)$,

(12.166) $$y(0) = A, \qquad |A| < \frac{\omega^2}{2\varepsilon}$$

where the inequality is required to assure *periodic* motion as noted in our phase-plane discussion.

Now inserting Eq. (12.163) for $\bar{\omega}$ and Eq. (12.146) for y, but now with *all* y_j having τ as *argument* into Eq. (12.145) we get, upon equating the coefficient of each power of ε to zero separately:

(12.167)
$$\begin{cases} y_0'' + y_0 = 0 \\ y_1'' + y_1 = [-2\omega f_1 y_0'' + y_0^3]\,\omega^{-2} \\ y_2'' + y_2 = [-2\omega f_1 y_1'' - (2\omega f_2 + f_1^2)y_0'' + 6y_0 y_1]\,\omega^{-2} \\ \quad \vdots \\ \vdots \end{cases}$$

Solving the first equation here subject to $y_0(0) = A$ gives

(12.168) $$y_0 = A \cos \tau + B_1 \sin \tau$$

where B_1 is *not* determined. This is then substituted in the right member of the next equation to give:

(12.169) $$y_1'' + y_1 = [-2\omega f_1(A \cos \tau + B_1 \sin \tau) + (A \cos \tau + B_1 \sin \tau)^3] \omega^{-2}$$

Here we expand the cubic terms and write all factors such as $\cos^3 \tau$ etc., in terms of *first* powers of $\cos 3\tau$ etc., as we did above. We obtain:

(12.170)
$$y_1'' + y_1 = \left[\left\{ \frac{A^3 - 3AB_1^2}{4} \cos 3\tau - \frac{B_1^3 - 3A^2 B_1}{4} \sin 3\tau \right\} + \{[\tfrac{3}{4}(A^3 - 3AB_1^2) - 2\omega f_1 A\} \cos \tau \right.$$
$$\left. + \{\tfrac{3}{4}(B_1^3 - 3A^2 B_1) - 2f_1 B_1\} \sin \tau + 3(AB_1^2 + A^2 B_1) \right] \omega^{-2}$$

At this point we impose our *periodicity* condition by noting that the terms in $\sin \tau$ and $\cos \tau$ will give rise to *nonperiodic* terms in the solution for y_1. This differential equation is just like that for a *forced oscillator* and, as we should recognize, an oscillator driven at its natural frequency gives rise to resonance, i.e., the terms in $\sin \tau$ and $\cos \tau$ *must* be eliminated to preserve periodicity. We do this by *choosing* B_1 and f_1, i.e., require the coefficients of $\sin \tau$ and $\cos \tau$ to vanish,

(12.171)
$$\begin{cases} \tfrac{3}{4}(A^3 - 3AB_1^2) - 2\omega f_1 A = 0 \\ \tfrac{3}{4}(B_1^3 - 3A^2 B_1) - 2\omega f_1 B_1 = 0 \end{cases}$$

This gives $B_1 = \pm A, 0$ and $f_1 = -3A^2/4\omega$ for any of the three values of B_1.

The process can be continued to any desired number of terms. Note that now y_0 is completely fixed as is f_1. When we integrate the y_1 equation, after eliminating the $\sin \tau$ and $\cos \tau$ terms on the right we will require the integral to be consistent with $y(0) = A$, and hence will again have another undetermined factor B_2, like B_1 above, enter the problem. This factor, along with f_2, will be fixed when we eliminate the resonance producing terms, $\sin \tau$ and $\cos \tau$ from the right number of the y_2 equation, etc.

Note that we now have to *first order* in ε,

(12.172) $$\bar{\omega} = \omega - \frac{3A^2}{4\omega} \varepsilon$$

so we see that the natural frequency depends on the *amplitude A*.

Example (12)

Forced motions of the soft spring. If a forcing term, say $F \sin \omega_f t$, is included in the equation of motion then the differential equation has the form

(12.173) $$m \frac{d^2 y}{dt^2} = -ay + by^3 + F \sin \omega_f t$$

Here the perturbation method for either periodic or nonperiodic motion is not materially altered. For the nonperiodic case the only significant change is in the zero-order differential equation, i.e., the first equation in Eq. (12.148). There the right member will now be $F/m \sin \omega_f t$ instead of zero but the procedure goes forward in the same manner. We obtain the solution to the now *inhomogeneous* equation for y_0, put this into the y_1 equation, etc. For the periodic case the changes are more elaborate and will not be developed here.

EXECUTE PROBLEM SET (12-5)

VARIATIONAL SOLUTION OF NONLINEAR DIFFERENTIAL EQUATIONS; AN APPROXIMATION METHOD

Earlier we have seen how a differential equation arises as the Euler–Lagrange equation of some variational condition on an integral. In the Rayleigh–Ritz method this fact was exploited to obtain estimates of eigenvalues of a differential operator. Although we did not emphasize the point there this same type of procedure generates an approximation to the solution of the differential equation.

We formulate the general idea here for emphasis although in our Example (5) of this chapter, Eq. (12.100) with the value of α we found, $\tilde{U}_0(r)$, is an approximate solution of the boundary value problem there.

Thus if we have some extremum condition,

$$(12.174) \qquad \delta I = \delta \iint F(U, U_x, U_y, x, y)\, dx\, dy$$

whose Euler–Lagrange equation is the differential equation whose solution we seek, subject to given boundary conditions, then we assume a trial function,

$$(12.175) \qquad \tilde{U} = \tilde{U}(\alpha, \beta, \ldots, x, y)$$

which *satisfies the boundary conditions* and contains some arbitrary parameters, α, β, etc. We put this function into the integrand of I and execute the integration over the independent variables, x and y in this case chosen as our example. Then we have $I(\alpha, \beta, \ldots)$ which we extremize as,

$$(12.176) \qquad \frac{\partial I}{\partial \alpha} = 0, \qquad \frac{\partial I}{\partial \beta} = 0, \quad \text{etc.}$$

thus providing a set of equations to solve simultaneously for the parameters α, β, δ, etc. If constraints are imposed we make the obvious modification of using say the Rayleigh–Ritz method.

PERTURBATION METHODS FOR THE EIGENVALUE PROBLEM

This problem is one that is basic to modern quantum theory, but also has wide applications in other areas. We use the notation and context of quantum mechanics[†] here.

We have given an operator H and seek the eigenfunctions ψ and their corresponding eigenvalues E, i.e.,

$$(12.177) \qquad H\psi = E\psi$$

where in our special case H is the *sum* of two operators H_0 and $\lambda \mathcal{H}$ where λ is a *small* parameter.

We then expand *both* ψ and E as power series in λ,

$$(12.178) \qquad \begin{cases} \psi = \psi_0 + \lambda\psi_1 + \lambda^2\psi_2 + \ldots \\ E = E_0 + \lambda E_1 + \lambda^2 E_2 + \ldots \end{cases}$$

[†] The operator H below is a *complex* differential operator in general but it is required to be *Hermitian* in that it must have *real* eigenvalues, although *complex* eigenfunctions. However, all of this discussion remains valid for H and all other quantities real.

(Note the similarity here to the treatment of the frequency in the perturbation treatment of the soft spring.) Putting these into Eq. (12.178) with $H = H_0 + \lambda \mathcal{H}$ as stated, and equating corresponding coefficients of λ^n yields

(12.179)
$$\begin{cases} H_0 \psi_0 = E_0 \psi_0 \\ H_0 \psi_1 + \mathcal{H} \psi_0 = E_0 \psi_1 + E_1 \psi_0 \\ H_0 \psi_2 + \mathcal{H} \psi_1 = E_0 \psi_2 + E_1 \psi_1 + E_2 \psi_0 \\ \vdots \end{cases}$$

We suppose H_0 to be a linear operator having a *complete set of orthonormal* solutions $U_n(\mathbf{r})$, corresponding to the *eigenvalues*, $\varepsilon_n, n = 1, 2, \ldots$

(12.180)
$$H_0 U_n = \varepsilon_n U_n, \qquad \mathbf{r} \text{ in } D$$

and†

(12.181)
$$\int_D U_n^* U_m \, d\tau = \delta_{nm} = \begin{cases} 1, & n = m \\ 0, & n \neq m \end{cases}$$

where $d\tau$ is the volume element in the space over which the $U_n(\mathbf{r})$ are defined.

Now we expand ψ_1 as a series in these U_n,

(12.182)
$$\psi_1 = \sum_n a_n^{(1)} U_n(\mathbf{r})$$

Here we have tacitly assumed that H_0, for the given domain, has a *discrete* spectrum of ε_n, and hence a discrete set of U_n. If such were not the case then we would have to form ψ_1 as an *integral* over all possible ε instead of the sum.

Thus if we now put

(12.183)
$$\psi_0 = U_m, \qquad E_0 = \varepsilon_m$$

and Eq. (12.181) for ψ_1 into the second member of Eq. (12.179) we obtain;

(12.184)
$$\sum_n a_n^{(1)} U_n \varepsilon_n + \mathcal{H} U_m = \varepsilon_m \sum_n a_n^{(1)} U_n + E_1 U_m$$

where we have made use of Eq. (12.180) in the first term here. If we simply multiply through by $U_k^* \, d\tau$ and integrate over the domain we obtain, by virtue of Eq. (12.181),

(12.185)
$$a_k^{(1)}(\varepsilon_k - \varepsilon_m) = E_1 \delta_{mk} - \int_D U_k^* \mathcal{H} U_m \, d\tau$$

where, if we take $k = m$, $\delta_{mk} = 1$ and thus

(12.186)
$$E_1 = \int_D U_m^* \mathcal{H} U_m \, d\tau$$

and for $k \neq m$,

(12.187)
$$a_k^{(1)} = \frac{\int_D U_k^* \mathcal{H} U_m \, d\tau}{\varepsilon_m - \varepsilon_k}$$

† Note that the orthogonality is between the *complex conjugate* U_n^*, and U_m.

Note that for $k = m$, $a_m^{(1)}$ is not determined. This situation is taken care of by the requirement that ψ be normalized, i.e.,

(12.188)
$$\int_D \psi^* \psi \, d\tau = 1$$

Thus, using Eq. (12.178) we see that the *first*-order terms in λ give,

(12.189)
$$\int_D (\psi_0 \psi_1^* + \psi_0^* \psi_1) \, d\tau = 0$$

and using $\psi_0 = U_m$ and the expansion, Eq. (12.182) for ψ_1, we get from the orthogonality stated in Eq. (12.181),

(12.190)
$$a_m^{(1)} + a_m^{*(1)} = 0$$

This fixes the *real* part of $a_m^{(1)}$ but not the phase, or imaginary part. In quantum mechanics this poses no problem since it is the eigenvalues and the products $\psi^* \psi$ which have *physical* meaning, thus the imaginary part of $a_m^{(1)}$ can be taken as zero with no loss whatever.

The procedure outlined here can be carried to higher order with essentially the same steps, but will not be given here.

EXECUTE PROBLEM SET (12-6)

CHANGE OF VARIABLES IN NONLINEAR, DIFFERENTIAL EQUATIONS

The student should not be led to think that only approximate solutions to nonlinear differential equations are possible. Actually, it is often possible to obtain an *exact* solution by a suitable change of variable. This is most dramatically illustrated in a reverse fashion.

The *linear* partial differential equation for diffusion, or heat flow,

(12.191)
$$\nabla^2 \rho = \frac{\partial \rho}{\partial t}$$

becomes under the substitution,

(12.192)
$$\rho = e^{cp}, \qquad c = \text{constant}$$

the *nonlinear* partial differential equation,

(12.193)
$$\frac{\partial p}{\partial t} = \nabla^2 p + c|\nabla p|^2$$

and of course the boundary conditions must be changed accordingly. Equation (12.191) is readily solved by our now familiar methods and hence by the substitution of Eq. (12.192) we also obtain the *exact* solution to Eq. (12.193).

Unfortunately no *general* rules can be given for generating the appropriate transformation for any given nonlinear equation.

PROBLEMS

Set (12-1)

(1) Find the extremum points for Example (1) of this chapter, the bead on a wire.

(2) Find the extremum points of the function

$$U = z - ax^2 - by^2$$

subject to the constraint,

$$dG = dz - z\,dx = 0$$

Is this constraint an exact differential? Is it integrable? Is there a unique extremum or a *family* of extremum points?

(3) Construct an integral having a function $y(x)$ and its first derivative in the integrand, such that the integral between two fixed points, a and b, has an extreme value if $y(x)$ is a solution of the equation,

$$\frac{d^2y}{dx^2} + Cy = W$$

where C and W are *given* functions of x.

(4) Consider a string of fixed length, L and mass per unit length ρ with ends attached at two fixed points in the x, y plane with y being vertical. Find the equation for the shape of the string in equilibrium when a uniform vertical gravitational field is present. [Hint: $\rho gy\,ds$ is the potential energy of the element of length ds and elevation y, g being the acceleration of gravity; and don't forget, the string has a *fixed* length, L. Use these facts in the condition that potential energy is a minimum in the equilibrium state.]

(5) Solve the famous problem first posed by John Bernoulli, the "brachistochrone" (Greek for "shortest time.") A bead slides without friction on a wire from fixed point A to fixed point B under a uniform gravitational field. What must be the shape of the wire, assumed in a vertical plane, such that the bead goes from A to B in the minimum time. [Hint: the speed of the bead is ds/dt, where ds is an element of length on the wire, so with v equal to the speed,

$$t = \int_A^B \frac{ds}{v}$$

is the travel time; also, by the conservation of energy,

$$v = \sqrt{2gy}$$

where g is the acceleration of gravity.

(6) Construct an integral having a function $\psi(x, y, z)$ and its first derivatives, $\partial\psi/\partial x$, $\partial\psi/\partial y$, and $\partial\psi/\partial z$ in the integrand such that the integral over a fixed domain of the x, y, z space is an extremum if ψ satisfies the Schrödinger equation,

$$-\frac{\hbar^2}{2m}\nabla^2\psi + V\psi = E\psi$$

where $\hbar^2/2m$ and E are constants and V is a function of x, y, z.

(7) Consider a string of mass per unit length, ρ and length L attached at two points to a rotating shaft, where the angular speed of rotation is ω radians per second and the distance between the points of attachment is less than L. If we neglect the earth's gravity the string *appears* to be in a nonuniform gravitational field of strength, or acceleration, $\omega^2 r$ at a distance r from the rotating shaft. Find the equilibrium shape of the string.

Set (12-2)

(1) In Problem (6) of Set (12.1) above the Schrödinger equation was formulated as the Euler–Lagrange equation of a variational problem *without* constraints. Show how this same equation results as the Euler–Lagrange equation for an integral which has an extremum subject to the integral constraint,

$$\int_D \psi^2 \, dx \, dy \, dz = \text{constant}$$

where D is the domain of definition of $\psi(x, y, z)$. Thus see E as an undetermined multiplier.

(2) In Problem (1) above no stipulation is required that ψ be complex, however we can show that the Schrödinger equation does result for ψ and its complex conjugate, ψ^*, as the *two* Euler–Lagrange equations corresponding to

$$\int_D \left(\frac{\hbar^2}{2m} \nabla \psi \cdot \nabla \psi^* + V \psi \psi^* \right) dx \, dy \, dz$$

being an extremum subject to the constraint.

$$\int_D \psi^* \psi \, dx \, dy \, dz = \text{constant}$$

(3) Following the pattern of Example (5) of this chapter use the Rayleigh–Ritz method to estimate the lowest frequency of vibration of a thin membrane clamped at the edge having the form of an ellipse with semimajor axis, a, and semiminor axis b.

Set (12-3)

(1) A uniform string of mass per unit length ρ and length L has one end attached to each of the two circles in the x, y plane (y vertical)

$$(x - 2R)^2 + y^2 = R^2, \qquad (x + 2R)^2 + y^2 = R^2$$

Assuming a uniform vertical gravitational field and $2R < L < 4R$ find the proper points of attachment and the shape of the string in the equilibrium configuration of minimum potential energy.

(2) Carry out the details of the analysis in Example (6) of this chapter and show that the air–liquid interface intersects the solid wall at an angle, θ, determined by:

$$\cos \theta = \frac{\gamma_{sl} - \gamma_{as}}{\gamma_{al}}$$

called the "contact angle."

Set (12-4)

(1) Determine the conditions for periodic solutions to exist for the "hard-spring problem," $b < 0$, as defined in Example (7) of this chapter.

(2) Construct the phase-plane trajectories for the equation

$$m \frac{d^2\theta}{dt^2} + \frac{mg}{l} \sin \theta = 0$$

describing the simple pendulum exactly.

(3) Construct the phase-plane diagram for the equation

$$m \frac{d^2\theta}{dt^2} + b\theta + C \sin^3\theta = 0$$

Under what conditions will solutions of the equation in Problem (2) above and those of this equation have nearly similar character?

Set (12-5)

(1) Use the perturbation method to approximate to the periodic solution of the van der Pol equation,

$$\frac{d^2y}{dt^2} - \varepsilon \frac{dy}{dt}(1 - y^2) + y = 0$$

for free oscillations with $y(0) = 0$. Show that,

$$y \approx 2 \sin \omega t + \frac{\varepsilon}{4}(\cos \omega t - \cos 3\omega t)$$

where

$$\omega = 1 - \frac{\varepsilon^2}{16}$$

(2) Show how one would set up the perturbation method for forced oscillation and analyze the case of forced oscillations of the van der Pol equation,

$$\frac{d^2y}{dt^2} - \varepsilon \frac{dy}{dt}(1 - y^2) + y = B \cos \omega t$$

with

$$y(0) = A$$

(3) Carry out the analysis of Example (12) of this chapter, forced oscillations of the soft spring.

Set (12-6)

(1) Solve the quantum mechanical problem of a particle in a one-dimensional square-well potential of infinite depth.

$$-\frac{\hbar^2}{2m} \frac{\partial^2\psi}{\partial x^2} = E\psi, \qquad 0 < x < L$$

where we require $\psi = 0$ for $x \le 0$ and $x \ge L$, then introduce the perturbation of a small "dimple" in the potential. Thus, ($\varepsilon = $ constant)

$$-\frac{\hbar^2}{2m} \frac{\partial^2\psi}{\partial x^2} = E\psi, \qquad 0 < x < \frac{L - \delta}{2}$$

$$-\frac{\hbar^2}{2m} \frac{\partial^2\psi}{\partial x^2} - \varepsilon\psi = E\psi, \qquad \frac{L - \delta}{2} < x < \frac{L + \delta}{2}$$

$$-\frac{\hbar^2}{2m} \frac{\partial^2\psi}{\partial x^2} = E\psi, \qquad \frac{L + \delta}{2} < x < L$$

where ε is the depth of the "dimple" in the potential and δ is its width. We require ψ and $\partial\psi/\partial x$ to be continuous at the edge of this square dimple of course, and we still require $\psi = 0$ at $x = 0, L$. Having found the eigenfunctions and eigenvalues for the unperturbed problem, use perturbation theory to find the perturbed eigenvalues and eigenfunctions considering small values for ε and δ.

REFERENCES

W. F. Ames, *Nonlinear Partial Differential Equations* (Academic Press, New York, 1956).

R. Courant and D. Hilbert, *Methods of Mathematical Physics* (Interscience Publishers, Inc., New York, 1953).

H. T. Davis, *Introduction to Non-linear Differential and Integral Equation* (Dover Publications, Inc., New York, 1962).

N. N. Lebedev, I. P. Skalskaya, and Y. S. Uflyand, *Problems of Mathematical Physics*, (translated by R. A. Silverman), (Prentice-Hall Inc., Englewood Cliffs, N.J., 1965).

H. Margeneau and G. M. Murphy, *The Mathematics of Physics and Chemistry* (D. Van Nostrand Company Inc., New York, 1961).

J. Mathews and R. L. Walker, *Mathematical Methods in Physics and Engineering* (W. A. Benjamin, New York, 1965).

P. M. Morse and H. Feshbach, *Methods of Theoretical Physics* (McGraw-Hill Book Company, Inc., New York, 1953).

R. A. Struble, *Nonlinear Differential Equations* (McGraw-Hill Book Company, Inc., New York, 1962).

CHAPTER THIRTEEN

Elements of probability theory

Probability becomes a necessary part of the description of any system when not every factor which may affect the system is known or controlled. Although rather loosely stated this admission of *lack of knowledge* is the argument most often given for introducing probability concepts into physical problems.

DEFINITION OF MATHEMATICAL PROBABILITY

"If, consistent with conditions, C, on a system there are exactly n distinct and *equally likely* states of the system, and among these precisely m are consistent with the configuration A, then the mathematical probability for A is defined to be $m/n = P$."

This definition is referred to as the "enumeration of cases definition." The other definition, sometimes called the "operational definition" is as follows.

"If consistent with conditions, C, we perform on a system a sequence of N observations of the state of the system and we find the configuration A to occur M times, then if our sequence is arbitrarily long, $N \to \infty$, we define

$$P = \lim_{N \to \infty} \frac{M}{N}$$

as the probability for the event A."

Generally most mathematical texts take the first definition above and show that the second definition given here, must be true as a mathematical consequence. This is sometimes called the *law of large numbers*. It is incorporated in *Poisson's theorem*† which states that the *probability of the inequality*,

$$(13.1) \qquad \left| \frac{m}{n} - \frac{M}{N} \right| \le \varepsilon, \qquad \varepsilon > 0$$

for fixed ε, no matter how small, can be made as near to unity (certainty) as we please, by taking N sufficiently large.

The simplest example which brings out all facets of these two ways of looking at probability is the toss of a coin. The conditions C are that the coin have two distinct sides and neither side is biased in any way. Then, since there are *two* states possible, head or tails, and the configuration, head, corresponds to A say, we see that if both sides are *equally likely* to show, then sure enough by the first definition $p = \frac{1}{2}$.

On the other hand if we continue *unbiased* tossing, say, 1000 times, we find that something like 496 times the head appeared and M/N is slightly less than $\frac{1}{2}$. But, as we continue we find M/N generally oscillating about $\frac{1}{2}$ with ever decreasing amplitude.

The application of the basic definition by enumeration leads to these rules.

RULES ON TOTAL AND COMPOUND PROBABILITY

(1) Exclusive events

If events $A_1, A_2, \ldots A_n$ are *mutually exclusive*, meaning that only one of them may occur at one observation, then the probability for A_1, *or* A_2, *or* $A_3, \ldots$ *or* A_n, to occur is,

$$(13.2) \qquad p = P_1(A_1) + P_2(A_2) + \ldots + P_n(A_n)$$

where $P_j(A_j)$ is the probability for A_j to occur.

(2) Exclusive and exhaustive events

If in the set of exclusive events $A_1, A_2, \ldots A_n$, *at least one must occur*, then we have the condition,

$$(13.3) \qquad p = 1 = P_1(A_1) + P_2(A_2) + \ldots + P_n(A_n)$$

The main difference, then, is that if it is *possible* for *no* A_j to occur then $p < 1$ is a possibility, but, if *some* A_j *must occur* then $p = 1$.

(3) Joint events

For two events A and B, which *may* occur together we formulate the mutually *exclusive* and *exhaustive* set. Letting A mean "A occur" and $\bar{A}$ mean "A does *not* occur," and similarly for B we see that from the above rules

$$(13.4) \qquad 1 = P(A, B) + P(\bar{A}, B) + P(A, \bar{B}) + P(\bar{A}, \bar{B})$$

and this could readily be generalized for the *joint occurrence* of any number of events. Here, $P(\bar{A}, B)$, for example, is the probability of the *joint* event, A does not occur and B does occur.

† See J. V. Uspensky, *Introduction to Mathematical Probability*. Dover Pub. Co. Inc., New York, 1967.

(4) Conditional probabilities

Consider two events, A, B which may or may not occur simultaneously, then we write

(13.5) $$P(A, B) = P_A(A)P_B(B|A)$$

where $P(A, B)$ is the probability for the joint event, A *and* B to occur; $P_A(A)$ is the *unconditional* probability for A to occur without reference to B, and $P_B(B|A)$ is the *conditional* probability for B to occur *given* that A actually does occur.

This definition can readily be generalized to multiple events, but, we defer this for the moment. If $P_B(B|A)$ turns out to be *independent* of A then we say the events A and B are *statistically independent*.

EXECUTE PROBLEM SET (13-1)

REPEATED TRIALS

Here, when we say a "trial" we mean a test, or observation, to see if an event in question has or has not occurred. When trials may be repeated, i.e., many tosses of a coin, we have a *series* of trials. The trials may be *independent*, as with the coin, or may be dependent, as for example, drawing colored beads from a jar *without* returning those drawn at each trial.

The most famous and by far most important problem of repeated trials is that called Bernoulli trials after the famed (there were many in that family) mathematician. We present it in a general context then point out some applications.

BERNOULLI TRIALS AND THE BINOMIAL DISTRIBUTION

The result of a trial is either, *yes*, the event occurred, or *no*, the event did not occur, as for example when we roll one die and ask "Did the six show?" Consider a sequence of $N = n + m$ trials in which the event A occurs n times and does not occur m times.

The probability for the n successes can be viewed in terms of the conditional probability above, i.e., Eq. (13.5). Take just the first two successes. If we let $p =$ *probability of* "*yes*," then event A is "yes" and event B is also "yes." But, in this case the fact that A was "yes" does not in any way affect the probability for B to be yes, i.e., *if the probability for B does not depend on the outcome of A* then

(13.6) $$P'_B(B|A) = P_B(B)$$

In this case we say B *is statistically independent of* A. Thus, for *two* "yeses" the probability is

$$p \cdot p$$

and for n occurrences of *yes*,

(13.7) $$p \cdot p \ldots p = p^n$$

is the probability.

Since the event either happens or does not happen we have, if q is the probability for "no,"

(13.8) $$p + q = 1$$

Then by exactly the argument above we see that

(13.9) $$(1 - p)^m = q^m$$

is the probability for m "noes" to occur.

Now the compound event of n yeses and m noes can occur in many different orders and according to the first rule on exclusive events we must *add* the probabilities for there to be n yeses *and* m noes in every possible order that can occur. First note that n yeses and m noes is a *compound* event with the probability

(13.10) $$p^n(1 - p)^m = p^n q^m$$

no matter in what *order* the yeses and noes occur. Thus, to obtain the total probability we just need to *multiply* this factor by the number of ways n yeses and m noes can occur in $n + m = N$ trials.

To see what this number is, write down a list of N symbols, nY's, and mN's, we have,

$$Y N N Y N Y Y \ldots$$

There are N choices for the first symbol, $N - 1$ for the second, etc., so there are

$$N! = N(N - 1)(N - 2) \ldots 1$$

ways of arranging these in a line. But, now since the Y's are all the same we have n choices for the first Y, $n - 1$ choices for the second, etc., or $n!$ permutations of the Y's that do *not* change the *order* of N's and Y's. Similarly, there are $m!$ permutations of N's which do not change the order of N's and Y's. Thus there are,

$$\frac{(n + m)!}{n!m!}$$

distinctly different orders in which the Y's and N's can occur.

Thus, the probability for n successes and m failures in N trials, with all trials being *independent* and the *order* of successes being immaterial is:

(13.11) $$P = \frac{N!}{n!(N - n)!} p^n(1 - p)^{N-n} = P(n, N, p)$$

where we have replaced $n + m$ by N and m by $N - n$. This is the *binomial distribution for independent trials*.

Here we should note that if we had not stipulated that the *order* of successes was to be ignored then for say n successes *followed* by $N - n$ failures the probability is just *one term* of the form $p^n(1 - p)^{N-n}$.

APPROXIMATIONS TO THE BINOMIAL DISTRIBUTION

(a) Poisson distribution

In many applications we deal with Bernoulli trials in which N is *very large* and p *extremely small*, but, the product

(13.12) $$\lambda = Np$$

is of moderate magnitude. For this case we can use an approximation to $P(n, N, p)$ due to Poisson. This is obtained by *induction* as follows.

For $n = 0$,

(13.13) $$P(0, N, p) = (1 - p)^N$$

or by Eq. (13.12)

(13.14)
$$P(0, N, p) = \left(1 - \frac{\lambda}{N}\right)^N$$

and as we know the familiar limit,

(13.15)
$$\lim_{N \to \infty}\left(1 - \frac{\lambda}{N}\right)^N = e^{-\lambda}$$

so that approximately,

(13.16)
$$P(0, N, p) \approx e^{-\lambda}$$

Also as can readily be seen from the form of $P(n, N, p)$,

(13.17)
$$P(1, N, p) = \frac{Np}{q}P(0, N, p) = \frac{\lambda}{1 - \lambda/N}e^{-\lambda}$$

or, approximately for *large N*,

(13.18)
$$P(1, N, p) \approx \lambda e^{-\lambda}$$

Furthermore, we have the exact relation and its approximation,

(13.19)
$$\frac{P(n + 1, N, p)}{P(n, N, p)} = \frac{p(N - n)}{n + 1}\frac{1}{1 - p} \approx \frac{\lambda}{n + 1}$$

for $p \ll 1$ and $n \ll N$. Thus, by induction it follows that,

(13.20)
$$P(n, N, p) \approx e^{-\lambda}\frac{\lambda^n}{n!}$$

which is called the *Poisson distribution*, applicable to *rare* events.

Example (1)

We consider a sequence of random events occurring in time, in particular say the emission of alpha particles by a radioactive nucleus, that is, from any one of many nuclei making up a large sample of material. Divide the time interval, t, into a great number, N, of intervals each of length t/N. During one of these intervals an α particle may, or may not appear. If one, or more, particles appear we call this a "success" in that "trial." Actually, for very *rare* events, and *very small time intervals, multiple* occurrences can essentially be ignored, so that essentially a success is *one* particle during the interval.

Here then, we have a sequence of N trials during the time interval t. These are independent trials and certainly the probability of success at one "trial" should be proportional to the duration of the elementary time interval, $p \propto \Delta t$, and hence, $Np = \lambda = \gamma t$. Thus, for rare events we see

(13.21)
$$P \approx e^{-\gamma t}\frac{(\gamma t)^n}{n!}$$

as the probability for n events, say n α particles, observed during a time t. Here, γ will be characteristic of the system in question. Actually, this does fit the data of radioactive decay rather well.

(b) Gaussian distribution

If we suppose n, N and $N - n$ all sufficiently large so that the Stirling† approximation is valid, i.e., we will let all, n, N and $N - n \to \infty$, so that for example,

$$(13.22) \qquad n! \approx \sqrt{2\pi n} \, n^n e^{-n}$$

then we have,

$$(13.23) \qquad P(n, N, p) = \frac{1}{\sqrt{2\pi n(1 - n/N)}} \left(\frac{Np}{n}\right)^n \left(\frac{Nq}{N - n}\right)^{N-n}$$

Here, we substitute the new variable

$$(13.24) \qquad x = n - Np$$

and have,

$$(13.25) \qquad P(n, N, p) = \left(\frac{N}{2\pi(Np + x)(Nq - x)}\right)^{\frac{1}{2}} \cdot \left\{ \frac{1}{\left(1 + \dfrac{x}{Np}\right)^{Np+x} \left(1 - \dfrac{x}{Nq}\right)^{Nq-x}} \right\}$$

To evaluate the limit of this last factor for *large* N and *small* x we write out the logarithm,

$$(13.26) \qquad -\ln\{\ \} = (Np + x)\ln\left(1 + \frac{x}{Np}\right) + (Nq - x)\ln\left(1 - \frac{x}{Nq}\right)$$

where $\{\ \}$ denotes the factor, and then expand the logarithms in series,

$$(13.27) \quad -\ln\{\ \} \approx (Np + x)\left(\frac{x}{Np} - \frac{x^2}{N^2 p^2} + \cdots\right) - (Nq - x)\left(\frac{x}{Nq} + \frac{x^2}{N^2 q^2} + \cdots\right)$$

or

$$(13.28) \qquad -\ln\{\ \} = \frac{x^2}{2N}\left(\frac{1}{p} + \frac{1}{q}\right) - \frac{x^3}{6N^3}\left(\frac{1}{p^2} - \frac{1}{q^2}\right) + \cdots$$

Now if n increases with N so that

$$(13.29) \qquad \frac{x^3}{N^2} \to 0$$

and $p \approx q$, we can neglect all terms but the first and our final approximate form for P can be written as,

$$(13.30) \qquad P \approx \frac{1}{\sqrt{2\pi Npq}} e^{-\frac{x^2}{2Npq}}$$

which is the *Gaussian Distribution* in x. Figure 13.1 illustrates the relationship of the Gaussian *curve* to the actual binomial distribution for one particular case.

Example (2)

As an important illustration of the Gaussian distribution as the limiting form of the binomial distribution for large N, which at the same time exhibits some other important ideas in probability theory, we consider the *random walk in one dimension*.

† See Chapter Eight.

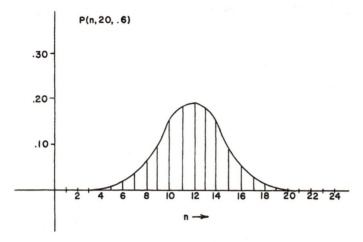

FIGURE 13.1.   Relation of the Gaussian Curve to the Binomial Distribution for One Set of Parameters.

A point moves along the ξ axis in unit steps starting from the origin, one step occurring in Δt units of time. Let p be the probability for a forward step and q the probability for a backward step at any point ξ. These probabilities are presumed to be fixed. Since, we have *unit* steps ξ will have one of the values $\ldots -2, -1, 0, 1, 2, \ldots k, \ldots$.

If the point has made n *plus* steps and m *minus* steps after N time increments its position will be,

$$(13.31) \qquad\qquad \xi = n - m$$

and since this is just a sequence of N Bernoulli trials with n successes we see that the probability for this situation to occur is just, $P(n, N, p)$ as constructed above. We also see that this takes the Gaussian form in the limit as $N \to \infty$ if p and q are of comparable magnitude, i.e., *neither* is near *zero*.

Now if we express the Gaussian form above in terms of ξ and t as used here,

$$(13.32) \qquad\qquad \begin{cases} N = \dfrac{t}{\Delta t} \\[2mm] n = \dfrac{\xi}{2} + \dfrac{N}{2} \end{cases}$$

we find

$$(13.33) \qquad\qquad \begin{cases} x^2 = \left[\dfrac{\xi}{2} - \left(p - \dfrac{1}{2}\right)\dfrac{1}{\Delta t}t\right]^2 \\[3mm] Npq = \left(\dfrac{pq}{\Delta t}\right)t \end{cases}$$

and, hence, the limiting form appears as,

$$(13.34) \qquad\qquad P \approx \frac{1}{\sqrt{\pi D t}}\, e^{-\frac{(\xi - vt)^2}{4Dt}}$$

where we have introduced the symbols,

(13.35)
$$\begin{cases} v = 2\left(p - \dfrac{1}{2}\right)\dfrac{1}{\Delta t} \\[3mm] D = \dfrac{2pq}{\Delta t} \end{cases}$$

This is the solution of the partial differential equation,†

(13.36)
$$D\frac{\partial^2 P}{\partial \xi^2} - v\frac{\partial P}{\partial \xi} = \frac{\partial P}{\partial t} + \delta(t)\,\delta(\xi)$$

and yields the physical interpretation of the random walk as follows. This is equivalent to a *diffusion* problem with a flow, or drift, of velocity v in the $+\xi$ direction with a *unit* quantity of material released at the origin, $\xi = 0$ at time $t = 0$. Observe, that if $p = q = \frac{1}{2}$, then $v = 0$, there is no drift and the peak of concentration (probability) remains at the origin but decreases in magnitude with time.

Later we will consider random-walk problems from another point of view.

EXECUTE PROBLEM SET (13-2)

RANDOM VARIABLES AND FUNCTIONS OF RANDOM VARIABLES

In our discussion of the simple random walk just above we let the position variable, x, take on discrete values, but in the limiting process it was treated as a continuous variable. We can define probability in terms of a continuous *random variable*, x, defined over some domain we call the *spectrum of* x, $a < x < b$, as follows. We simply let

(13.37)
$$dP = \rho(x)\,dx$$

be the probability of finding x between x and $x + dx$ at a single observation. We call $\rho(x)$ the probability *density* for x. From this we have

(13.38)
$$P(x_1 \le x \le x_2) = \int_{x_1}^{x_2} \rho(x)\,dx$$

as the probability for finding x in the interval $x_1 \le x \le x_2$.

Since, *some* value of x in (a, b) must occur we see

(13.39)
$$\int_a^b \rho(x)\,dx = 1$$

Equations (13.38) and (13.39) are just the natural extensions of our rules for exclusive and exhaustive events.

If we have *two* random variables, x, y, then the joint probability for x in the interval x to $x + dx$ and y in y to $y + dy$ will appear as,

(13.40)
$$dP = \rho(x, y)\,dx\,dy$$

† Compare this to the problem of finding a Green's function.

and again

(13.41)
$$\int_a^b \int_c^d \rho(x, y)\, dx\, dy = 1$$

if $a \le x \le b$, $c \le y \le d$ are the domains of x and y.

We may wish to consider some functions of random variables such as say $F(x, y)$, and then we may ask what is the probability for F to have a value between F and $F + dF$, where $F + dF$ corresponds to replacing x by $x + dx$ and y by $y + dy$. If F is a *single valued function* of x and y then the probability is just dP in Eq. (13.40), but if this particular value of F can occur for the pairs of values, or points, $(x, y), (x', y'), (x'', y'')$ say, then the probability is by our rules of exclusive events,

(13.42)
$$dP = [\rho(x, y) + \rho(x', y') + \rho(x'', y'')]\, dx\, dy$$

and one can readily see how to *generalize* this expression.

Thus, what is the probability for (x, y) to be on the surface, (to within an infinitesimal distance),

(13.43)
$$F(x, y) = C$$

where C is some constant. We must give this probability as the integral,

(13.44)
$$\iint_{F(x,y)=C} \rho(x, y)\, dx\, dy = P(F = C)$$

where the integration is over *all* points of the x, y space consistent with Eq. (13.43).

MEANS, DISPERSIONS, AND CORRELATIONS

Whether we consider discrete or continuous random variables the *concepts* are essentially the same. Thus, in all of the following discussion we will use, say $P(x)$, to denote the probability distribution for the variable x, and if x is *discrete* as indicated then we must have

(13.45)
$$\sum_x P(x) = 1$$

where the sum is over all possible values of x, i.e., the *spectrum* of x. However, the reader may equally well take this to be an *integral* with $P(x)$ being read as $\rho(x)\, dx$. This is the *only* difference we note between continuous and discrete variables.

MEAN VALUE

The *mean value* of a random variable with distribution $P(x)$ is defined as:

(13.46)
$$\bar{x} = \sum_x xP(x)$$

MEAN-SQUARE VALUE

The *mean-square value* of x in the same circumstances is,

(13.47)
$$\bar{x}^2 = \sum_x x^2 P(x)$$

DISPERSION

The *dispersion* of x is defined as:

(13.48)
$$\sigma = \langle(x - \bar{x})^2\rangle_{\text{av}} = \sum_x (x - \bar{x})^2 P(x)$$

This is also called the mean square *deviation* of x.

CORRELATION COEFFICIENT

When we have the *joint* probability $P(x, y)$ for the joint event, x, y, we call

(13.49)
$$\langle xy \rangle_{\text{av}} = \sum_x \sum_y xy P(x, y)$$

the correlation of x and y. (*Note:* This, divided by the product $\bar{x} \cdot \bar{y}$, is called the correlation coefficient.) Here, the *individual means*,

(13.50)
$$\bar{x} = \sum_x \sum_y x P(x, y)$$

and

(13.51)
$$\bar{y} = \sum_x \sum_y y P(x, y)$$

are defined exactly as above except we must sum over the spectra of *both* variables.

MODE OF A DISTRIBUTION

The *mode* of a distribution of probability is just the point in the spectrum of the variables at which P has a maximum value. For a simple distribution such as the Gaussian distribution this is a unique point, but many distributions have multiple "peaks" and the mode loses its simple meaning.

EXECUTE PROBLEM SET (13-3)

RANDOM VARIABLES AND CONDITIONAL PROBABILITIES

Suppose we are given the probability $P(x, y)$ for the joint occurrence of a definite value of x and a definite value of y, these of course being functionally independent variables, i.e., x may change arbitrarily holding y fixed and vice versa.

If we sum over all possible values of y we obtain the probability for a definite value of x to occur without regard to what value of y occurs, i.e., x may occur with $y = y_1$, with probability $P(x, y_1)$, or x with y_2 with probability $P(x, y_2)$, etc., and by our rules for compound events,

(13.52)
$$P_x(x) = \sum_y P(x, y)$$

is then the probability for x to occur with *some* y. Of course, we have in the same way

(13.53)
$$P_y(y) = \sum_x P(x, y)$$

as the probability for a value of y without regard to the x value. Here P_x and P_y are *unconditional probabilities*.

Now if we form the quotient,

(13.54) $$P_y(y|x) = \frac{P(x, y)}{\sum_y P(x, y)} = \frac{P(x, y)}{P_x(x)}$$

we form the *conditional probability* for a certain value of y to occur, *given* the value of x. Note that

(13.55) $$\sum_y P_y(y|x) = 1$$

by virtue of the definition of $P_x(x)$ given above. Of course, we can in the same way define the conditional probability for a value of x, *given* y, as:

(13.56) $$P_x(x|y) = \frac{P(x, y)}{P_y(y)}$$

where also,

(13.57) $$\sum_x P_x(x|y) = 1$$

automatically follows.

Here, we see that the joint probability, $P(x, y)$ has two equally acceptable representations:

(13.58) $$P(x, y) = \begin{cases} P_x(x)P_y(y|x) \\ P_y(y)P_x(x|y) \end{cases}$$

We also note what at first appears as an alarming feature of Eqs. (13.55) and (13.57). At face value these seem to indicate that we have a function of x equal to a *constant* in the first case and a function of y equal to a *constant* in the second case. These would imply only certain "roots" as acceptable values of x and y, respectively. However, this difficulty is obviated by noting that conditional probabilities *must* have the functional structure,

(13.59) $$P_y(y|x) = A(x)B(y, x)$$

for example then Eq. (13.55) places no constraint on the possible x values, i.e., putting Eq. (13.59) into Eq. (13.55) we get

(13.60) $$\frac{1}{A(x)} = \sum_y B(y, x)$$

so that we simply have a rule for forming a function $A(x)$ out of the *set* of functions $B(y, x)$, i.e., one $B(y, x)$ for each value of y.

The ideas outlined, here, can readily be extended to many random variables.

We call attention to the fact that if x and y are *statistically independent* then $P_x(x|y)$ does *not* depend on y and $P_y(y|x)$ does not depend on x. In this case the *two* representations of $P(x, y)$ become the *single* representation,

(13.61) $$P(x, y) = P_x(x)P_y(y)$$

This is the basic form for *statistically independent* random variables and can also be extended to any number of variables.

Example (3)

To illustrate a few important points about the above discussion of conditional probabilities, recall the limiting form of the random walk for one dimension, Eq. (13.34). Here, we have, (now in proper notation for a continuous variable).

$$(13.62) \qquad dP = \frac{1}{\sqrt{\pi D(t - t')}} e^{-\frac{[\xi - v(t - t')]^2}{4D(t - t')}} d\xi, \qquad t - t' > 0$$

as the probability for the "particle" to have coordinate between $\xi(t)$ and $\xi + d\xi$ at time t *if* the "particle" appeared at the origin at time t'. This is a *conditional probability* because of the "if" with regard to the *time* the particle has been wandering. Here, the functions corresponding to "A" and "B", respectively, in Eq. (13.59) are

$$(13.63) \qquad \begin{cases} A(t - t') = \dfrac{1}{\sqrt{\pi D(t - t')}} \\ B(\xi, t - t') = e^{-\frac{[\xi - v(t - t')]^2}{4D(t - t')}} d\xi \end{cases}$$

and we note that the integral,

$$(13.64) \qquad \int_{-\infty}^{+\infty} e^{-\frac{[\xi - v(t - t')]^2}{4D(t - t')}} d\xi = \sqrt{\pi D(t - t')}$$

is the continuous analog of Eq. (13.60).

If we also had given the *unconditional probability* for the "particle" to appear at the origin at time between t' and $t' + dt'$ then we would have the *joint* probability for $\xi(t)$ and t' as:

$$(13.65) \qquad dP(\xi, t') = f(t') \frac{e^{-\frac{[\xi - v(t - t')]^2}{4D(t - t')}}}{\sqrt{\pi D(t - t')}} dt' d\xi$$

where $[f(t') dt']$ is the just named probability for t' and we must have

$$(13.66) \qquad \int_0^\infty f(t') dt' = 1$$

i.e., the range of t is $0 \le t \le \infty$. Equation (13.65) is the continuous analog of one member of the pair of equations in Eq. (13.58).

In terms of the diffusion analog of the limiting form of the random walk we note that the integral,

$$(13.67) \qquad C(\xi, t) = \int_0^t f(t') \frac{e^{-\frac{[\xi - v(t - t')]^2}{4D(t - t')}}}{\sqrt{\pi D(t - t')}} dt'$$

corresponds to the "concentration" of diffusing material at ξ at time t when material is liberated from the origin at the rate $f(t)$.

GAUSSIAN DISTRIBUTION IN SEVERAL VARIABLES; STATISTICAL DEPENDENCE

To further illustrate and emphasize certain aspects of statistical dependence or independence among random variables, i.e., as mentioned in regard to Eq. (13.61) above, we consider the following.

We wish to consider a set of random variables $x_1, x_2, \ldots x_n$ with the probability distribution,

$$(13.68) \qquad dP = \alpha e^{-\sum_{ij} \beta_{ij} x_i x_j} dx_1 \, dx_2 \ldots dx_n$$

Since this does *not* factor in a *product* of functions of the individual variables x_j as required in Eq. (13.61), we see that these x_j are satistically dependent. However, an appropriate *change of variables* to a set $x_i', i = 1, 2, \ldots n$ can be shown to yield a statistically independent set. However, before we examine this we wish to point out *why* the particular problem considered here is so important.

Consider *any* set of random variables $y_j, j = 1, 2, \ldots n$ having a probability distribution,

$$(13.69) \qquad d\bar{P} = \rho(y_1, y_2, \ldots y_n) \, dy_1 \, dy_2, \ldots dy_n$$

which has some rather *sharp maximum*, i.e., the function, $\rho(y_1, y_2, \ldots y_n)$ has a "peak" defined by

$$(13.70) \qquad \frac{\partial \rho}{\partial y_j}(y_1', y_2', \ldots y_n') = 0, \qquad j = 1, 2, \ldots n$$

We note that, since $\rho \geq 0$, we can write

$$(13.71) \qquad \rho(y_i) = e^{U(y_i)}$$

and we then expand U in a Taylor series about the point $y_i = y_i', i = 1, 2, \ldots n$. In view of Eq. (13.70) we then have,

$$(13.72) \qquad d\bar{P} = e^{U(y_i') - \sum_i \sum_j \beta_{ij}(y_i - y_i')(y_j - y_j')} dy_1 \, dy_2 \ldots dy_n$$

and with the change of notation, $x_i = y_i - y_i'$, this is exactly of the same form as Eq. (13.68) above. Here, $\beta_{ij} = -\frac{1}{2} \partial^2 \ln \rho / \partial y_i \, \partial y_j$ evaluated at the peak, and only terms of second order are retained.

Thus, we see that any probability distribution in several variables will be approximated by the multivariate Gaussian distribution if it has a clear cut maximum.

Now, we return to our original point, namely that we can transform from the *statistically dependent* set of variables $x_i, i = 1, 2, \ldots n$, to a *statistically independent* set, $x_i', i = 1, 2, \ldots n$. This transformation also makes possible the evaluation of the factor, α, in Eq. (13.68), i.e., we must have,

$$(13.73) \qquad \alpha \int_{-\infty}^{+\infty} \int_{-\infty}^{+\infty} \cdots \int_{-\infty}^{+\infty} e^{-\sum_i \sum_j \beta_{ij} x_i x_j} dx_1 \, dx_2 \ldots dx_n = 1$$

We require a *linear* transformation matrix,† $a_{ij}, i = 1, 2, \ldots n, j = 1, 2, \ldots n$

$$(13.74) \qquad x_i' = \sum_{j=1}^{n} a_{ij} x_j, \qquad i = 1, 2, \ldots n$$

or

$$(13.75) \qquad X' = AX$$

such that the quadratic form in the exponent of Eq. (13.6) or Eq. (13.72) is simply a sum of

† Consult Chapter Two for details of these matrix operations.

squares. This quadratic form can be written in matrix form as,

$$(13.76) \qquad \sum_{i=1}^{n} \sum_{j=1}^{n} \beta_{ij} x_i x_j = \tilde{X} B X$$

where the elements of B are the β_{ij}. (See Chapter Two.)

Here, then we substitute for X and $\tilde{X}$ from Eq. (13.74) to obtain (see Chapter Two)

$$(13.77) \qquad \overbrace{A^{-1} X' B A^{-1} X'}^{} = \tilde{X} B X$$

If we require that A be an *orthogonal* transformation, $A^{-1} = \tilde{A}$, this becomes,

$$(13.78) \qquad \tilde{X}'(A B \tilde{A}) X' = \tilde{X} B X$$

and we want the left member to reduce to a form such as

$$(13.79) \qquad \sum_{i=1}^{n} b_i x_i'^2 = \tilde{X}'(A B \tilde{A}) X'$$

Hence, we see that $A B \tilde{A}$ must be a diagonal matrix with elements $b_1, b_2, \ldots b_n$ on the main diagonal. As we saw in the exercises of Chapter Two an orthogonal transformation, A, can *always* be found to diagonalize a matrix B in this way *if B is a symmetric matrix*, $\beta_{ij} = \beta_{ji}$.[†]

Assuming this to be the case we can replace the exponent in Eq. (13.68) by the left member of Eq. (13.79), but we must also express the "volume element" $dx_1 \, dx_2 \ldots dx_n$ in terms of $dx_1' \, dx_2' \ldots dx_n'$. Note that from Chapter One,

$$(13.80) \qquad dx_1 \, dx_2 \ldots dx_n = \left| \frac{\partial x_i}{\partial x_j} \right| dx_1' \, dx_2' \ldots dx_n'$$

or

$$(13.81) \qquad dx_1 \, dx_2 \ldots dx_n = |\tilde{A}| \, dx_1' \, dx_2' \ldots dx_n'$$

and from the above requirement,

$$(13.82) \qquad |b| = |A| \, |B| \, |\tilde{A}|$$

but

$$(13.83) \qquad |A| = |\tilde{A}|$$

so we have

$$(13.84) \qquad |\tilde{A}| = \pm \sqrt{\frac{|b|}{|B|}}$$

If in particular we *choose* all the b_i to be unity then we have a simple means of evaluating α in Eq. (13.68), i.e., Eq. (13.80) then reads in the new variables,

$$(13.85) \qquad \frac{\alpha}{\sqrt{|B|}} \int_{-\infty}^{+\infty} \int_{-\infty}^{+\infty} \cdots \int_{-\infty}^{+\infty} e^{-\sum_{i=1}^{n} x_i'^2} \, dx_1' \, dx_2' \ldots dx_n' = 1$$

But this is a product of integrals, each with the value $\sqrt{\pi}$, thus,

$$(13.86) \qquad \alpha = \pi^{\frac{n}{2}} \sqrt{|B|}$$

which gives α in terms of the determinant of the matrix of the β_{ij}, $|B|$.

[†]And the determinant of B is not zero.

From the physical point of view the important point here is that in one *reference frame* the random variables, x_i, are correlated, or *statistically dependent*, but in another *reference frame*, obtained by a rotation of axes, the variables, x_i', are *uncorrelated*, or statistically independent.

EXECUTE PROBLEM SET (13-4)

STATIONARY TIME SERIES AND ENSEMBLES

Here, we consider a collection of similar physical systems, for example a collection, or *ensemble*, of strings, each of the same length, mass density, etc., mounted between supports with the same tension in identical vessels filled with air. We suppose the collection to have always existed and to continue to exist indefinitely.

Due to the random impacts of air molecules on the strings they are all undergoing some random vibrations. We then consider the *vertical* displacement of the *midpoint* of the *k*th string at time *t* as $y_k(t)$, a *random function of t.*

Now at any instant, *t*, we can compute the *average* value of this *y* over the whole ensemble as

(13.87)
$$\langle y(t) \rangle = \frac{1}{N} \sum_{k=1}^{N} y_k(t)$$

or we could select any *one* of the strings, say the *k*th string, and compute the time average as

(13.88)
$$\bar{y}_k = \frac{1}{T} \int_0^T y_k(t)\, dt$$

These are the *ensemble* and *time* averages, respectively, of the same random variable.

Note in the ensemble average that if there are M_j of the systems having $y_k(t)$ equal to $y_j(t)$ then we can write the *ensemble average* for an infinite ensemble, $N \to \infty$, as,

(13.89)
$$\langle y(t) \rangle = \lim_{N \to \infty} \sum_{j=1}^{K} \left(\frac{M_j}{N} \right) y_j(t) = \sum_y P(y) y$$

where, in the limit, M_j/N is just the probability, $P(y_j)$, for y_j to occur, and K is the number of distinct values in the spectrum of *y*. Generally $P(y)$ could depend on *t*.

We call a system, such as one of our strings, a *stationary system* if there is no preferred origin in time for the system and $P(y)$ does not depend on *t*. Thus, our terminology of stationary time series for the random function $y_k(t)$ which persists forever, $T \to \infty$. If we also assume an infinite collection for our ensemble and assume that *each* system will in the course of a sufficiently long time pass through *every* state accessible to it,† then the ensemble average and the *time average* of $y_k(t)$ will be *equal*.

We define the integral

(13.90)
$$K_{jk}(\tau) = \frac{1}{T} \int_t^{t+T} y_j(t) y_k(t + \tau)\, dt$$

as the cross correlation between two random time series over the interval *t* to $t + T$ with "step out," *τ*. In general this quantity would depend on the choice of time interval and the choice of the two systems, and will be an explicit function of *τ*. In many *numerical* evaluations of correlations of time series a finite interval of correlation, *T*, is used, but for purposes of

† This is the *ergodic hypothesis*.

theoretical discussion we define

(13.91)
$$K_{jk}(\tau) = \lim_{T \to \infty} \frac{1}{T} \int_0^T y_j(t) y_k(t + \tau) \, dt$$

as the *cross correlation* of $y_j(t)$ and $y_k(t)$.

When we "correlate" a time function with itself, $y_j(t)$ with $y_j(t)$, we call the result the *autocorrelation*, or just the *correlation* of y_j. Thus, we introduce the symbol

(13.92)
$$A_j(\tau) = \lim_{T \to \infty} \frac{1}{T} \int_0^T y_j(t) y_j(t + \tau) \, dt$$

Note that if the system giving rise to $y_j(t)$, i.e., the string, is an *ergodic* system; a member of an ensemble of identical systems, we could *delete the subscript j*.

For an *ergodic* system we can also write $A(\tau)$ as,†

(13.93)
$$A(\tau) = \lim_{N \to \infty} \sum_{i=1}^{N} \frac{y_i(t) y_i(t + \tau)}{N}$$

and this must be *independent of t*. Thus, for ergodic systems

(13.94)
$$A(\tau) = \overline{y(t) y(t + \tau)} = \langle y(t) y(t + \tau) \rangle$$

indicating equality of the time average and ensemble average of the indicated product. Note the deletion of subscripts. We illustrate these ideas with an example.

Example (4)

For *small* displacements of the stretched strings described above the actual vertical displacement at the midpoint at any time t can be shown to be a sum of the form

(13.95)
$$y(t) = \sum_{n=0}^{\infty} a_n \sin(n\omega t + \phi_n)$$

where ω is the fundamental angular frequency of the string c/L, and a_n and ϕ_n are random variables having some probabilities $P_a(a_n)$ and $P_\phi(\phi_n)$. Here, we considered only vertical impacts on the string.

Thus, as a *simple* example we compute $A(\tau)$ for the single sinusoidal term,

(13.96)
$$y = \sin(\omega' t + \phi), \qquad -\infty < t < +\infty$$

where ϕ is a random variable with probability $P(\phi)$. We have,

(13.97)
$$A(\tau) = \lim_{T \to \infty} \int_0^T \sin(\omega' t + \phi) \sin(\omega' t + \omega' \tau + \phi) \, dt$$

for the correlation. Using the identity

(13.98)
$$\sin(\omega' t + \phi) \sin(\omega' t + \omega' \tau + \phi) = \tfrac{1}{2}[\cos \omega' \tau - \cos(2\omega' t + 2\phi)]$$

we have as $A_T(\tau)$ *before* the limit is taken,

(13.99)
$$A_T(\tau) = \frac{1}{2} \cos \omega' \tau - \frac{1}{4\omega' T} \sin(2\omega' T + 2\phi)$$

† See Example (4) for the expression of this in terms of a probability distribution.

Thus, as $T \to \infty$,

(13.100)
$$A(\tau) = \tfrac{1}{2} \cos \omega' \tau$$

Now consider the ensemble way of calculating,

(13.101)
$$A(\tau) = \lim_{N \to \infty} \frac{1}{N} \sum_{k=1}^{N} y_k(t) y_k(t + \tau)$$

we write this in terms of the probability, just as in Eq. (13.89),

(13.102)
$$A(\tau) = \sum_{\phi} P(\phi) \sin(\omega' t + \phi) \sin(\omega' t + \omega' \tau + \phi)$$

Now here we have no information about $P(\phi)$ but we do know that Eq. (13.102) should yield the same result as found above *if* the system is ergodic. *Assume*, here, that ϕ is *bounded*, $-\pi < \phi < \pi$ and all values are *equally likely*,

(13.103)
$$P(\phi) \Rightarrow \frac{d\phi}{2\pi}$$

then in Eq. (13.102) our sum over the spectrum of ϕ becomes the integral,

(13.104)
$$A(\tau) = \int_{-\pi}^{\pi} \sin(\omega' t + \phi) \sin(\omega' t + \omega' \tau + \phi) \frac{d\phi}{2\pi}$$

By the same trigonometric identity used above we find that indeed

(13.105)
$$A(\tau) = \tfrac{1}{2} \cos \omega' \tau$$

so that in this case the system *is* ergodic.

PROPERTIES OF THE CORRELATION FUNCTION AND ITS FOURIER TRANSFORM FOR ERGODIC SYSTEMS

Here, we list a few important properties of the correlation $A(\tau)$ for a stationary time series:
    (a) $A(0)$ is the mean-square value of $y(t)$

$$\overline{y^2} = \langle y^2 \rangle = A(0)$$

    (b) $A(\tau)$ is an *even* function,

$$A(\tau) = A(-\tau)$$

    (c) $A(0)$ is an upper bound for $A(\tau)$,

$$A(\tau) \leq A(0), \qquad |\tau| > 0$$

These can be readily shown and will be left to the exercises.
    The sufficient condition for the Fourier transform of $A(\tau)$ to exist is

(13.106)
$$\int_{-\infty}^{+\infty} |A(\tau)| \, d\tau < \infty$$

as for *any* function of τ. Assuming this condition fulfilled we define

(13.107)
$$\Omega(\omega) = \int_{-\infty}^{+\infty} e^{-i\omega\tau} A(\tau) \, d\tau$$

as the Fourier transform of $A(\tau)$. The inverse is, of course,

$$(13.108) \qquad A(\tau) = \frac{1}{2\pi} \int_{-\infty}^{+\infty} e^{+i\omega\tau}\Omega(\omega)\,d\omega$$

From the definition of $A(\tau)$ one can readily deduce the following properties of $\Omega(\omega)$:

(a) $\Omega(\omega)$ is an *even* function of ω.

$$\Omega(\omega) = \Omega(-\omega)$$

(b) $\Omega(\omega)$ is a *real* function of ω,

$$\Omega^*(\omega) = \Omega(\omega)$$

(c) $\Omega(\omega)$ is always positive,

$$\Omega(\omega) \geq 0$$

(d) The mean-square value of $y(t)$ is given in terms of $\Omega(\omega)$ by:

$$\overline{y^2} = \frac{1}{2\pi} \int_{-\infty}^{+\infty} \Omega(\omega)\,d\omega$$

We also leave the demonstration of these properties to the exercises.

EXECUTE PROBLEM SET (13-5)

LINEAR SYSTEMS AND RANDOM NOISE

At several points in this text we have considered the response of a linear system to a given input. Here, we again consider the output (reading) of a linear instrument, $I(t)$, due to an input $S(t)$ and write,

$$(13.109) \qquad I(t) = \int_{-\infty}^{+\infty} R(t - \tau)S(\tau)\,d\tau$$

for a system as discussed in Chapter Nine. Here, $R(t)$ is the response function of the system, as defined in Chapter Nine.

Now we consider the input, $S(t)$, to the system to consist of a "signal," $S_0(t)$ *plus* some random "noise," $S_N(t)$, which we *assume* to be a stationary time series and to satisfy *ergodic* conditions. If we assume the instrument to be *pure*, i.e., it introduces no additional noise, then

$$(13.110) \qquad I(t) = I_0(t) + I_N(t)$$

as can readily be verified by substituting

$$(13.111) \qquad S(t) = S_0(t) + S_N(t)$$

into Eq. (13.109). Here,

$$(13.112) \qquad I_0(t) = \int_{-\infty}^{+\infty} R(t - \tau)S_0(\tau)\,d\tau$$

and

$$(13.113) \qquad I_N(t) = \int_{-\infty}^{+\infty} R(t - \tau)S_N(\tau)\,d\tau$$

Now we can form the correlation of the *output* function as:

(13.114) $\overline{I(t)I(t + \tau)} = \overline{I_0(t)I_0(t + \tau)} + \overline{I_0(t)I_N(t + \tau)} + \overline{I_N(t)I_0(t + \tau)} + \overline{I_N(t)I_N(t + \tau)}$

and we can do the same thing with the *input function*.

(13.115) $\overline{S(t)S(t + \tau)} = \overline{S_0(t)S_0(t + \tau)} + \overline{S_0(t)S_N(t + \tau)} + \overline{S_0(t + \tau)S_N(t)} + \overline{S_N(t)S_N(t + \tau)}$

Now the "power" of a signal is defined as proportional to the square of its magnitude and, hence, the *mean power* is just the mean square value of the time function.

Thus, for the *input* we have, in view of this definition, by putting $\tau = 0$ in Eq. (13.113),

(13.116) $$\overline{S^2} = \overline{S_0^2} + \overline{2S_0(t)S_N(t)} + \overline{S_N^2}$$

and similarly for the output. We saw in Chapter Nine† that,

(13.117) $$\int_{-\infty}^{+\infty} F(t)G(t)\, dt = \frac{1}{2\pi} \int_{-\infty}^{+\infty} f(\omega)g(-\omega)\, d\omega$$

for *any* two functions $F(t)$, $G(t)$, having Fourier transforms $f(\omega)$ and $g(\omega)$ (the "Parseval relation"). From this we can readily show that if the frequency distributions of $S_0(t)$ and $S_N(t)$ do not significantly overlap then $\overline{S_0(t)S_N(t)}$ will be negligible compared to the other two terms. Then, we define the *input signal to noise ratio* as,

(13.118) $$\mathcal{R}_S = \sqrt{\frac{\overline{S_0^2}}{\overline{S_N^2}}}$$

Similarly, and on the same basis of neglecting, $\overline{I_0(t)I_n(t)}$, we define the *output signal to noise ratio* as,

(13.119) $$\mathcal{R}_I = \sqrt{\frac{\overline{I_0^2}}{\overline{I_N^2}}}$$

Thus, since I_0 and I_N can be computed from S_0 and S_N, for a known response function $R(t)$, we can examine the effect of the "instrument" on the signal to noise ratio.

To pursue this topic further would take us far beyond the scope of the present text. These ideas lead to studies of filter design in electrical circuits, telephone line transmissions, etc. The student can readily see that a knowledge of *Fourier transforms* and *probability theory* is essential in these areas. The references should be consulted for further study.

GENERAL RANDOM WALK IN THREE DIMENSIONS AND DIFFUSION

Here, we show how the general rules for combining probabilities can be used to formulate a *difference equation* whose solution is the desired probability in a problem, and then we show how this leads, via a limiting process to the diffusion equation.

We consider a point that is free to move along any one of the rectangular cartesian axes, x, y, z by unit steps of size Δl during a time Δt. We let p_x, p_y, p_z and q_x, q_y, q_z be the probabilities for a positive step along x, y, or z, respectively, and a *negative* step along x, y, or z, respectively, at time $N\Delta t = t$ during Δt. We also allow a probability γ that the point remains fixed during this time interval. Since these form a mutually exclusive and exhaustive set of events (we do not here, consider *absorption* or *sources*) we have,

(13.120) $$1 = \gamma + p_x + p_y + p_z + q_x + q_y + q_z$$

† Problem (5), set (9-2).

Now letting $P(i, j, k, N) = P_{i,j,k,N}$ denote the probability for the point to be at $x = i\,\Delta l$, $y = j\,\Delta l$, $z = k\,\Delta l$ at time $t = N\,\Delta t$, we see that this can occur in the following exclusive ways.

$$(i - 1, j, k, N - 1) \to (i, j, k, N)$$

$$(i, j - 1, k, N - 1) \to (i, j, k, N)$$

$$(i, j, k - 1, N - 1) \to (i, j, k, N)$$

$$(i + 1, j, k, N - 1) \to (i, j, k, N)$$

$$(i, j + 1, k, N - 1) \to (i, j, k, N)$$

$$(i, j, k + 1, N - 1) \to (i, j, k, N)$$

$$(i, j, k, N - 1) \to (i, j, k, N)$$

Thus, by the rules for combining probabilities the probability for the first process above is just $p_x P(i - 1, j, k, N - 1)$, where $P(i - 1, j, k, N - 1)$ is the probability for $i - 1, j, k, N - 1$ to occur and p_x is the *conditional* probability for the plus step in x *if* the said event does exist at $N - 1$. To obtain $P(i, j, k, N)$ then we sum all such probabilities over all the exclusive cases above, making use of Eq. (13.118) to eliminate γ. We obtain

$$P_{i,j,k,N} = [P_{i,j,k} + p_x(P_{i-1,j,k} - P_{i,j,k}) + q_x(P_{i+1,j,k} - P_{i,j,k}) + p_y(P_{i,j-1,k} - P_{i,j,k})$$
(13.121)
$$+ q_y(P_{i,j+1,k} - P_{i,j,k}) + p_z(q_{i,j,k-1} - P_{i,j,k}) + q_z(P_{i,j,k-1} - P_{i,j,k})]$$

where all P's on the right carry a subscript $N - 1$ which has been deleted here to avoid confusion.

Here, if we were given some initial condition on $P(i, j, k, N)$ at $N = 0$ and some boundary conditions, if the domain is not infinite, we could, in principle, solve this equation to determine $P(i, j, k, N)$. However, we will not undertake this here, instead we show how this leads to the ordinary diffusion equation if a certain limiting process is applied.

We assume, here, $p = p_x = q_x = p_y \ldots$ etc., i.e., all the "transition probabilities" are equal to p. Then we rearrange and write with division by Δt,

$$\frac{P_{i,j,k,N} - P_{i,j,k,N-1}}{\Delta t} = \frac{(\Delta l)^2 p}{\Delta t}$$
(13.122)

$$\bullet \left\{ \frac{P_{i+1,j,k} + P_{i-1,j,k} - 2P_{i,j,k}}{(\Delta l)^2} + \frac{P_{i,j+1,k} + P_{i,j-1,k} - 2P_{i,j,k}}{(\Delta l)^2} + \frac{P_{i,j,k+1} + P_{i,j,k-1} - 2P_{i,j,k}}{(\Delta l)^2} \right\}$$

Now in the limit the term on the left approaches $\partial P/\partial t$ evaluated at $x = i\,\Delta l$, $y = j\,\Delta l$, $z = k\,\Delta l$ and $t = (N - 1)\,\Delta t$, similarly,

$$\text{(13.123)} \qquad \lim_{\Delta l \to 0} \frac{P_{i+1,j,k} + P_{i-1,j,k} - 2P_{i,j,k}}{(\Delta l)^2} = \frac{\partial^2 P}{\partial x^2}$$

evaluated at the same point and time. The other terms yield $\partial^2 P/\partial y^2$ and $\partial^2 P/\partial z^2$, respectively. Thus, if we assume that the limits $\Delta t \to 0$, $\Delta l \to 0$ are taken in such a manner that

$$\text{(13.124)} \qquad \lim_{\substack{\Delta t \to 0 \\ \Delta l \to 0}} \left[\frac{(\Delta l)^2 p}{\Delta t} \right] = D$$

is *finite*, then we obtain,

(13.125)
$$\frac{\partial P}{\partial t} = D\nabla^2 P$$

which is just the diffusion equation for $P(x, y, z, t)$.

EXECUTE PROBLEM SET (13-6)

MARKOV PROCESSES

In the random-walk process just above the probabilities, p_x, q_x, etc., did *not* depend on what step, or steps that had preceded a given step. That is, the *transition probability* did not depend on the history of the past events. The simplest generalization of such a sequence of *independent* steps, or trials, is to make the probability for a step or trial depend on the outcome of the immediately preceeding trial. Such a process is called a *Markov process*. We give a *general* symbolic formulation of such a process as follows.

We have a sequence of trials with the outcome of any trial being one of the numbers $E_1, E_2, \ldots E_j, \ldots$, which may be an infinite or finite set. For *independent* trials we would have our old system in which a fixed probability p_j would exist for E_j at any trial. But now we define a set of probabilities, p_{ij}, which represent the probabilities for outcome of *a* trial to be E_j if E_i occurred at the previous trial. Thus, the p_{ij} are a set of *conditional probabilities*.

By our usual rules then for combining probabilities we have the probability for the sequence of outcomes, $E_1, E_2, E_3 \ldots E_k$,

(13.126)
$$P(E_1, E_2, \ldots E_k) = a_1 p_{1,2} p_{2,3} p_{3,4} \cdots p_{k-1,k}$$

where we have inserted a_1 as the *unconditional probability* for the starting event of the *Markov chain* to be E_1.

If one sets up a "random-walk" problem in which the p_x, p_y, etc., depend on position, then, in essence we have a Markov process. The limiting case, then, is one in which the diffusion constant, or constants for anisotropic cases depend on position.

Higher-order transition probabilities can be formulated by building on the $p_{i,j}$ already defined. Thus, note that in Eq. (13.126) we see the transition from E_1 to E_k in exactly k unique steps. However, there are other sequences that also lead from E_1 to E_k, in k steps. But let us look first at a simpler case.

$p_{j,k}^{(1)}$ is the probability of the step $i \to j$, *if i* occurs; this is in *one step*. We could have the sequence $i \to l$ then $l \to j$; this would have the probability $p_{i,l} p_{l,j}$. Of course, there are *many* possibilities for the intermediate state l for this *two-step* process. Thus, we see

(13.127)
$$p_{i,j}^{(2)} = \sum_l p_{i,l} p_{l,j}$$

as the probability of the transition $i \to j$ in *some* two-step process, *if i* does occur.

The subject of *Markov processes* leads to *many* important applications in many areas of science, but here, we can do no more than essentially *define* such processes. The student must consult the references for a complete discussion of these topics.

CENTRAL LIMIT THEOREM

Since so much of physics and engineering deals with *linear* processes it is appropriate that we close this chapter with a *statement* of an important theorem about the probability for occurrence of a particular value of a *linear* combination,

(13.128) $$u = C_1 y_1 + C_2 y_2 + \ldots + C_n y_n$$

of statistically independent random variables. Here, the y_n are these random variables. We can multiply u by the probability $P_1(y_1)P_2(y_2) \ldots P_n(y_n)$ and compute by summation,

(13.129) $$\bar{u} = C_1 \bar{y}_1 + C_2 \bar{y}_2 + \ldots + C_n \bar{y}_n$$

the *mean*, and then define a set of new random variables, $x_i = C_i(y_i - \bar{y}_i)$, all of which have mean $\bar{x}_i = 0$.

In terms of these we define the *moment of order* k as,

(13.130) $$\sum_{x_i} x_i^k P(x_i) = \mu_k^{(i)}$$

i.e., this is the kth moment of the probability distribution in the variable x_i.† We also define the dispersion of the sum as

(13.131) $$\sigma_n = \sum_{x_1} \sum_{x_2} \ldots \sum_{x_n} (x_1 + x_2 + \ldots + x_n)^2 P_1(x_1) P_2(x_2) \ldots P_n(x_n)$$

Then if the limit,

(13.132) $$\underset{n \to \infty}{\text{limit}} \frac{\mu_{2+\delta}^{(1)} + \mu_{2+\delta}^{(2)} + \ldots + \mu_{2+\delta}^{(n)}}{\sigma_n^{1+\frac{1}{2}\delta}} = 0$$

for all $\delta > 0$, the probability for

(13.133) $$v \le \frac{x_1 + x_2 + \ldots + x_n}{\sqrt{\sigma_n}} \le v + dv$$

is:

(13.134) $$dP = \frac{1}{\sqrt{2\pi}} e^{-\frac{v^2}{2}} \, dv$$

This is the Laplace–Liapounoff form of the Central Limit Theorem.

This means that *in the limit of large n*, the probability distribution for the linear combination, u in Eq. (13.126) is,

(13.135) $$dP(u) = \frac{1}{\sqrt{2\pi}\sigma} e^{-\frac{(u - \bar{u})^2}{2\sigma}} \, du$$

where

(13.136) $$\sigma = \underset{n \to \infty}{\text{limit}} (\sigma_1 + \sigma_2 + \ldots + \sigma_n)$$

and the σ_j are defined by:

(13.137) $$\sigma_j = \sum_{y_j} C_j^2 (y_j - \bar{y}_j)^2 P_j(y_j)$$

Thus, in the limit of large n every *linear combination of statistically independent random variables has a probability distribution approximating to a normal or Gaussian distribution.*

EXECUTE PROBLEM SET (13-7)

---

† In the Appendix an important connection between these moments and a Fourier transform is pointed out.

PROBLEMS

Set (13-1)

(1) Show from the definition of probability, in terms of enumeration of cases that Eq. (13.2), (13.3), and (13.4) are true.

(2) Consider the following problem. An urn is filled with red and white balls and every ball is marked with a number, either *one* or *two*. Thus for example, let there be N_{W1} *white* balls marked with a *one*, N_{W2} *white* balls marked with a *two*, N_{R1} red balls marked with a *one*, and N_{R2} red balls marked with a *two*. Drawing a ball from the urn is a trial for a *joint event*, namely observing a color, W or R, as say event A, and observing a number, 1 or 2, as event B. Use the enumeration of cases to define: $P(W)$, $P(R)$, $P(1)$, $P(2)$; $P(W, 2)$, $P(W, 1)$ $P(R, 1)$, $P(R, 2)$, and also the conditional probabilities $P(1|R)$, $P(2|R)$, etc. Thus in terms of this example, show that Eq. (13.5) holds true for all of these various probabilities which can be defined.

(3) Consider the problem of drawing cards from a well shuffled deck of 52 cards. Define the conditional probability function for drawing an ace on the *second* draw from the deck if the first card drawn is *not* returned to the deck. What does this become if the first card drawn is returned and the deck shuffled before the second card is drawn?

Set (13-2)

(1) What is the probability in 10 consecutive rolls of a pair of dice that a seven will show on each of the first five rolls and will not show again until the tenth roll? Compare this to the probability that the same number of sevens will show in *any* order whatever during the 10 rolls of the dice.

(2) Assuming all sex distributions to be equally probable, what proportion of families with exactly six children should be expected to have three boys and three girls?

(3) Consider the probability, $P(n, \gamma, t)$ for n occurrences of a rare event during a "time" period t as in Eq. (13.21). Now consider *repeating* periods of observation of duration t *many* times, say M periods altogether. Then suppose that in M_0 of these periods the event does not happen at all, in M_1 of these periods it happens *one* time, in M_2 it happens *two* times, etc., so that

$$M_0 + M_1 + M_2 + \ldots = M$$

The total number of events observed is

$$M_1 + 2M_2 + 3M_3 + \ldots = T$$

Now according to the "law of large numbers" we should expect

$$\frac{M_k}{M} = P(k, \gamma, t)$$

Thus

$$T \approx M[P(1, \gamma, t) + 2P(2, \gamma, t) + \ldots]$$

Show that this yields

$$\gamma t = \frac{T}{M}$$

and hence offers a way of estimating γ from observational data. For example, consider the ringing of your home telephone during any 10-min period as a rare event with some parameter γ. Thus simply by keeping a record of total calls during one week you can estimate γ for $t = 10$.

(4) The "peak," or mode, or point of maximum probability in the Gaussian distribution occurs at $x = 0$, i.e., in Eq. (13.30). With what speed does this peak move in the diffusion approximation to the random-walk problem, and how does the amplitude of the peak change with time?

Set (13-3)

(1) Show for the Gaussian distribution,

$$dP(x) = \frac{1}{\sqrt{2\pi\sigma}} e^{-\frac{x^2}{2\sigma}} \, dx$$

that:

$$\int_{-\infty}^{+\infty} dP(x) = 1$$

$$\int_{-\infty}^{+\infty} x \, dP(x) = 0$$

and,

$$\int_{-\infty}^{+\infty} x^2 \, dP(x) = \sigma$$

(2) Show for the Poisson distribution,

$$P(n, \gamma, t) = e^{-\gamma t} \frac{(\gamma t)^n}{n!}$$

that,

$$\sum_{n=0}^{N} P(n, \gamma, t) = 1$$

only if we allow $N \to \infty$. Compute the mean value of n, $\bar{n}$, for this case.

(3) Show that if two random variables are statistically independent, and each has mean-value zero, then their correlation is zero. Thus, if we consider two variables x and y which are *not* statistically independent, and whose means are *not* zero then the quantity,

$$\rho = \frac{\overline{(x - \bar{x})(y - \bar{y})}}{\bar{x}\bar{y}}$$

is a measure of the statistical dependence between x and y, being zero for statistical independence and greater than zero (in magnitude) for statistical dependence.

(4) Compute the expectation value or mean value of n for the binomial distribution. Show that it is precisely $n = Np$.

Set (13-4)

(1) Show how the probabilities deduced in Problem (2), Set (13–1), fit into the format of Eq. (13.58).

(2) Consider drawing a card from a standard deck of 52 cards that has been well shuffled as a joint event, i.e., the card value is one random variable, x, which has the spectrum A, 2, 3, ... 10, J, Q, K; while the suit of the card is another random variable, y, having the spectrum: diamond, club, heart or spade. For simplicity we can say x has the spectrum 1, 2, 3 ... 13, while y has the spectrum, 1, 2, 3, 4. What is the probability of drawing any specified hand of five cards after a previous hand of five cards has been drawn? Write a general formula then write down the probability of drawing a full house, three cards of one value and two of another, *if* the first hand drawn was a flush, i.e., all in the same suit.

(3) Show that the probability distribution function,

$$dP = B[x(1 - x)y(1 - y)]^m \, dx \, dy$$

approximates to a *product* of a Gaussian distribution in x and a Gaussian distribution in y, with $\sigma = \frac{1}{4}m$ being the mean-square variance, if m is very large. [Hint: Follow the format of Eq. (13.72). In this approximation find B.]

(4) Find the transformation matrix, $[a_{ij}] = A$, which transforms the bivariate Gaussian distribution,

$$dP = \alpha e^{-\frac{x^2 + y^2 + \rho \, x \, y}{2\sigma}} dx \, dy$$

to variables x', y' which have a Gaussian distribution but are statistically independent. Also find the value of α.

Set (13-5)

(1) Prove properties (a), (b), and (c) of the autocorrelation function, $A(\tau)$ for an ergodic system. These are listed below Eq. (13.105).

(2) Prove properties (a), (b), (c), and (d) for the Fourier transform, $\Omega(\omega)$, of the autocorrelation function of an ergodic system. [These are listed following Eq. (13.108).

Set (13-6)

(1) Show that if all the transition probabilities are *not* taken as being equal in the general random-walk problem then the limiting form can be shown to lead to the general diffusion equation with "drift,"

$$D\nabla^2 P - \mathbf{v} \cdot \nabla P = \frac{\partial P}{\partial t}$$

where D and $\mathbf{v}$ are suitably defined.

Set (13-7)

(1) If we consider a Markov process as a sequence of spatial transitions of a point in time then we can define an *unconditional* probability for a "state" represented by a "position vector," $\mathbf{r}$ at time t as $P(\mathbf{r}, t)$ and a *conditional* probability for a transition from a "point," $\mathbf{r} - \mathbf{u}$ in this space during a *period* Δt to the point $\mathbf{r}$ as $C(\mathbf{u}|\mathbf{r} - \mathbf{u}, t) \, du_1 \, du_2 \, du_3$ (for a three dimensional Cartesian space). Show then that

$$P(\mathbf{r}, t + \Delta t) = \int P(\mathbf{r} - \mathbf{u}, t)C(\mathbf{u}|\mathbf{r} - \mathbf{u}, t) \, du_1 \, du_2 \, du_3$$

Also show that if C is independent of $\mathbf{r} - \mathbf{u}$ and has mean of $\mathbf{u}$ equal to zero then expansion of $P(\mathbf{r} - \mathbf{u}, t)$ in a Taylor series in u_1, u_2, u_3 yields the diffusion equation, to second order approximation in the u_i.

(2) Show that if the outcome of an experiment is the value of a variable x, and if the experiment is repeated n times, then assuming x to be a random variable with standard variance, or root-mean-square error $\sqrt{\sigma}$ indicates that the *mean* value x of the outcome of many experiments has root-mean-square error $\sqrt{\sigma}/\sqrt{n}$.

REFERENCES

H. Cramer, *Mathematical Methods of Statistics* (John Wiley and Sons, Inc., New York, 1950).

W. Feller, *An Introduction to Probability Theory and its Applications* (John Wiley and Sons, Inc., New York, 1961).

J. V. Misyurkeyev, *Problems in Mathematical Physics with Solutions* (McGraw-Hill Book Company, Inc., New York, 1966).

J. V. Uspensky, *Introduction to Probability Theory* (Dover Publishing Co. Inc., New York, 1967).

J. Walker and R. L. Mathews, *Mathematical Methods of Physics* (W. A. Benjamin, Inc., New York, 1965).

A. M. Yaglom, *An Introduction to the Theory of Stationary Random Functions* (translated by R. A. Silverman), (Prentice-Hall, Inc., Englewood Cliffs, N.J., 1962).

CHAPTER FOURTEEN

Miscellaneous topics: evaluation of integrals, summation of series, curve fitting, transcendental equations

EVALUATION OF INTEGRALS

One of the most difficult steps encountered in the solution of many problems in applied mathematics is often the evaluation of some integral in *closed* form, either exactly or approximately. Some of the elementary "tricks" normally used are:

(a) change of variables,
(b) integration by parts,
(c) expansion by partial fractions,

but these are often inadequate for accomplishing the integration. In such cases one of the techniques already treated in this text may prove suitable for the task. We list these again, here in summary for quick reference.

(d) The Laplace transform method, treated in Chapter Six and applicable to a wide range of integral types:

(e) Integration around the unit circle in the complex plane, treated in Chapter Eight and applicable to integrals of the type,

$$\int_0^{2\pi} F(\sin \theta, \cos \theta) \, d\theta$$

with certain restrictions on the acceptable forms for the function, F.

(f) Integration around the infinite half-circle and along the real axis in the complex plane treated in Chapter Eight also, and suitable for evaluating infinite integrals of the types,

$$\int_{-\infty}^{+\infty} f(x)\, dx, \qquad \int_{-\infty}^{+\infty} e^{imx} g(x)\, dx \quad \text{or} \quad \int_{0}^{\infty} h(x)\, dx$$

provided the integrands meet certain specifications.

(g) Integration on the contour of an infinite circle with a branch line cut to a branch point in the complex plane, treated in Chapter Eight and suitable for infinite integrals of the type

$$\int_{0}^{\infty} F(x)\, dx$$

provided $F(x)$ meets certain conditions.

(h) Method of steepest descent, treated in Chapter Eight and suitable for the approximate evaluation of integrals of the type

$$\int_{C} f(z) e^{tg(z)}\, dz$$

on some contour in the complex plane. Here, t, $g(z)$, and $f(z)$ must meet certain specifications. This is also applicable to *real* integrals in some circumstances.

(i) Method of orthogonal transformations, used in the problem of the multivariate Gaussian distribution in Chapter Thirteen, suitable for multiple integrals of the form treated there and some extensions.

There are a few other "tricks," which can sometimes be applied to advantage, which we describe below. These are all based on the following general ideas.

LINEAR OPERATIONS ON INTEGRALS

The process of integration is itself a *linear* operation, i.e., a limit of a sum. Here, we consider some other *linear operation*, indicated by O which has an inverse O^{-1} and *commutes* with integration. Thus, say to a function $f(x)$ we may apply the operation,

$$\int dx$$

(14.1)
$$I = \int_{a}^{b} f(x)\, dx$$

to define the integral I. We can also apply O to f to produce

(14.2)
$$Of = \bar{f}(x)$$

and since the inverse is required to exist,

(14.3)
$$O^{-1}\bar{f}(x) = f(x)$$

Also, since O and O^{-1}, commute with

$$\int dx$$

we have,

(14.4)
$$O \int_{a}^{b} f(x)\, dx = \int_{a}^{b} Of(x)\, dx = \int_{a}^{b} \bar{f}(x)\, dx$$

then,

$$(14.5) \qquad O^{-1} \int_a^b \bar{f}(x)\, dx = \int_a^b O^{-1}\bar{f}(x)\, dx = \int_a^b f(x)\, dx$$

The student should recognize that these two equations, Eqs. (14.4) and (14.5), represent, in a general symbolic way, exactly the processes applied in using the Laplace transform to evaluate an integral, i.e., in that case O corresponds to taking the Laplace transform of $f(x)$ *with respect to a parameter*, say t, in $f(x)$; then this transform, $\bar{f}(x)$ is integrated with respect to x, and finally the inverse Laplace transform, corresponding to O^{-1} here, is applied to this result, to yield the integral in closed form.

There are a wide variety of linear operations, O, which can be applied in this manner to yield the value of an integral in closed form. The essential element is that the integrand, $f(x, t)$, contains a *parameter*, t here, on which O acts, and that suitable conditions of uniform convergence with respect to t, etc. be met. We illustrate such an operation by the following example.

Example (1)

The operation, $O = \partial/\partial t$ has the inverse,

$$O^{-1} = \int dt$$

thus, consider the integral,

$$(14.6) \qquad I = \int_0^\infty \frac{e^{-tx} \sin bx}{x}\, dx = I(t, b), \qquad b \neq 0$$

which *converges uniformly* for all $t \geq 0$. Thus, we have the legitimate operation,

$$(14.7) \qquad \frac{\partial I}{\partial t} = -\int_0^\infty e^{-tx} \sin bx\, dx$$

differentiating *under* the integral sign. This integral can be evaluated by elementary means to yield,

$$(14.8) \qquad \frac{\partial I}{\partial t} = -\frac{b}{b^2 + t^2}$$

Then we apply the *inverse* operation of integrating with respect to t to yield,

$$(14.9) \qquad I(t, b) = -\tan^{-1}\frac{t}{b} + C$$

where C is a constant of integration. Next, we note that in the original expression for I, Eq. (14.6), if we let $t \to \infty$ then $I \to 0$. Thus,

$$(14.10) \qquad 0 = -\tan^{-1}(\infty) + C$$

or, $C = \pi/2$. Hence, we have the result

$$(14.11) \qquad \int_0^\infty \frac{e^{-tx} \sin bx}{x}\, dx = \frac{\pi}{2} - \tan^{-1}\frac{t}{b}, \qquad b \neq 0$$

We also note that had we been confronted with say,

(14.12)
$$I' = \int_0^\infty \frac{e^{-x} \sin x}{x} \, dx$$

we could readily insert the parameter t into the exponent, carry out the above operations and *then* put $t = 1$ in the final result along with $b = 1$ for this special case.

EXECUTE PROBLEM SET (14-1)

SUMMATION OF SERIES

The summation of series, finite or infinite, in *closed form* is *not* a highly developed area in applied mathematics, although there do exist a few methods for series of certain types. Before we consider these special types, however we note that the same ideas with regard to the use of *linear operators* applies to the evaluation of sums as to integrals. The general idea is this: Given a sum

(14.13)
$$S(x) = \sum_n a_n(x)$$

in which a parameter x appears (or is appropriately inserted if it is not originally present), we operate on $S(x)$ with some linear operator, O, acting on x, to produce

(14.14)
$$O\{S(x)\} = \bar{S} = \sum_n O\{a_n(x)\} = \sum_n \bar{a}_n$$

where the series must be such that O and

$$\sum_n$$

commute. If O has been properly chosen we should *recognize the sum*, $\bar{S}$, and be able to write it in a *closed form*. Then the inverse operation, O^{-1} is applied to $\bar{S}$ to yield $S(x)$ in closed form, i.e.,

(14.15)
$$S(x) = O^{-1}\{\bar{S}\}$$

In many cases the Laplace transform can be used in this way as well as various differential operators, just as for integrals, as shown above. Here, as with integrals the student must be cautioned that existence criteria, uniform convergence, etc. must be carefully scrutinized in each individual case. For *finite* sums these conditions are not at all stringent and are easily met for nearly *all* linear operations, O.

A key point in this method for summing series is that the application of O to $S(x)$ yield a series which we *know* already in *closed* form. This means that it would be nice to have readily available a *table* of useful sums. We present a limited table of this sort as Table 14–1.

Now we illustrate with a few examples.

Example (2)

Find the value of the sum

(14.16)
$$S = \sum_{n=1}^\infty \frac{1}{n2^n}$$

TABLE 14-1. Sums of Series

(1) $(1 - x)^{-1} = 1 + x + x^2 + \ldots + x^n + \ldots,$ $\qquad [x^2 < 1]$

(2) $(1 + x)^n = 1 + nx + \dfrac{n(n-1)}{2!}x^2 + \ldots + \dfrac{n!x^k}{k!(n-k)!} + \ldots + x^n$ $\qquad [x^2 < 1]$

(3) $(1 + x)^{\frac{1}{2}} = 1 + \dfrac{1}{2}x - \dfrac{1}{2\cdot 4}x^2 + \dfrac{3}{2\cdot 4\cdot 6}x^3 - \dfrac{3\cdot 5}{2\cdot 4\cdot 6\cdot 8}x^4 + \ldots$ $\qquad [x^2 < 1]$

(4) $e^x = 1 + x + \dfrac{x^2}{2!} + \dfrac{x^3}{3!} + \ldots + \dfrac{x^n}{n!} + \ldots,$ $\qquad [x^2 < \infty]$

(5) $\ln x = (x-1) - \dfrac{1}{2}(x-1)^2 + \dfrac{1}{3}(x-1)^3 - \ldots,$ $\qquad [0 < x < 2]$

(6) $\ln x = \dfrac{x-1}{x} + \dfrac{1}{2}\left(\dfrac{x-1}{x}\right)^2 + \dfrac{1}{3}\left(\dfrac{x-1}{x}\right)^3 + \ldots,$ $\qquad [x > \frac{1}{2}]$

(7) $\ln(1 + x) = x - \dfrac{1}{2}x^2 + \dfrac{1}{3}x^3 - \dfrac{1}{4}x^4 + \ldots,$ $\qquad [x^2 < 1]$

(8) $\sin x = x - \dfrac{x^3}{3!} + \dfrac{x^5}{5!} - \dfrac{x^7}{7!}\ldots,$ $\qquad [x^2 < \infty]$

(9) $\cos x = 1 - \dfrac{x^2}{2!} + \dfrac{x^4}{4!} - \dfrac{x^6}{6!}\ldots,$ $\qquad [x^2 < \infty]$

(10) $\tan^{-1} x = x - \dfrac{1}{3}x^3 + \dfrac{1}{5}x^5 - \dfrac{1}{7}x^7 + \ldots,$ $\qquad [x^2 < 1]$

(11) $1 = \dfrac{4}{\pi}\left[\sin\dfrac{\pi x}{L} + \dfrac{1}{3}\sin\dfrac{3\pi x}{L} + \dfrac{1}{5}\sin\dfrac{5\pi x}{L} + \ldots\right],$ $\qquad [0 < x < L]$

(12) $x = \dfrac{2L}{\pi}\left[\sin\dfrac{\pi x}{L} - \dfrac{1}{2}\sin\dfrac{2\pi x}{L} + \dfrac{1}{3}\sin\dfrac{3\pi x}{L} - \ldots\right],$ $\qquad [-L < x < L]$

(13) $x^2 = \dfrac{L^2}{3} - \dfrac{4L^2}{\pi^2}\left[\cos\dfrac{\pi x}{L} - \dfrac{1}{2^2}\cos\dfrac{2\pi x}{L} + \dfrac{1}{3^2}\cos\dfrac{3\pi x}{L} - \ldots\right],$ $\qquad [-L < x < L]$

(14) $\ln\sin\dfrac{x}{2} = -\ln 2 - \cos x - \dfrac{1}{2}\cos 2x - \dfrac{1}{3}\cos 3x - \ldots,$ $\qquad \left[0 < x < \dfrac{\pi}{2}\right]$

Here, except for the $1/n$ factor in each term this looks like

$$\sum_{n=1}^{\infty} (\tfrac{1}{2})^n$$

which is essentially the infinite geometric series, Item Number (1) of Table 14–1, with $x = \frac{1}{2}$. Thus, we start with

(14.17) $$G(x) = \dfrac{1}{1-x} = \sum_{k=0}^{\infty} x^k$$

then integrate term by term to get

$$(14.18) \qquad \int_0^x \frac{dx}{1-x} = -\ln(1-x) = \sum_{k=0}^{\infty} \frac{x^{k+1}}{k+1}$$

Then, with the change of notation $n = k + 1$, and putting $x = \frac{1}{2}$, we have

$$(14.19) \qquad \ln 2 = \sum_{n=1}^{\infty} \frac{1}{n2^n}$$

Checking back we note that all the operations we have used are legitimate since $G(x)$ converges *uniformly* for $0 < x^2 < 1$.

As a by-product of this problem we see that we have also shown the general result,

$$(14.20) \qquad \ln(1-x) = -\sum_{n=1}^{\infty} \frac{x^n}{n}$$

valid for $x^2 < 1$. (See Item Number (7) of Table 14–1.)

Example (3)

Find the value of the sum

$$(14.21) \qquad S = \sum_{n=2}^{\infty} \frac{e^{-nt}}{n^2 - 1}, \qquad t > 0$$

Here, several methods might be used. Note that if we let $x = e^{-t}$ this appears as

$$(14.22) \qquad S = \sum_{n=2}^{\infty} \frac{x^n}{n^2 - 1}, \qquad x^2 < 1$$

for $t > 0$, and again a connection to the geometric series is suggested. Thus, we use *partial fractions*,

$$(14.23) \qquad \frac{1}{n^2 - 1} = \frac{1}{2}\left(\frac{1}{n-1} - \frac{1}{n+1}\right)$$

and have

$$(14.24) \qquad S = S_1 - S_2 = \sum_{n=2}^{\infty} \frac{1}{2}\frac{x^n}{n-1} - \sum_{n=2}^{\infty} \frac{1}{2}\frac{x^n}{n+1}$$

Here, note that, with $G(x) = (1-x)^{-1}$, and Item (1) of Table (14-1),

$$(14.25) \qquad \frac{1}{x}\int_0^x G(x)\,dx = \sum_{n=0}^{\infty} \frac{x^n}{n+1}$$

and, therefore,

$$(14.26) \qquad 2S_2 = \frac{1}{x}\int_0^x G(x)\,dx - 1 - \frac{x}{2}$$

or, evaluating the integral

$$(14.27) \qquad 2S_2 = -\frac{1}{x}\ln(1-x) - 1 - \frac{x}{2}$$

Similarly, if we change the summation index in S_1, $n = n' + 2$,

(14.28)
$$2S_1 = \sum_{n'=0}^{\infty} \frac{x^{n'+2}}{n'+1}$$

we see that,

(14.29)
$$2S_1 = x \int_0^x G(x)\,dx = -x\ln(1-x)$$

Thus, collecting these results we have,

(14.30)
$$S = \frac{1}{2} + \frac{x}{4} + \frac{1}{2}\frac{1-x^2}{x}\ln(1-x)$$

or putting back in the form $x = e^{-t}$, we have,

(14.31)
$$\sum_{n=2}^{\infty} \frac{e^{-nt}}{n^2-1} = \frac{1}{2}\left(1 + \frac{e^{-t}}{2}\right) + \sinh t \ln(1 - e^{-t})$$

Here, as in our previous example all the operations we have used in the series are legitimate because of the uniform convergence for $x^2 < 1$ in all the series encountered.

EXECUTE PROBLEM SET (14-2)

SUMMATION OF SERIES BY CONTOUR INTEGRATION

The basis of this method of summing a series can be outlined as follows. Consider, the contour integral of the *product* of two functions, $F(z)G(z)$, around some contour C on which the product is single valued and analytic, and inside which $F(z)$ has a *small finite* number of poles while $G(z)$ may have a very large number, N, of *simple poles*. Let none of the poles of G and F coincide and let the poles of G be at points,

(14.32)
$$\text{Poles of } G(z) = \xi_1, \xi_2, \ldots \xi_N$$

with residues

(14.33)
$$\text{Res}\{G(\xi_j)\} = a_j, \qquad j = 1, 2, \ldots N$$

Also, let the poles of $F(z)$ be,

(14.34)
$$\text{Poles of } F(z) = z_1, z_2, \ldots z_m, \qquad (m \ll N)$$

Then by the Cauchy–Residue Theorem of Chapter Eight we have

(14.35)
$$\oint_C F(z)G(z)\,dz = 2\pi i \left\{ \sum_{j=1}^{N} a_j F(\xi_j) + \sum_{z_i} \text{Res}[G(z)F(z)] \right\}$$

The first sum, here, takes this simple form because the poles of G are *simple poles*, whereas, the second sum may be more complex in form because some of the poles of $F(z)$ may be high-order poles. Here, then, *if* the contour integral on the left vanishes, or can by other means be written in a simple closed form, say I, we have:

(14.36)
$$\sum_{j=1}^{N} a_j F(\xi_j) = \frac{I}{2\pi i} - \sum_{z_i} \text{Res}[G(z)F(z)]$$

which gives us a formula for the sum on the left. We can now use this approach to develop some special cases.

Case (1)

$G(z) = \pi \csc \pi z$; then the poles are at $z = \xi_n = 0, \pm 1, \pm 2, \ldots$, and the residues are

$$a_n = (-1)^n$$

If $F(z)$ has poles only in the *finite part* of the z plane, we call it a *meromorphic function*. Then, if $|zF(z)| \to 0$ on the infinite circle $|z| \to \infty$, we take the contour to be the infinite circle not passing through any singularity of the integrand. $F(z)$ must also be a single valued function of z everywhere. For these conditions the contour integral,

$$\oint = I$$

in our formula above, vanishes and we have:

(14.37)
$$\sum_{n=-\infty}^{+\infty} (-1)^n F(n) = - \sum_{z_i} \text{Res}[F(z)\pi \csc \pi z]$$

Case (2)

$G(z) = \pi \cot \pi z$; then the poles are again at $z = \xi_n = 0, \pm 1, \pm 2, \ldots$, but the residues $a_n = +1$ for all n. For the *same* conditions on $F(z)$ and the same choice of contour C, I again is zero and we obtain,

(14.38)
$$\sum_{n=-\infty}^{+\infty} F(n) = - \sum_{z_i} \text{Res}[F(z)\pi \cot \pi z]$$

so that we have two useful formulas for evaluating certain types of sums.

Quite obviously, there is a wide variety of functions we could choose for $G(z)$, $1/J_0(\beta z)$ for example, which has simple poles at $\beta z_j = x_j$, where the x_j are roots of $J_0(x) = 0$. This is valuable in summing series in which the sum is over the roots of J_0. We now illustrate by examples.

Example (4)

Find the sum,

(14.39)
$$S = \sum_{n=0}^{\infty} \frac{1}{a^2 + n^2}$$

We must first put this into the proper form of a sum on n from $-\infty$ to $+\infty$, thus,

(14.40)
$$S = \frac{1}{2} \sum_{-\infty}^{+\infty} \frac{1}{a^2 + n^2} + \frac{1}{2a^2}$$

and we then consider,

(14.41)
$$S' = \sum_{-\infty}^{+\infty} \frac{1}{a^2 + n^2}$$

which fits Case (2) above with,

(14.42)
$$F(z) = \frac{1}{a^2 + z^2}$$

This has simple poles at $\pm ia$ and hence, even if a is an integer, these do *not* coincide with the ξ_j. Evaluating the residues

$$(14.43) \qquad \operatorname*{Res}_{z=-ia} \left\{ \frac{\pi \cot \pi z}{a^2 + z^2} \right\} = \frac{\pi \cot i\pi a}{2ia} = -\frac{\pi}{2a} \coth \pi a$$

and

$$(14.44) \qquad \operatorname*{Res}_{z=-ia} \left\{ \frac{\pi \cot \pi z}{a^2 + z^2} \right\} = -\frac{\pi}{2a} \coth \pi a$$

Thus according to Eq. (14.38) we can write,

$$(14.45) \qquad S' = \frac{\pi}{a} \coth \pi a$$

and hence,

$$(14.46) \qquad S = \sum_{n=0}^{\infty} \frac{1}{a^2 + n^2} = \frac{\pi}{2a} \coth \pi a + \frac{1}{2a^2}$$

and we have the sum S in closed form.

Example (5)

Find the sum of the Fourier series,

$$(14.47) \qquad y(x) = \sum_{n=0}^{\infty} (-1)^n \frac{\cos nx}{n^2 + a^2}, \qquad -\pi < x < \pi$$

where x must be *real*, as must a also. Again we must put this into the form of a sum on n from $-\infty$ to $+\infty$, so we write,

$$(14.48) \qquad y(x) = \frac{1}{2} \sum_{-\infty}^{+\infty} (-1)^n \frac{\cos nx}{n^2 + a^2} + \frac{1}{2a^2}$$

as we did above. The sum

$$(14.49) \qquad S'(x) = \sum_{-\infty}^{+\infty} (-1)^n \frac{\cos nx}{n^2 + a^2}$$

now fits Case (1) above with $F(z)$ being,

$$(14.50) \qquad F(z) = \frac{\cos xz}{z^2 + a^2}$$

and here, we must not confuse this x with the real part of z. The poles of $F(z)$ are again at $z = \pm ia$ and this time the residues are:

$$(14.51) \qquad \operatorname*{Res}_{z=ia} \left\{ \frac{\pi \csc \pi z \cos xz}{z^2 + a^2} \right\} = -\frac{\pi}{2a} \frac{\cosh ax}{\sinh a\pi}$$

and

$$(14.52) \qquad \operatorname*{Res}_{z=-ia} \left\{ \frac{\pi \csc \pi z \cos xz}{z^2 + a^2} \right\} = -\frac{\pi}{2a} \frac{\cosh ax}{\sinh a\pi}$$

Thus, we have for S' according to Eq. (14.37),

(14.53)
$$S' = \frac{\pi}{a} \frac{\cosh ax}{\sinh ax}$$

and therefore,

(14.54)
$$y(x) = \sum_0^\infty \frac{(-1)^n \cos nx}{n^2 + a^2} = \frac{\pi}{2a} \frac{\cosh ax}{\sinh a\pi} + \frac{1}{2a^2}$$

is the sum of the Fourier series.

EXECUTE PROBLEM SET (14-3)

CURVE FITTING, GENERAL REMARKS

One of the most useful and practical tools of applied mathematics is *curve fitting*, but yet this topic is often slighted in texts on applied mathematics. Here, we mean by curve fitting the construction of a mathematical function which takes on, or approximates to, specified values at a finite set of discrete points in the space of its arguments. For the simplest case we may have *data values* y_i, $i = 1, 2, \ldots N$ corresponding to values x_i, $i = 1, 2, \ldots N$ of a variable x. We desire a function $y = f(x)$, usually continuous in x over some interval $a < x < b$ which includes all the data points x_i, such that *either* $y_i = f(x_i)$, *or* $y_i - f(x_i)$ is "small" by some criterion. Building such functions is our concern here. Quite obviously, we readily generalize our concern to functions $U = U(x, y, z)$ of many variables. However, we here point out certain generalities about the whole process.

There are *three factors* to be concerned with in curve fitting, these are:

(1) The use to which we will put the constructed function, i.e., what is to be done with the function once we have it?

(2) The criterion, or criteria, we apply to the "goodness of fit," i.e., how well does the constructed function have to match the "data points"?

(3) What mathematical procedures are to be used to determine the constants, or parameters, in the function being constructed?

However, these three factors are closely interrelated and in all cases must be considered together. Usually, this leads to a compromise of some sort.

With regard to the *purposes* of curve fitting there are three main applications. One application is the *comparison of theory and experiment*, i.e., a theory says the equation describing experimental data should be of a certain *form*, thus, we try to fit such an equation to the data.

Another purpose of curve fitting is simply for *interpolation between data points* and the last most common use of curve fitting is to facilitate *subsequent calculation or analysis*.

For example, we may wish to construct an approximate form for some integral. In this case, we fit a function that is easily integrated in closed form to a set of points computed from the integrand.

Thus, in general, the *purpose* determines the *form* of the function that we select to fit to the data, although we may have to adjust this form for convenience in the fitting process.

With regard to the criterion for the "goodness of fit" the simplest is that applied in *interpolation formulas*. Here, the function must be reasonably smooth over the whole range of available data points and must *match the data* exactly, within the numerical precession being used, at *every available data* point. Thus, *no* restrictions apply to the form of the function *outside* the range of available data.

In the case of some theory or "partial theory" which tells us that certain restrictions apply to the *form* of the function we almost always relax the requirement of perfect matching at data points, but, insist upon proper *form*. Also, as in the approximate evaluation of an integral, or other further calculation we choose the *form* to make the subsequent calculation as easy as possible and yet maintain reasonable accuracy.

We note the obvious fact that for example if we are to fit a function $y = f(\alpha_1, \alpha_2 \ldots \alpha_n, x)$ to data of y and x we must have *at least n* distinct data points in order to solve for the unknown parameters, i.e., in general, if N is *the number of parameters* in the fitting function and M is *the number of data points* we *must have*;

$$(14.55) \qquad\qquad M \geq N$$

in every case.

INTERPOLATION (REPEATED DIFFERENCE METHOD)

We consider, here, first the construction of an interpolation formula based on *repeated differences* for data of y at *equal intervals* $x_{i+1} - x_i = \Delta x$. We form the tabular scheme: $y_i = y(x_i)$

$$
\begin{array}{lllll}
y_1 \\
y_2 & \Delta^1 y_2 \\
y_3 & \Delta^1 y_3 & \Delta^2 y_2 \\
y_4 & \Delta^1 y_4 & \Delta^2 y_3 & \Delta^3 y_3 \\
y_5 & \Delta^1 y_5 & \Delta^2 y_4 & \Delta^3 y_4 & \Delta^4 y_3 \quad \cdots \\
y_6 & \Delta^1 y_6 & \Delta^2 y_5 & \Delta^3 y_5 & \Delta^4 y_4 \quad \text{etc.} \\
y_7 & \Delta^1 y_7 & \Delta^2 y_6 & \Delta^3 y_6 & \Delta^4 y_5 \\
y_8 & \Delta^1 y_8 & \Delta^2 y_7 & \Delta^3 y_7 & \Delta^4 y_6
\end{array}
$$

where,

$$(14.56) \qquad \begin{cases} \Delta^1 y_k = y_k - y_{k-1} \\[4pt] \Delta^{n+1} y_k = \Delta^n y_{k+1} - \Delta^n y_k, & n+1 \quad \text{even} \\[4pt] \Delta^{n+1} y_k = \Delta^n y_k - \Delta^n y_{k-1}, & n+1 \quad \text{odd} \end{cases}$$

These are the repeated differences.

Now we can show that if the Nth repeated difference,

$$(14.57) \qquad\qquad \Delta^N y_k = C = \text{constant}, \qquad \text{all } k$$

then the data can be fitted *exactly* with a polynomial in x of order N, i.e.,

$$(14.58) \qquad\qquad y = \sum_{n=0}^{N} a_n x^n$$

can be made to match every data point over the range of points for which Eq. (14.57) holds. We leave the proof as an exercise for the student.

In practice this may be applied by fitting a polynomial, Eq. (14.57), of *low* order over a few data points for which Eq. (14.57) holds approximately. This is then applicable for interpolation of y values between *these* data points. As we move down the range of data we then must fit a different polynomial to each successive "segment of data."

This segmental curve fitting or interpolation with polynomials of *low order* is generally preferred for many purposes. One such application is the determination of the *derivative* dy/dx at, or between data points, or in constructing integration formulas for

$$\int y \, dx$$

for *numerical evaluation.*

Observe, that to fit a polynomial *exactly* to the data we have, for $N + 1$ data points and a polynomial of order N the set of $N + 1$ simultaneous equations.

(14.59) $$y_i = a_0 + a_1 x_i + a_2 x_i^2 + \ldots + a_N x_i^N, (i = 1, 2, \ldots N + 1)$$

or in the matrix form,

$$\begin{pmatrix} y_1 \\ y_2 \\ \vdots \\ y_{N+1} \end{pmatrix} = \begin{pmatrix} 1 & x_1 & x_1^2 & x_1^3 \ldots & x_1^N \\ 1 & x_2 & x_2^2 & x_2^3 \ldots & x_2^N \\ \vdots & \vdots & & & \vdots \\ 1 & \ldots & \ldots & \ldots & x_{N+1}^N \end{pmatrix} \begin{pmatrix} a_0 \\ a_1 \\ \vdots \\ a_N \end{pmatrix}$$

so in general we need to find the *inverse* of the square X matrix here to compute the a_n, i.e., as in Chapter Two.

However, we *caution against using high-order polynomials* unless Eq. (14.57) is fulfilled *exactly* to high precision, because such polynomials generally introduce extraneous "wiggles" between data points.

FOURIER INTERPOLATION FUNCTION (USE OF DELTA FUNCTION)

This method of interpolation has a most interesting basis. The development given here, is interesting in other contexts also, because it is a revealing exploitation of the Dirac delta function.

The Dirac delta function is defined such that

(14.61) $$\delta(x - x') = \delta(x' - x) = \begin{cases} 0, & x \neq x' \\ \infty, & x = x' \end{cases}$$

and

(14.62) $$F(x) = \int_{-\infty}^{+\infty} F(x') \, \delta(x - x') \, dx'$$

Now take the *Fourier transform* of $\delta(x - x')$ and obtain,

(14.63) $$\int_{-\infty}^{+\infty} e^{-i\omega x} \delta(x - x') \, dx' = e^{-i\omega x'}$$

in view of Eq. (14.62); thus, the *inverse Fourier transform* gives,

(14.64) $$\delta(x - x') = \frac{1}{2\pi} \int_{-\infty}^{+\infty} e^{+i\omega(x - x')} \, d\omega$$

But here suppose we look at this as,

(14.65)
$$\delta(x - x') = \lim_{w \to \infty} \int_{-w}^{w} \frac{e^{+i\omega(x-x')}}{2\pi} \, d\omega$$

The integral appearing here, can be evaluated as,

(14.66)
$$\frac{1}{2\pi} \int_{-w}^{w} e^{+i\omega(x-x')} \, d\omega = \frac{\sin w(x - x')}{\pi(x - x')}$$

Let us call this function $\bar{G}_w(x - x')$, then

(14.67)
$$\delta(x - x') = \lim_{w \to \infty} \bar{G}_w(x - x')$$

This function is sketched in Fig. 14–1.

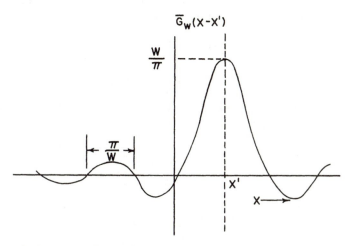

FIGURE 14.1. **Sketch of the Function** $[\sin w(x - x')]/\pi(x - x')$.

Observe that there is a central peak of height w/π at $x = x'$ and zeros are equally spaced at intervals of π/w to each side of this peak, continuing out to infinity in both directions. Thus, the function formed from $\bar{G}_w$ by multiplying by π/w has unit *height* for the peak and this function,

(14.68)
$$G_w(x - x') = \frac{\sin w(x - x')}{w(x - x')}$$

is the *Fourier interpolation function*.

To see how this is used suppose we have data of a function, $F(x)$, at *equal intervals*,

(14.69)
$$\Delta x = \frac{\pi}{w}$$

about the origin, i.e., we have

(14.70)
$$F_n = F(x_n), \qquad x_n = n\frac{\pi}{w}, \qquad n = 0, \pm 1, \pm 2, \ldots$$

Then erect a G_w function, centered at each data point, x_n, but multiplied by the value F_n corresponding to that point. The family of such curves appears as in Fig. 14–2.

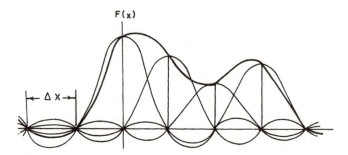

FIGURE 14.2.   Illustrating Interpolation with the Fourier Interpolation Function.

If we next form the *sum* of these curves we construct the "envelope" indicated as the heavy curve in Fig. 14–2. Analytically, the equation for this envelope is then,

$$(14.71) \qquad \tilde{F}(x) = \sum_n F_n G_w(x - x_n)$$

This defines a continuous function of x which takes on *exactly* the values of $F(x)$ at $x = x_1, x_2, \ldots$, i.e., *all* the data points. However, outside the interval over which data points are given $\tilde{F}(x)$ approaches zero asymptotically in an oscillatory fashion. Thus, $\tilde{F}(x)$ is a good *smooth* interpolation function in the data interval.

Now note that we have,

$$(14.72) \qquad \tilde{F}(x) = \sum_n F(x_n) \frac{\sin w(x - x_n)}{\pi(x - x_n)} \Delta x$$

where we have put back into Eq. (14.71) the definition of G_w and Δx from above. Here, we see that as $w \to \infty$ we have $\Delta x = \pi/w \to 0$ and this approaches a limit,

$$(14.73) \qquad F(x) = \int_{x-\varepsilon}^{x+\varepsilon} F(x') \, \delta(x - x') \, dx', \qquad \varepsilon > 0$$

where by this construction the "data interval" is $x - \varepsilon < x < x + \varepsilon$. In this way we see this important integral relation as a limiting form of an interpolation equation.

EXECUTE PROBLEM SET (14-4)

LEAST-SQUARES CRITERION AND CURVE FITTING

At the end of Chapter Eight we stated the central limit theorem of probability theory which stated *in essence* that if a quantity u is the sum,

$$(14.74) \qquad u = \sum_{i=1}^N (\Delta x)_i$$

of N independent random variables then in the limit of large N the probability distribution for u would approximate a normal, or Gaussian, distribution. Furthermore, if the $(\Delta x)_i$ are each distributed with mean zero then u will have mean equal zero.

Suppose then we look at some *measured* variable, say y_m. This value differs from the "true" value, y, by some amount, $\Delta y = (y_m - y)$, which we attribute to "random experimental errors" in the measuring process. If we adopt the *hypothesis* that this experimental error is the *sum* of a very great number, N, of elemental random errors, $(\delta y)_i, i = 1, 2, \ldots N$, each arising from *independent causes* and having mean zero, then we see by the central limit theorem that Δy would have a *normal distribution* of probability. That such is often actually the case can be verified experimentally.

If we make a great many repeated measurements of y_m, then plot the frequency distribution of y_m we do often find an essentially Gaussian shaped curve about some mean $\bar{y}$ (at the mode or peak). In such case we can see the justification for assuming that this mean, $\bar{y}$, is a good approximation to the "true" value, y.

Now suppose that this measured value of y is some function of x. Then we might have a measured value y_m at many different values of x, x_m, $m = 1, 2, \ldots M$. These data points (x_m, y_m) might appear on a graph as in Fig. 14–3, where we have also indicated by the smooth curve the "true" values of y at every value of x.

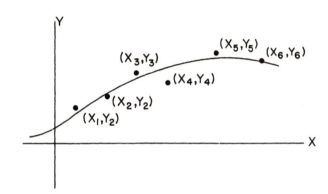

FIGURE 14.3. Sketch Indicating Relation of Data Points to "True" Curve.

According to our hypothesis about the elemental errors each of the errors,

(14.75)
$$\Delta y_m = u_m = [y_m - y(x_m)]$$

is a normally distributed random variable with mean zero. Thus, if we *assume* the mean square deviation, σ, to be the *same for each* of these random variables we have

(14.76)
$$P = \int\int\int_{(D)} \cdots \int \frac{e^{-\frac{u_1^2 + u_2^2 + \cdots + u_M^2}{2\sigma}}}{(2\pi\sigma)^{\frac{m}{2}}} \, du_1 \, du_2 \ldots du_M$$

as the *probability* for the sum,

(14.77)
$$0 \le u_1^2 + u_2^2 + \ldots + u_M^2 < t$$

where the domain (D) of integration in Eq. (14.76) is defined by Eq. (14.77). This multiple integral can be transformed to the single integral,[†]

(14.78)
$$P(t) = \frac{1}{(\sqrt{2\sigma})^M \Gamma(M/2)} \int_0^t e^{-\frac{\xi}{2\sigma}} \xi^{\frac{M}{2}-1} \, d\xi$$

Certainly then, under our present hypotheses each $u_m = \Delta y_m$ would be exactly zero *if* each y_m had its true *value* and, therefore, the *sum of the squares of the errors, t* here, would be

---

† See J. V. Uspensky, *Introduction to Mathematical Probability*, Dover Pub. Co. Inc., New York, 1967, page 331.

zero if each y_m had the true value. *But*, we don't *know* the *form* of the true function $y(x)$ so we proceed as follows.

Assume the true function $y(x)$ has the form,

(14.79) $$y = f(\alpha_1, \alpha_2, \ldots \alpha_K, x)$$

where we *choose* some form of $f(x)$ containing K parameters, $\alpha_1, \alpha_2, \ldots \alpha_K$. Then to make this as *close to the true function as possible* we require that the parameters, α_j, be selected so that,

(14.80) $$\Delta^2 = \sum_{m=1}^{M} [y_m - f(\alpha_1, \alpha_2, \ldots \alpha_K, x_m)]^2 \approx t$$

which approximates to t above, is made as *small as possible*, i.e., we vary the α_j to minimize Δ^2. Thus, we require,

(14.81) $$\frac{\partial \Delta^2}{\partial \alpha_j} = 0, \qquad j = 1, 2, \ldots K$$

which yields K simultaneous equations to be solved for the α_j. This is the *method of least squares* for fitting the *chosen* function, $f(x)$ to the data. Note that here the number, M, of data points will in general greatly exceed the number K, of parameters being determined.

Note that once the α_j are thus determined we can insert them back into Eq. (14.80) and get an estimate of t. Then, if we also have an estimate for σ we can use Eq. (14.78) to compute the probability that this value of t would occur "due to chance alone." Should this probability turn out to be very *small*, this would mean that a t, or Δ^2, of this size is very unlikely, due to random errors *only*. Consequently, a large part of the "error" Δ^2, is due to the wrong choice of *form* for $f(x)$.

In practice, however, one rarely goes through this elaborate procedure of testing the validity of the choice of *form* for $f(x)$. The form for $f(x)$ is almost always fixed by other considerations to be discussed in more detail shortly. In this regard, note that the set of *simultaneous equations*, Eqs. (14.81), for determining the α_j may be quite awkward to deal with. We will explore this point in more detail shortly also.

In all of our discussion above there is no feature which restricts y to be a function of a *single* variable x for we only used y values explicitly. Thus, we readily generalize to $y = y(x, z, \xi)$, or any set of variables and, hence, generalize the least-squares criterion to fit a function $f(\alpha_1, \alpha_2, \ldots \alpha_K, x, z, \xi)$ to measured data simply as:

(14.82) $$\Delta^2 = \sum_{m=1}^{M} [y_m - f(\alpha_1, \alpha_2, \ldots \alpha_K, x, z, \xi)]^2 = \text{minimum}$$

and Eqs. (14.81), then are used here to be solved for the α_j.

In *summary* then the criterion of Δ^2 equals a minimum for curve fitting is based on the hypothesis that the errors in y_m are normally distributed about the true curve. However, we often use the method for *convenience* even when this criterion is not met.

Example (6)

The least-squares method for functions linear in the free parameters. If we wish to fit a polynomial in x to data of y and x,

(14.83) $$y = a_0 + a_1 x + a_2 x^2 + \ldots + a_n x^n$$

or if we are fitting data of U as a function of *several* variables, x, y, z, again of polynomial type,

(14.84) $\quad U = a_0 + a_1 x + a_2 x^2 + b_1 y + b_2 y^2 + c_1 z + h_1 xy + j_1 yz + \ldots$ etc.

we see the parameters, a_i, b_j, h_l, etc. all enter our equations *linearly*. Thus, we consider the general *type*

(14.85) $$U = a_0 + a_1 \xi_1 + a_2 \xi_2 + \ldots + a_n \xi_n$$

of functions linear in the parameters. Here, $\xi_1, \xi_2 \ldots$, etc. play the role of x, x^2, x^3 etc. if we have the situation as in Eq. (14.83), or x, x^2, y, xy, etc. as in Eq. (14.84).

To determine the α_j we first note that the average value of U over M data points is:

(14.86) $$\bar{U} = \sum_{m=1}^{M} \frac{U_m}{M} = a_0 + a_1 \bar{\xi}_1 + a_2 \bar{\xi}_2 + \ldots + a_n \bar{\xi}_n$$

where

(14.87) $$\bar{\xi}_j = \frac{1}{M} \sum_{m=1}^{M} \xi_{jm}$$

is the mean value of ξ_j over the data values, $\xi_{j1}, \xi_{j2}, \ldots \xi_{jm}$ of ξ_j. Thus, define

(14.88) $$\begin{cases} V = U - \bar{U} \\ \eta_j = \xi_j - \bar{\xi}_j \end{cases}$$

and have

(14.89) $$V = \sum_{j=1}^{n} a_j \eta_j$$

eliminating a_0 for the moment.

Now require

(14.90) $$\Delta^2 = \sum_{m=1}^{M} \left(V_m - \sum_{j=1}^{n} a_j \eta_{jm} \right)^2 = \text{minimum}$$

Taking the partial derivative with respect to a_k, one of the a_j, we get the equation,

(14.91) $$\sum_{m=1}^{M} \eta_{km} V_m = \sum_{j=1}^{n} a_j \left(\sum_{m=1}^{M} \eta_{jm} \eta_{km} \right)$$

and this form applies for $k = 1, 2, \ldots n$. This then is a set of simultaneous *linear* equations to be solved† for the $a_j, j = 1, 2, \ldots n$. To determine a_0 we return to Eq. (14.88) as,

(14.92) $$a_0 = \bar{U} - \sum_{j=1}^{n} a_j \bar{\xi}_j$$

where all quantities on the right are known.

CHOICE OF "FORM" IN CURVE FITTING AND GRAPHICAL METHODS

In our discussion above we saw that if the "least-squares criterion" is applied then functions linear in the free parameters are easy to deal with, i.e., we have very well developed methods for solving simultaneous linear equations. However, this imposes somewhat stringent

† See Chapter Two.

restrictions on the possible forms we might choose for our fitting functions. Quite often other theoretical considerations dictate that the function should be of a specific type.

For example theory may require that y and x be related as,

(14.93) $$y = Ae^{-\beta x}$$

and, here, while A enters linearly, β does not. However, taking the natural logarithm,

(14.94) $$U = \ln y = \ln A - \beta x = a_0 - a_1 x$$

shows that if we consider our function in terms of $U = \ln y$ and x we have the two parameters $a_0 = \ln A$ and $a_1 = \beta$ appearing *linearly* and we can proceed as in Example (6).

However, we point out the obvious fact that if the errors in y are normally distributed, then those in $\ln y$ are not and vice versa, and strictly speaking we can only legitimately use the least squares criterion when the error distribution *is* normal. Even so, we usually ignore this fact and proceed to use $U = \ln y$, and x, as here indicated, because of *convenience*.

Here, we call attention to the fact that one should *always* make some preliminary graphical estimates, or tests, of the functional form *before* applying any detailed calculation of parameters *if possible*. Here, for example, we should plot $\ln y$ versus x from the data to see if a straight line results. In fact, we may be content with estimating "by eye" the best straight line through the data and read off the intercept $\ln A$ at $x = 0$ and compute the slope $-\beta$ for the line by using two points. Such *graphical methods* are extremely valuable and should always be used, if practical, for estimates.

Strictly speaking curve fitting is an *art* that is developed with experience. However, we list, here, a few guide lines for choosing forms.

(1) Be aware that the equation for any curve changes drastically with rotation and translation of the coordinate axes, i.e.,

$$xy = \text{constant}$$

and

$$\frac{x^2}{a^2} - \frac{y^2}{b^2} = 1$$

describe the same *kind* of curve with axes rotated through 90°.

(2) Make use of all theoretical constraints such as location of zeros and asymptotes, or constraints on form in asymptotic domains.

(3) When feasible, examine and approximately fit segments of the data, i.e., for small x or large x, as a guide to the general form, use graphical methods.

These ideas are illustrated in the following example.

Example (7)

We have data given as indicated in Fig. 14-4 of y versus t and we also have the theoretical constraints that y has a horizontal asymptote as $y \to \infty$, and also that $dy/dx = 0$ at $x = 0$. Now we could try a *rational function* such as

(14.95) $$y = \frac{bx^2}{1 + cx + dx^2}$$

because this has the desired asymptote and satisfies the condition at the origin. It also has

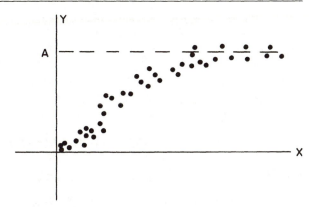

**FIGURE 14.4.   Data Points and
the Asymptote Line.**

an additional advantage. If we multiply through by the denominator we get

(14.96) $$y = bx^2 - c(xy) - d(x^2y)$$

where all the parameters, b, c, d enter *linearly*. Thus, if we treated x^2, xy, and x^2y here, as separate variables the *least-squares* method would be easy to apply. However, this process sometimes introduces extraneous *zeros* in the denominator of Eq. (14.95) unless constraints are imposed on c and d.

Another possible form is of the type

(14.97) $$y = Bx^2e^{-\gamma x} + A(1 - e^{-\beta x})$$

and this becomes for *large* x,

(14.98) $$y \approx A(1 - e^{-\beta x})$$

thus, we see A as the value of the asymptote and we can estimate this value from the given data, i.e., the dotted line in Fig. 14-4. Also, then by rearranging this we see

(14.99) $$U = \ln\left(1 - \frac{y}{A}\right) \approx -\beta x$$

for *large* x, and having estimated A we plot, with the data values, the left member versus x. The data for large x should approach a straight line with slope $-\beta$ passing through the origin, i.e., as in Fig. 14-5. Thus, we estimate the value of β.

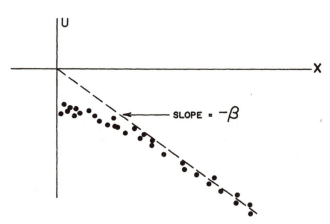

SLOPE $= -\beta$

**FIGURE 14.5.   Plot of $U(y, x)$
versus x.**

Next, since A and β are now estimated, note that

(14.100) $$V = \ln\left[\frac{y - A(1 - e^{-\beta x})}{x^2}\right] = \ln B - \gamma x$$

and we can plot V versus x and the result should be a straight line with intercept $\ln B$ at $x = 0$ and slope $-\gamma$. Thus, all constants are estimated by *graphical means*. Of course, if an electronic digital computer is available these estimates can be made *numerically* by the computer.

EXECUTE PROBLEM SET (14-5)

NEWTONIAN ITERATION FOR SOLUTION OF TRANSCENDENTAL EQUATIONS

In the problem of least-squares curve fitting in which we have the set of simultaneous equations, Eq. (14.81), to solve, we often encounter nonlinear, or even transcendental equations to solve. Such transcendental equations also arise in the study of special functions; for example, we have not yet in this text described how one might actually determine the numerical values of say, the roots, λ_i, of $J_0(\lambda_i) = 0$. Here, we describe *one* method of accomplishing this task.

Suppose we have given a function,

(14.101) $$y = F(x)$$

and we desire to determine the root, or *roots*, specified by

(14.102) $$F(x) = 0$$

First of all we may select a set of values of x, say $x_1^{(0)}, x_2^{(0)}, \ldots$, etc. and evaluate y at these points. If y *changes sign* in going from $x_j^{(0)}$ to $x_{j+1}^{(0)}$ then the product,

(14.103) $$F(x_j^{(0)})F(x_{j+1}^{(0)}) < 0$$

and this is the criterion for *at least one*, or some *odd* number of roots, between $x_j^{(0)}$ and $x_{j+1}^{(0)}$. If the product in Eq. (14.104) is *positive* then there are no roots, or an *even* number of roots between $x_j^{(0)}$ and $x_{j+1}^{(0)}$. We could proceed with *halving* the intervals and continuing with this testing procedure until we determine one or more roots to any desired precision, but instead we now turn to the following *Newtonian iteration* scheme.

Knowing that there is a root of $F(x)$ in the interval $x_j^{(0)}$ to $x_{j+1}^{(0)}$ we define

(14.104) $$x_0 = \frac{x_{j+1}^{(0)} + x_j^{(0)}}{2}$$

as a *first guess* for the root. Then we look at the Taylor expansion of the function $F(x)$ about a point x_n. Thus,

(14.105) $$F(x_{n+1}) = F(x_n) + \frac{\partial F(x_n)}{\partial x}(x_{n+1} - x_n) + \frac{1}{2}\frac{\partial^2 F(x_n)}{\partial x^2}(x_{n+1} - x_n)^2 + \cdots$$

and we will assume $x_{n+1} - x_n$ small enough to neglect second and higher-order terms. Now *if* x_{n+1} were the desired root then the left member would be zero, and we would have the relation,

(14.106) $$x_{n+1} = x_n - \frac{F(x_n)}{\dfrac{\partial F(x_n)}{\partial x}}$$

giving the value of x_{n+1} in terms of x_n. Thus, we insert x_0 into this relation as our first guess and compute a value x_1 as the next best guess, then insert x_1 and compute an improved value, x_2, etc. continuing until there ceases to be any *significant* change in the values obtained. When this occurs we have determined a root of $F(x)$. The process can be repeated in other intervals until we have determined all the roots, or as many as desired.

The nature of this process is shown graphically in Fig. 14-6. As we see here the tangent line at x_1 intersects $y = 0$, the x axis, at x_2. Thus, since $-\tan \theta = dF(x_1)/dx$ as indicated, we see Eq. (14.108) as the equation describing this situation.

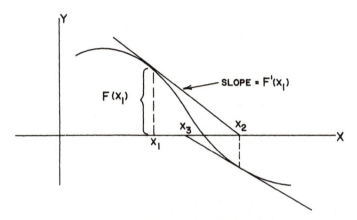

FIGURE 14.6.   Graphical Presentation of Newtonian Iteration.

At $x = x_2$, the new tangent is drawn intersecting the x-axis at x_3, *closer* to the root, i.e., the point where the curve crosses the axis. Thus, it is immediately obvious that if the roots are not *too* close together the process should converge to the desired root.

Example (8)

Set up the iteration scheme to construct a table of roots of,

(14.107) $J_0(x) = 0$

By the Newtonian scheme we have

(14.108) $$x_{n+1} = x_n - \frac{J_0(x_n)}{\dfrac{dJ_0(x_n)}{dx}}$$

but we know that $J_0' = -J_1$ so,

(14.109) $$x_{n+1} = x_n + \frac{J_0(x_n)}{J_1(x_n)}$$

Hence, if we choose a starting value, x_0 *near* some root, call it λ_k now, then all we need is a good table of $J_0(x)$ and $J_1(x)$ to iterate to the root λ_k. In actual practice we *rarely* use tables of Bessel functions any more, since the advent of modern high-speed computers. Instead we use some high-order approximation function for these functions constructed by curve fitting procedures.† Otherwise, if we do use tables we find the necessity for *interpolation* between table entries to carry out this root finding process, this too, can be set up in a computer, i.e., a table of values in the computer memory and one of our interpolation schemes programmed into the machine.

EXECUTE PROBLEM SET (14-6)

APPROXIMATION TO A FUNCTION BY A FINITE SERIES OF ORTHOGONAL FUNCTIONS

In Chapter Three we showed that the *proper* solutions of a Sturm–Liouville equation formed a *complete set* on some interval of the independent variable, $a < x < b$, and, therefore, that a function $F(x)$, satisfying the *same* end conditions as the member functions, $y_j(x), j = 1, 2, \ldots$, of the complete set could be represented *exactly* by an *infinite series* of such functions, provided that $F(x)$ was sectionally continuous in the interval, and bounded, i.e., Eq. (3.79). We also showed there that this representation of $F(x)$ was the "best fit" to a *discrete* set of values of $F(x)$, using a *finite series*, in a *least-squares sense*, i.e., Eq. (3.84).

Thus, we see that if we wish to construct an approximation,

$$(14.110) \qquad F(x) = \sum_{j=1}^{N} A_j y_j(x), \qquad a < x < b$$

as the equation to *fit* a set of *data* values of F, say F_i at $x = x_i$, $i = 1, 2, \ldots K$, then, we need to determine the appropriate A_j's so that Eq. (3.85), now shown as,

$$(14.111) \qquad \sum_{i=1}^{K} \left[F_i - \sum_{j=1}^{N} A_j y_j(x_i) \right]^2 \omega(x_i) = \text{minimum}$$

is satisfied. Here, we restrict the x_i to *equal spacing*,

$$x_{i+1} - x_i = \Delta x = \text{constant}$$

and, of course $\omega(x)$ is the weight function appropriate to the y_j set.

This leads directly, by differentiating with respect to the A_j, to the set of simultaneous equations:

$$(14.112) \qquad \sum_{i=1}^{K} F_i y_k(x_i) \omega(x_i) = \sum_{j=1}^{N} A_j \left[\sum_{i=1}^{K} y_k(x_i) y_j(x_i) \omega(x_i) \right]$$

for determination of the A_j. The method can easily be extended to nonequally spaced data points.

It is important to note that the "kind" of functions chosen is rather arbitrary, the only criterion is actually that of the end conditions we wish $F(x)$ to satisfy, and the domain, $a < x < b$, on which $F(x)$ is defined.

EXECUTE PROBLEM SET (14-7)

---

† See C. Hastings, *Approximations for Digital Computers* (Princeton University Press, Princeton, New Jersey, 1955).

PROBLEMS

Set (14-1)

(1) Evaluate

$$I(t) = \int_0^\infty \frac{\sin xt \, dx}{x(1 + x^2)}$$

[Hint: Show that

$$\frac{d^2 I}{dt^2} - I + \frac{\pi}{2} = 0$$

then deduce from the original expression for I the values of $I(0)$ and $dI(0)/dt$. Solve this differential equation with these conditions to obtain $I(t)$ in closed form. For comparison evaluate this integral by the Laplace transform method of Chapter Six.]

(2) Evaluate,

$$I = \int_{-\infty}^{+\infty} \frac{e^{i\omega t} \, d\omega}{a^2 + \omega^2}$$

using a technique similar to that in Problem (1) above.

(3) Evaluate,

$$I = \int_0^\infty \frac{e^{-ax} J_0(\beta x) x \, dx}{b^2 + x^2}$$

Set (14-2)

(1) Show that the sum, $S(x)$, of Example (3),

$$S(x) = \sum_{n=2}^\infty \frac{x^n}{n^2 - 1}, \qquad x^2 < 1$$

is a solution of the equation,

$$\frac{d}{dx}\left(x\frac{dS}{dx}\right) - S = \frac{1}{1-x} - 1 - x$$

subject to suitable boundary conditions.

(2) Show from the series representation that,

$$\frac{1}{2} \ln \frac{1+x}{1-x} = \tan^{-1} x, \qquad x^2 < 1$$

(3) Find the sum of the series,

$$S(x) = 1 + 2x + \frac{3x^2}{2!} + \frac{4x^3}{3!} + \cdots$$

(4) Find the sum of the series,

$$S(x) = \frac{2L}{\pi} \sum_{n=1}^\infty \frac{(-1)^n}{n} \frac{\sin \frac{n\pi x}{L}}{a^2 - \frac{n^2 \pi^2}{L^2}}, \qquad |x| < L,$$

[Hint: Reduce to a simple differential equation and obtain the solution in closed form.]

Set (14-3)

(1) Use the residue method to show that,

$$\sum_1^\infty (-1)^n \frac{n \sin nx}{(-r)^{n+1} n^2 - t^2} = \frac{\pi \sin xt}{2 \sin \pi t}$$

(2) Use the residue method to show that

$$\sum_{n=1}^\infty \frac{\sin an \sin bn}{(n^2 + \alpha^2)(n^2 + \beta^2)} = \frac{\pi}{2(\alpha^2 - \beta^2)} \left[\frac{\sinh \alpha \cosh \pi\beta \sinh b\beta}{\beta \sinh(\pi\beta)} - \frac{\sinh a\infty \cos l \pi\alpha \sinh b\alpha}{\alpha \sinh(\pi\alpha)} \right]$$

What restrictions apply to a and b?

(3) Use the residue method to sum the series

$$S(x) = \sum_{n=1}^\infty \frac{2x}{x^2 + n^2\pi^2}$$

(4) Construct a general formula for summing series of the type,

$$\sum_{j=1}^\infty F(\xi_j) \frac{J_0(\xi_j r)}{J_1(\xi_j)}$$

where the ξ_j are roots of $J_0(\xi_j) = 0$. Stipulate what restrictions apply to the form of the function $F(\xi)$.

Set (14-4)

(1) Show that if

$$y(x) = \sum_{n=0}^N a_n x^n$$

then the Nth repeated difference, $\Delta^N y_k$ (for equal intervals of x, $\Delta x = x_{j+1} - x_j$) is a constant.

(2) Consider the data values of y and x in the x, y pairs: $(0, 1)$, $(1, 4)$, $(2, 1)$, $(3, -11)$, $(4, -39)$, $(5, -89)$, $(6, -167)$, $(7, -279)$, $(8, -431)$. Can these data be interpolated by a polynomial in x? If so construct the polynomial.

(3) In Chapter Nine, in our consideration of the response of a linear instrument to an input signal, $S(t)$, this signal was approximated by a series of delta functions, i.e., Eq. (9.132),

$$S(t) = \sum_{n=0}^\infty S(nT) \delta(t - nT)$$

In the present chapter we have shown that the delta function can be approximated by Eq. (14.66). Combine these ideas into an expression for the response of the linear instrument in terms of our Gaussian interpolation function.

(4) A more useful approach than that indicated in Problem (3) above is to retain the stated representation of $S(t)$ but represent the response function of the instrument by the Gaussian interpolation function as in Eq. (14.71). Use this approach to construct a useful expression for the response of the linear instrument.

Set (14-5)

(1) Given the data of x, y pairs: (10, 0.053), (20, 0.140), (30, 0.222), (40, 0.287), (50, 0.333), (60, 0.356), (70, 0.356), (80, 0.355), (90, 0.331), (100, 0.302), (110, 0.265), (120, 0.224), (130, 0.182), (140, 0.139), (150, 0.100), and the required asymptotic form of the relation between x and y,

$$\lim_{x \to \infty} y(x) \approx e^{-\frac{x}{100}} \sin x$$

(for x in degrees as in the above data), find a function to fit these data and suitable for extrapolation in the interval $0 \le x \le \infty$. Use a *graphical method*.

(2) Having ascertained a suitable *form* for the function $y(x)$ in Problem (1) just above set up a *linear least-squares* calculation procedure to determine the parameters in your chosen functions. How *small* is Δ^2 for your solution?

Set (14-6)

(1) Set up and carry out a Newtonian iteration for solution of

$$\tfrac{1}{3}x - \ln x = F(x) = 0$$

to determine the root x to three significant digits.

(2) Consider the data pairs of x and y values : (0.87, 5.1), (1.75, 11), (2.68, 14), (3.64, 20), (4.65, 25), (5.78, 29), (7.00, 35), (8.40, 41). First show *graphically* that these data are fitted approximately by

$$y = \tan^{-1} bx$$

then set up an iteration scheme to carry out the *nonlinear least squares* determination of b as :

$$\sum_i (y_i - \tan^{-1} bx_i)^2 = \Delta^2 = \text{minimum}$$

Set (14-7)

(1) Set up the least squares method to represent the function $y(x) = x$ by a series of 3 sine functions using 10 equally spaced points on the interval $0 \le x \le 1$. Carry out the calculation and compare the coefficients you obtain to those of the exact series appearing as Entry Number (12) in Table 14-1.

(2) Set up the *same* problem as Problem (1) above but use cosine functions (3 terms) to fit $y = x^2$. Compare the result to Entry Number (13) of Table 14-2.

REFERENCES

P. Dennery and A. Krzywicki, *Mathematics for Physicists* (Harper and Row, Publishers, Inc., New York, 1967).

C. Hastings, *Approximations for Digital Computers* (Princeton University Press, Princeton, New Jersey, 1955).

J. Irving and N. Mullineux, *Mathematics in Physics and Engineering* (Academic Press Inc., New York, 1959).

P. M. Morse and H. Feshbach, *Methods of Theoretical Physics* (McGraw-Hill Book Company, Inc., New York, 1953).

A. Ralston and H. S. Wilf, *Mathematical Methods for Digital Computers* (John Wiley & Sons, Inc., New York, 1960).

J. Walker and R. L. Mathews, *Mathematical Methods of Physics* (W. A. Benjamin, Inc., New York, 1965).

Appendix I

CAUCHY'S PROBLEM AND THE CLASSIFICATION OF PARTIAL DIFFERENTIAL EQUATIONS: CHARACTERISTICS†

The problem considered by Cauchy can be simply stated as follows: *given* that a function $\psi(x, y)$ must satisfy the second-order partial differential equation,

(I-1)
$$A\frac{\partial^2\psi}{\partial x^2} + 2B\frac{\partial^2\psi}{\partial x\,\partial y} + C\frac{\partial^2\psi}{\partial y^2} = F$$

where A, B, C, and F are *specified* functions of x, y, $\partial\psi/\partial x$, $\partial\psi/\partial y$, in some domain R, bounded by a curve S, *is it possible* to determine $\psi(x, y)$ at interior points of R given only some information about $\psi(x, y)$, or its derivative, $\partial\psi/\partial n$, normal to S, on the curve S?

In particular we make the three classifications of *boundary conditions* on S:

(a) *Dirichlet* conditions, ψ specified at each point of S,
(b) *Neumann* conditions, $\partial\psi/\partial n$ specified at each point of S,
(c) *Cauchy* conditions, ψ and $\partial\psi/\partial n$ specified at each point of S.

The analysis required to answer the question posed by Cauchy is outlined as follows: Let the curve S bounding R have the parametric representation,

(I-2)
$$x = x(s), \qquad y = y(s)$$

then the components of the unit vector normal to S are: $-dy/ds$ and dx/ds so that

(I-3)
$$\frac{\partial\psi}{\partial n} = -\left(\frac{\partial\psi}{\partial x}\right)\frac{dy}{ds} + \left(\frac{\partial\psi}{\partial y}\right)\frac{dy}{ds}$$

Also, the change in ψ *along* the curve S is

(I-4)
$$\frac{d\psi}{ds} = \left(\frac{\partial\psi}{\partial x}\right)\frac{dx}{ds} + \left(\frac{\partial\psi}{\partial y}\right)\frac{dy}{ds}$$

Now the plan of attack is simply this; we will expand $\psi(x, y)$ in a Taylor series about some point on the boundary to find ψ at an interior point. If we can show that the given boundary information, plus the partial differential equation, allows us to determine *all* the coefficients of this expansion then the answer to Cauchy's question is in the affirmative. This expansion requires that we be able to determine *all* the derivatives of ψ, $\partial\psi/\partial x$, $\partial\psi/\partial y$, $\partial^2\psi/\partial x^2$, $\partial^2\psi/\partial x\,\partial y$, $\partial^2\psi/\partial y^2$, $\partial^3\psi/\partial x^3$, etc. up to *all* orders.

To begin with we can solve Eqs. (I-3) and (I-4) simultaneously for $\partial\psi/\partial x$ and $\partial\psi/\partial y$ in terms of $\partial\psi/\partial n$ and $\partial\psi/ds$, i.e.,

(I-5)
$$\begin{cases} \left(\dfrac{\partial\psi}{\partial x}\right)_s = -\left(\dfrac{\partial\psi}{\partial n}\right)_s\dfrac{dy}{ds} + \left(\dfrac{\partial\psi}{\partial s}\right)_s\dfrac{dx}{ds} \\[2mm] \left(\dfrac{\partial\psi}{\partial y}\right)_s = \left(\dfrac{\partial\psi}{\partial n}\right)_s\dfrac{dx}{ds} + \left(\dfrac{\partial\psi}{\partial s}\right)_s\dfrac{dy}{ds} \end{cases}$$

Next we seek the second derivatives, $(\partial^2\psi/\partial x^2)_s$, $(\partial^2\psi/\partial x\,\partial y)_s$, $(\partial^2\psi/\partial y^2)_s$. We find these by

† For more extensive discussions consult the references.

differentiating Eqs. (I-5) with respect to s along the *known* curve S

(I-6)
$$
\begin{cases}
\dfrac{d}{ds}\left(\dfrac{\partial \psi}{\partial x}\right)_s = \left(\dfrac{\partial^2 \psi}{\partial x^2}\right)_s \dfrac{dx}{ds} + \left(\dfrac{\partial^2 \psi}{\partial x\, \partial y}\right)_s \dfrac{dy}{ds} \\[2mm]
\dfrac{d}{ds}\left(\dfrac{\partial \psi}{\partial y}\right)_s = \left(\dfrac{\partial^2 \psi}{\partial x\, \partial y}\right)_s \dfrac{dx}{ds} + \left(\dfrac{\partial^2 \psi}{\partial y^2}\right)_s \dfrac{dy}{ds}
\end{cases}
$$

and using Eq. (I-1) as our third equation. We have three equations (inhomogeneous) in three unknowns (the sought for second derivatives). These have a solution *if* the determinant of coefficients is *not* zero, i.e.,

(I-7)
$$
\begin{vmatrix}
\dfrac{dx}{ds} & \dfrac{dy}{ds} & 0 \\[2mm]
0 & \dfrac{dx}{ds} & \dfrac{dy}{ds} \\[2mm]
A & 2B & C
\end{vmatrix} \neq 0
$$

Thus it is *possible* to find these second derivatives (and actually all higher derivatives by continued differentiation) provided that the boundary curve S is *not* one on which the above determinant *is zero*, i.e., if the curve S is such that

(I-8)
$$
A\left(\frac{dy}{ds}\right)^2 - 2B\frac{dx}{ds}\frac{dy}{ds} + C\left(\frac{dx}{ds}\right)^2 = 0
$$

then the second and higher derivatives *cannot* be found.

At each point this equation determines *two directions*, at that point. Curves whose tangents are everywhere tangent to these characteristic directions are called *characteristics* of the partial differential equation. Thus Cauchy conditions will determine the solution if S is nowhere tangent to a characteristic.

We see that if the characteristics are to be *real* curves we must have $B^2 > AC$. If $B^2 > AC$ the partial differential equation is called *hyperbolic*. If $B^2 = AC$ the equation is called *parabolic* and if $B^2 < AC$ the equation is elliptic.

We state, without proof, that for these three basic types of partial differential equations the types of regions, R (i.e., S is either an open or closed curve) and the *kind* of boundary conditions required in order to solve the Cauchy problem are as follows:

Equation	Example	Condition	Boundary
Hyperbolic	$\dfrac{\partial^2 \psi}{\partial x^2} - \dfrac{1}{c^2}\dfrac{\partial^2 \psi}{\partial t^2} = 0$	Cauchy	open
Elliptic	$\dfrac{\partial^2 \psi}{\partial x^2} + \dfrac{\partial^2 \psi}{\partial y^2} = 0$	Dirichlet or Neumann	closed
Parabolic	$\dfrac{\partial^2 \psi}{\partial x^2} = \dfrac{\partial \psi}{\partial t}$	Dirichlet or Neumann	open

That the stated conditions allow the solution of Cauchy's problem can be readily demonstrated. These example forms are the simplest cases of the special forms.

To see how the characteristics play a major role in determining the general form of the solution of a partial differential equation note that for the simplest hyperbolic form (c = constant)

(I-9)
$$\frac{\partial^2 \psi}{\partial x^2} - \frac{1}{c^2} \frac{\partial^2 \psi}{\partial t^2} = 0$$

[where t replaces y], the characteristics are given by:

(I-10)
$$\left(\frac{dt}{ds}\right)^2 - \frac{1}{c^2}\left(\frac{dx}{ds}\right)^2 = 0$$

or,

(I-11)
$$\left(\frac{dx}{dt}\right)^2 = c^2$$

Thus the characteristics are the straight lines,

(I-12)
$$\begin{cases} x - ct = \xi = \text{constant} \\ x + ct = \eta = \text{constant} \end{cases}$$

If we transform the partial differential equation to be expressed in terms of ξ and η we get, the *normal form* of the hyperbolic equation,

(I-13)
$$\frac{\partial^2 \psi}{\partial \eta\, \partial \xi} = 0$$

so that the general solution is

(I-14)
$$\psi = f(\xi) + g(\eta)$$

where f and g are *arbitrary functions*.

Appendix II
MOVING BOUNDARY PROBLEMS

One class of boundary value problems which has not been discussed in the text is that in which one or more boundaries of the domain of definition of a solution change with time or *move*. The classic example is the melting-ice problem, i.e., within the ice, $T < 0$, and the equation of heat conduction

(II-1)
$$\frac{k}{C\rho}\nabla^2 T = \frac{\partial T}{\partial t}, \quad \text{in } D$$

applies.

However at the surface $T = 0°C$ the ice *melts* and in the region outside of this surface, $T > 0$, water exists and here

(II-2)
$$\frac{k'}{C'\rho'}\nabla^2 T = \frac{\partial T'}{\partial t}, \quad \text{in } D'$$

applies. As the temperature distribution changes the surface,

(II-3)
$$T(x, y, z, t) = 0 = T_0$$

changes shape or *moves*. Thus we have,

$$\text{(II-4)} \qquad \frac{dT_0}{dt} = 0 = \frac{\partial T_0}{\partial t} + \mathbf{v} \cdot \nabla T_0 = 0$$

describing the *moving surface*. Here we define the vector, $\mathbf{v}$, *normal* to the surface, and hence *parallel*† to $\pm \nabla T$, as the velocity of advance of the surface.

Thus in such a problem there are *three* partial differential equations to be solved simultaneously. Of course appropriate boundary conditions have to be assigned on all the fixed boundaries *and* the moving boundary.

Analytical solutions to such problems are generally only possible for cases of extreme symmetry, for example complete spherical symmetry when $T_0 = 0$ on a *simple surface* $r = R(t)$, i.e. a sphere.

Appendix III
LAURENT'S SERIES

Theorem: If $f(z)$ is an analytic function of the complex variable z on two concentric circles, C_1 and C_2, and throughout the region between them, then at each point z between these circles $f(z)$ is represented by the sum of two convergent series as,

$$\text{(III-1)} \qquad f(z) = \sum_{n=0}^{\infty} a_n (z - z_0)^n + \sum_{n=1}^{\infty} \frac{b_n}{(z - z_0)^n}$$

where

$$\text{(III-2)} \qquad a_n = \frac{1}{2\pi i} \oint_{C_1} \frac{f(z')\, dz'}{(z' - z_0)^{n+1}}, \qquad n = 0, 1, 2, \ldots$$

and

$$\text{(III-3)} \qquad b_n = \frac{1}{2\pi i} \oint_{C_2} \frac{f(z')\, dz'}{(z' - z_0)^{-n+1}}, \qquad n = 1, 2, \ldots$$

each integral being counterclockwise and C_1 enclosing C_2 with common center z_0.

The outline of the proof is indicated as follows. By Cauchy's integral formula

$$\text{(III-4)} \qquad f(z) = \frac{1}{2\pi i} \oint_{C_1} \frac{f(z')\, dz'}{z' - z} - \frac{1}{2\pi i} \oint_{C_2} \frac{f(z')\, dz'}{z' - z}$$

simply by connecting the two circles with two straight lines, making a "cut" of infinitesimal width across the annular region. In the first integral use the identity,

$$\frac{1}{z' - z} = \frac{1}{(z' - z_0) - (z - z_0)} = \frac{1}{z' - z_0} + \frac{z - z_0}{(z' - z_0)^2} + \ldots + \frac{(z - z_0)^{n-1}}{(z' - z_0)^n} + \frac{(z - z_0)^n}{(z' - z_0)^n (z' - z)}$$

(III-5)

In the second, put

$$\text{(III-6)} \qquad -\frac{1}{z' - z} = \frac{1}{(z - z_0) - (z' - z_0)}$$

† The plus or minus sign will be determined by whether the ice is freezing or melting.

and factor $(z - z_0)$ out of the denominator to write

(III-7) $$-\frac{1}{z' - z} = \frac{1}{z - z_0} + \frac{(z' - z_0)}{(z - z_0)^2} + \dots + \frac{(z' - z_0)^{n-1}}{(z - z_0)^n} + \frac{(z' - z_0)^n}{(z - z_0)^n(z - z')}$$

Inserting these into Eq. (III-4) then yields,

(III-8) $$f(z) = a_0 + a_1(z - z_0) + a_2(z - z_0)^2 + \dots + a_{n-1}(z - z_0)^{n-1} + R_n$$

$$+ \frac{b_1}{z - z_0} + \frac{b_2}{(z - z_0)^2} + \dots + \frac{b_n}{(z - z_0)^n} + S_n$$

where

(III-9) $$\begin{cases} R_n = \frac{(z - z_0)^n}{2\pi i} \oint_{C_1} \frac{f(z') \, dz'}{(z' - z_0)^n(z' - z)} \\[2mm] S_n = \frac{1}{2\pi i(z - z_0)^n} \oint_{C_2} \frac{(z' - z_0)^n f(z') \, dz'}{z - z'} \end{cases}$$

Let $r = |z - z_0|$ then $r_1 > r > r_2$. If M is the maximum value of $|f(z)|$ on C_2 then

(III-10) $$|S_n| \le \left(\frac{r_2}{r}\right)^n \frac{M r_2}{r - r_2}$$

and $S_n \to 0$ as $n \to \infty$. A similar argument applies for R_n, i.e.,

(III-11) $$|R_n| \le \frac{r_1 M'}{r_1 - r}\left(\frac{r}{r_1}\right)^n$$

where M' is the maximum value of $|f(z)|$ on C_1. Thus $R_n \to 0$ as $n \to \infty$.

Appendix IV
ANALYTIC CONTINUATION

Theorem: If a function of the complex variable z is analytic throughout a region, then it is uniquely determined by its values over an arc, or throughout a subregion, of the given region.

As a consequence if $f_1(z)$ and $f_2(z)$ were analytic throughout the given region and *assume the same value* on the given arc, A_0, then their difference would be an analytic function in R that vanishes on A_0 and therefore

(IV-1) $$f_1(z) - f_2(z) = 0$$

everywhere in the region.

Thus consider the region R composed of two parts, R_1 and R_2, adjoining each other on a curve, A_0. Let $f_1(z)$ be analytic throughout R_1 and suppose $f_2(z)$ is analytic in R_2 and equal to $f_1(z)$ on A_0. Then these *two* functions are used to define *one* function

(IV-2) $$F(z) = \begin{cases} f_1(z) & \text{in } R_1 \\ f_2(z) & \text{in } R_2 \end{cases}$$

that is analytic in the composite region, $R = R_1 + R_2$.

For example the function,

(IV-3) $$f_1(z) = \sum_{n=0}^{\infty} z^n$$

is analytic in the region $R_1 \triangleq |z| < 1$, but is undefined for $|z| \geq 1$. However the function

(IV-4)
$$F(z) = \frac{1}{1 - z}$$

is defined and analytic everywhere except at $z = 1$. According to the theorem stated above the $F(z)$ is *unique*.

Appendix V
WIENER–HOPF METHOD

Originally this extension of the Fourier integral transform was developed as a method for solving certain types of integral equations but lends itself to the solution of certain boundary value problems which are difficult to solve by other methods.

The Fourier transform of $F(x)$ was originally defined as

(V-1)
$$f(k) = \int_{-\infty}^{+\infty} F(x) e^{-ikx} dx$$

for *real* k. The Wiener–Hopf technique extends k to be complex, $k = \omega + i\gamma$. Then,

(V-2)
$$f(x) = \int_{-\infty}^{+\infty} [F(x) e^{+\gamma x}] e^{-i\omega x} dx$$

and the factor $e^{+\gamma x}$ may cause the integral to diverge at either $x \to +\infty$ or $x \to -\infty$ depending upon the sign of γ and the asymptotic form of $F(x)$ in these regions. The *form* of $F(x)$ may prevent the divergence for a *limited* range of γ. Thus, if,

(V-3)
$$\begin{cases} F(x)e^{+\beta x} \to 0 & \text{as } x \to -\infty \\ F(x)e^{+\alpha x} \to 0 & \text{as } x \to +\infty \end{cases}$$

then $F(k)$ is analytic in the strip $\beta < \gamma < \alpha$.

Another alternative is $F(x) = 0$ for all $x < 0$, then clearly $f(k)$ is analytic for $\gamma < 0$, i.e., $\mathscr{Im}(k) < 0$. In this case not only is $f(k)$ analytic for $\mathscr{Im}(k) < 0$, but also $f(k) \to 0$ uniformly as $\mathscr{Im}(k) \to -\infty$ since now,

(V-4)
$$f(k) = \int_0^\infty [F(x) e^{+\gamma x}] e^{-i\omega x} dx$$

Thus this alternative is a much *stronger* condition on $f(k)$ than that above.

It also follows that just as the behavior of $F(x)$ at $x = \pm\infty$ determines the region of analyticity of $f(k)$, so does the behavior of $f(k)$ at $\pm\infty$ (actually the real part of k at $\pm\infty$) determine the region of analyticity of $F(x)$.

In addition to extending k to be complex the Wiener–Hopf technique depends on a decomposition of $f(k)$ in a *unique* manner. Thus suppose $f(k)$ is analytic in a strip $\beta < \mathscr{Im}(k) < \alpha$ and goes to zero at the ends of the strip $[\mathscr{Re}(k) \to \pm\infty]$ then we write

(V-5)
$$f(k) = f_+(k) + f_-(k)$$

where $f_+(k)$ is analytic for $\mathscr{Im}(k) > \beta$ and goes to zero as $\mathscr{Im}(k) \to +\infty$, and $f_-(k)$ is analytic for $\mathscr{Im}(k) < \alpha$ and goes to zero as $\mathscr{Im}(k) \to -\infty$. Then by Cauchy's integral theorem

(V-6)
$$f(k) = -\frac{1}{2\pi i} \int_{\infty + i\alpha}^{-\infty + i\alpha} \frac{f(k') \, dk'}{k' - k} + \frac{1}{2\pi i} \int_{-\infty + i\beta}^{\infty + i\beta} \frac{f(k') \, dk'}{k' - k}$$

with k in the strip, $\alpha > \mathscr{Im}(k) > \beta$.

Appendix VI

CONDITIONS OF MAXIMUM OR MINIMUM IN EXTREMUM PROBLEMS

We confine this discussion to a function $z(x, y)$ of two variables with no constraints. The approach provides some insight into the problem.

For an extremum at $x = x'$, $y = y'$ we have

(VI-1)
$$\frac{\partial z}{\partial x} = \frac{\partial z}{\partial y} = 0$$

as *necessary* for any sort of extremum. In the neighborhood of x', y' let a particular straight path through x', y' be chosen, i.e.,

(VI-2)
$$x = x' + lr, \qquad y = y' + mr$$

where $l = \cos\theta$, $m = \sin\theta$, and r is the distance from the extremum point, i.e., l and m are fixed. Then we have, since now z is a function of r only,

(VI-3)
$$\frac{dz}{dr} = \frac{\partial z}{\partial x}\frac{dx}{dr} + \frac{\partial z}{\partial y}\frac{dy}{dr} = l\frac{\partial z}{\partial x} + m\frac{\partial z}{\partial y}$$

and we must therefore have,

(VI-4)
$$\frac{dz}{dr} = 0$$

for *any* θ, according to Eq. (VI-1).

Now taking the second derivative we get

(VI-5)
$$\frac{d^2 z}{dr^2} = l^2\frac{\partial^2 z}{\partial x^2} + 2lm\frac{\partial^2 z}{\partial x\,\partial y} + m^2\frac{\partial^2 z}{\partial y^2}$$

and for $x = x'$, $y = y'$ we let

(VI-6)
$$\frac{\partial^2 z}{\partial x^2} = A, \qquad \frac{\partial^2 z}{\partial x\,\partial y} = B, \qquad \frac{\partial^2 z}{\partial y^2} = C$$

to obtain,

(VI-7)
$$\frac{d^2 z}{dr^2} = l^2 A + 2Blm + m^2 C$$

This can be rearranged by completing the square, either in terms of l, l^2 or m, m^2 to obtain,

(VI-8)
$$\frac{d^2 z}{dr^2} = \begin{cases} \dfrac{1}{A}\{(Al + Bm)^2 + m^2(AC - B^2)\} \\[2ex] \dfrac{1}{B}\{(Bl + Cm)^2 + l^2(AC - B^2)\} \end{cases}$$

Thus the whole discussion of the *sign* of d^2z/dr^2 at x', y' hinges on the nature of $AC - B^2$. Examining the various POSSIBILITIES leads to the theorem.

Theorem:
If for $x = x'$, $y = y'$

$$\frac{\partial z}{\partial x} = 0, \qquad \frac{\partial z}{\partial y} = 0$$

and

$$\Delta(x', y') = \frac{\partial^2 z}{\partial x^2} \frac{\partial^2 z}{\partial y^2} - \left(\frac{\partial^2 z}{\partial x \partial y}\right)^2 > 0$$

then the function $z(x, y)$ will have a maximum or minimum at x', y' according as

$$\frac{\partial^2 z}{\partial x^2}, \qquad \left(\text{or } \frac{\partial^2 z}{\partial y^2}\right)$$

is negative or positive at x', y'. If the first derivatives vanish at x', y' but $\Delta < 0$ at x', y' then the point x', y' is neither a maximum nor a minimum. If $\Delta(x', y') = 0$ the question is undecided.

Actually the case $\Delta < 0$ corresponds to x', y' being a saddle point of the function $z(x, y)$.

Appendix VII
CHARACTERISTIC FUNCTION OF A PROBABILITY DISTRIBUTION

A probability distribution $\rho(x)$ defined such that the probability for the value of x to lie in the interval x to $x + dx$ is,

(VII-1) $$dP(x) = \rho(x)\,dx, \qquad -\infty < x < +\infty$$

can be *uniquely* characterized by its moments, the nth moment, $M^{(n)}$, being defined as,

(VII-2) $$M^{(n)} = \int_{-\infty}^{+\infty} x^n \rho(x)\,dx$$

This is evident since the Fourier transform of $\rho(x)$, called the *characteristic function* of the distribution, is uniquely related to the complete set of moments. Thus,

(VII-3) $$\phi(k) = \int_{-\omega}^{+\omega} \rho(x)\, e^{-ikx}\,dx$$

and expanding the exponential in its power series given

(VII-4) $$\phi(k) = \int_{-\infty}^{+\infty} \rho(x)\,dx - ik\int_{-\infty}^{+\infty} x\rho(x)\,dx - k^2\int_{-\infty}^{+\infty} x^2\rho(x)\,dx + ik^3\int_{-\infty}^{+\alpha} x^3\rho(x)\,dx + \ldots$$

or,

(VII-5) $$\phi(k) = M^{(0)} - ikM^{(1)} - k^2 M^{(2)} + ik^3 M^{(3)} + \ldots$$

Hence if *all* the moments are *known* then $\phi(k)$ is determined and hence $\rho(x)$ can be constructed uniquely as the inverse Fourier transform,

(VII-6) $$\rho(x) = \frac{1}{2\pi}\int_{-\infty}^{+\infty} e^{+ikx}(M^{(0)} - ikM^{(1)} - k^2 M^{(2)} + ik^3 M^{(3)} + \ldots)\,dk$$

Appendix VIII
SOLID ANGLE

The element of solid angle, $d\Omega$, which has so many applications in physical problems is defined as the ratio of the area on a sphere, concentric to the point of origin of the angle, to the square of the radius of the sphere. Thus since the area of an element on the surface of a sphere of radius r is

(VIII-1)
$$dA = r^2 \sin \theta \, d\theta \, d\phi$$

where θ and ϕ are the usual polar and azimuthal angles of the spherical coordinate system with center at $r = 0$, we have

(VIII-2)
$$d\Omega = \frac{dA}{r^2} = \sin \theta \, d\theta \, d\phi$$

The total solid angle of a sphere is then 4π.

Appendix IX
LINEAR DEPENDENCE OF FUNCTIONS AND THE WRONSKIAN

A set of functions, $f_i(x), i = 1, 2, \ldots N$, of the variable x are said to be *linearly independent* if the only set of constants, C_i, which renders

(IX-1)
$$C_1 f_1(x) + C_2 f_2(x) + \ldots + C_N f_N(x) = 0$$

is the set $C_i = 0, i = 1, 2, \ldots N$, for *all* x in the internal $a < x < b$ over which the functions are defined. The case of *two* functions is especially important because we are generally concerned with the existance of two linearly independent solutions of a second-order ordinary differential equation. Thus if $f_1(x)$ and $f_2(x)$ *are* linearly dependent on $a < x < b$ then

(IX-2)
$$C_1 f_1(x) + C_2 f_2(x) = 0$$

holds for nonzero C_1 and C_2. Also, differentiating gives

(IX-3)
$$C_1 f'_1(x) + C_2 f'_2(x) = 0$$

For a fixed point x we look upon these as two equations to be solved simultaneously for C_1 and C_2, we know that a solution exists only if the determinant of coefficients vanishes, since the equations are homogeneous. Or,

(IX-4)
$$\frac{C_1}{C_2} = -\frac{f_2(x)}{f_1(x)} = -\frac{f'_2(x)}{f'_1(x)}$$

from the above two equations. In either way we see

(IX-5)
$$W(x) = [f_1(x)f'_2(x) - f_2(x)f'_1(x)] = 0$$

as the condition for the C's to exist. This is called the Wronskian of the two solutions and its vanishing tells us that f_1 and f_2 are linearly dependent functions.

Appendix X
THE SCHMIDT ORTHONORMALIZATION PROCESS AND BASIS VECTORS

Linear vector spaces are not treated in their most general form in this text but the manner in which a *basis set*, or set of mutually orthogonal unit vectors, can be built on a set of linearly independent vectors is fundamental to the concept of a vector space and is exhibited here. This is called the Schmidt orthonormalization process.

We define a vector as an ordered set of elements,

(X-1) $$\mathbf{y} = (y, y_2, \ldots, y_n)$$

Also we define *addition, multiplication by a scalar* and the *inner product,* i.e., multiplication by a scalar a is defined as,

(X-2) $$a\mathbf{y} = (ay_1, ay_2, \ldots, ay_n),$$

addition is defined as,

(X-3) $$\mathbf{y}_a + \mathbf{y}_b = (y_{a1} + y_{b1}, y_{a2} + y_{b2}, \ldots, y_{an} + y_{bn})$$

and the inner product is defined as,

(X-4) $$\mathbf{y}_a \cdot \mathbf{y}_b = \sum_{i=1}^{n} y_{ai} y_{bi}$$

In addition to these operations we have the *length* of a vector given by

(X-5) $$y = \{\mathbf{y}^* \cdot \mathbf{y}\}^{\frac{1}{2}}$$

where the asterisk denotes the complex conjugate. For *real vectors* $\mathbf{y}^* = \mathbf{y}$ and we see this as our usual definition of length, just as the inner product corresponds to the familiar dot product.

A set of m vectors $\mathbf{y}_i, i = 1, 2, \ldots m$ is defined to be *linearly independent* if the only set of constants a_i for which,

(X-6) $$a_1\mathbf{y}_1 + a_2\mathbf{y}_2 + \ldots + a_m\mathbf{y}_m = 0$$

is true is the trivial set $a_i = 0, i = 1, 2, \ldots m.$

Now assuming we have a set of m linearly independent vectors, as just defined, we construct a *unit* vector as,

(X-7) $$\mathbf{1}_1 = \frac{\mathbf{y}_1}{N_1}$$

where N_1 is just the length y_1 of $\mathbf{y}_1$. Then we form a linear combination of this vector with $\mathbf{y}_2$ as

(X-8) $$\mathbf{1}_2 = a_1\mathbf{1}_1 + a_2\mathbf{y}_2$$

which we require to satisfy,

(X-9) $$\mathbf{1}_2 \cdot \mathbf{1}_2 = 1 \quad \text{and} \quad \mathbf{1}_1 \cdot \mathbf{1}_2 = 0$$

These give two equations for determining a_1 and a_2; we find then,

(X-10) $$\mathbf{1}_2 = \frac{\mathbf{y}_2 - (\mathbf{1}_1 \cdot \mathbf{y}_2)\mathbf{1}_1}{\{y_2^2 - (\mathbf{1}_1 \cdot \mathbf{y}_2)^2\}^{\frac{1}{2}}}$$

We continue this procedure with the general form,

(X-11) $$\mathbf{1}_k = a_1\mathbf{1}_1 + a_2\mathbf{1}_2 + \ldots + a_{k-1}\mathbf{1}_{k-1} + a_k\mathbf{y}_k$$

requiring

(X-12) $$\mathbf{1}_i \cdot \mathbf{1}_k = \delta_{ik} = \begin{cases} 1, i = k, i = 1, 2, \ldots, k \\ 0, i \neq k \end{cases}$$

which determine the a_i. The general result is then,

(X-13)
$$1_k = \frac{1}{N_k}\left[\mathbf{y}_k - \sum_{i=1}^{k-1}(\mathbf{1}_i \cdot \mathbf{y}_k)\mathbf{1}_i\right]$$

where

(X-14)
$$N_k = \left\{y_k^2 - \sum_{i=1}^{k-1}(\mathbf{1}_i \cdot \mathbf{y}_k)^2\right\}^{\frac{1}{2}}$$

The process terminates with $k = m$ since there are m linearly independent vectors.

Thus a basic set of m mutually orthogonal unit vectors is constructed. It is not unique, however, i.e., taking the $\mathbf{y}_k$ in a different order yields again m unit vectors which are mutually orthogonal but not identical to the first set.

INDEX

THE HISTORICAL BACKGROUND OF CHEMISTRY, Henry M. Leicester. Evolution of ideas, not individual biography. Concentrates on formulation of a coherent set of chemical laws. 260pp. 5⅜ x 8½. 61053-5

A SHORT HISTORY OF CHEMISTRY, J. R. Partington. Classic exposition explores origins of chemistry, alchemy, early medical chemistry, nature of atmosphere, theory of valency, laws and structure of atomic theory, much more. 428pp. 5⅜ x 8½. (Available in U.S. only.) 65977-1

GENERAL CHEMISTRY, Linus Pauling. Revised 3rd edition of classic first-year text by Nobel laureate. Atomic and molecular structure, quantum mechanics, statistical mechanics, thermodynamics correlated with descriptive chemistry. Problems. 992pp. 5⅜ x 8½. 65622-5

Engineering

DE RE METALLICA, Georgius Agricola. The famous Hoover translation of greatest treatise on technological chemistry, engineering, geology, mining of early modern times (1556). All 289 original woodcuts. 638pp. 6¾ x 11. 60006-8

FUNDAMENTALS OF ASTRODYNAMICS, Roger Bate et al. Modern approach developed by U.S. Air Force Academy. Designed as a first course. Problems, exercises. Numerous illustrations. 455pp. 5⅜ x 8½. 60061-0

DYNAMICS OF FLUIDS IN POROUS MEDIA, Jacob Bear. For advanced students of ground water hydrology, soil mechanics and physics, drainage and irrigation engineering and more. 335 illustrations. Exercises, with answers. 784pp. 6⅛ x 9¼. 65675-6

ANALYTICAL MECHANICS OF GEARS, Earle Buckingham. Indispensable reference for modern gear manufacture covers conjugate gear-tooth action, gear-tooth profiles of various gears, many other topics. 263 figures. 102 tables. 546pp. 5⅜ x 8½. 65712-4

MECHANICS, J. P. Den Hartog. A classic introductory text or refresher. Hundreds of applications and design problems illuminate fundamentals of trusses, loaded beams and cables, etc. 334 answered problems. 462pp. 5⅜ x 8½. 60754-2

MECHANICAL VIBRATIONS, J. P. Den Hartog. Classic textbook offers lucid explanations and illustrative models, applying theories of vibrations to a variety of practical industrial engineering problems. Numerous figures. 233 problems, solutions. Appendix. Index. Preface. 436pp. 5⅜ x 8½. 64785-4

STRENGTH OF MATERIALS, J. P. Den Hartog. Full, clear treatment of basic material (tension, torsion, bending, etc.) plus advanced material on engineering methods, applications. 350 answered problems. 323pp. 5⅜ x 8½. 60755-0

A HISTORY OF MECHANICS, René Dugas. Monumental study of mechanical principles from antiquity to quantum mechanics. Contributions of ancient Greeks, Galileo, Leonardo, Kepler, Lagrange, many others. 671pp. 5⅜ x 8½. 65632-2

A CATALOG OF SELECTED
DOVER BOOKS
IN SCIENCE AND MATHEMATICS

Astronomy

BURNHAM'S CELESTIAL HANDBOOK, Robert Burnham, Jr. Thorough guide to the stars beyond our solar system. Exhaustive treatment. Alphabetical by constellation: Andromeda to Cetus in Vol. 1; Chamaeleon to Orion in Vol. 2; and Pavo to Vulpecula in Vol. 3. Hundreds of illustrations. Index in Vol. 3. 2,000pp. 6⅛ x 9¼.
23567-X, 23568-8, 23673-0 Three-vol. set

THE EXTRATERRESTRIAL LIFE DEBATE, 1750–1900, Michael J. Crowe. First detailed, scholarly study in English of the many ideas that developed from 1750 to 1900 regarding the existence of intelligent extraterrestrial life. Examines ideas of Kant, Herschel, Voltaire, Percival Lowell, many other scientists and thinkers. 16 illustrations. 704pp. 5⅜ x 8½.
40675-X

A HISTORY OF ASTRONOMY, A. Pannekoek. Well-balanced, carefully reasoned study covers such topics as Ptolemaic theory, work of Copernicus, Kepler, Newton, Eddington's work on stars, much more. Illustrated. References. 521pp. 5⅜ x 8½.
65994-1

AMATEUR ASTRONOMER'S HANDBOOK, J. B. Sidgwick. Timeless, comprehensive coverage of telescopes, mirrors, lenses, mountings, telescope drives, micrometers, spectroscopes, more. 189 illustrations. 576pp. 5⅜ x 8¼. (Available in U.S. only.)
24034-7

STARS AND RELATIVITY, Ya. B. Zel'dovich and I. D. Novikov. Vol. 1 of *Relativistic Astrophysics* by famed Russian scientists. General relativity, properties of matter under astrophysical conditions, stars, and stellar systems. Deep physical insights, clear presentation. 1971 edition. References. 544pp. 5⅜ x 8¼.   69424-0

Chemistry

CHEMICAL MAGIC, Leonard A. Ford. Second Edition, Revised by E. Winston Grundmeier. Over 100 unusual stunts demonstrating cold fire, dust explosions, much more. Text explains scientific principles and stresses safety precautions. 128pp. 5⅜ x 8½.
67628-5

THE DEVELOPMENT OF MODERN CHEMISTRY, Aaron J. Ihde. Authoritative history of chemistry from ancient Greek theory to 20th-century innovation. Covers major chemists and their discoveries. 209 illustrations. 14 tables. Bibliographies. Indices. Appendices. 851pp. 5⅜ x 8½.
64235-6

CATALYSIS IN CHEMISTRY AND ENZYMOLOGY, William P. Jencks. Exceptionally clear coverage of mechanisms for catalysis, forces in aqueous solution, carbonyl- and acyl-group reactions, practical kinetics, more. 864pp. 5⅜ x 8½.
65460-5

Physics

OPTICAL RESONANCE AND TWO-LEVEL ATOMS, L. Allen and J. H. Eberly. Clear, comprehensive introduction to basic principles behind all quantum optical resonance phenomena. 53 illustrations. Preface. Index. 256pp. 5⅜ x 8½.   65533-4

ULTRASONIC ABSORPTION: An Introduction to the Theory of Sound Absorption and Dispersion in Gases, Liquids and Solids, A. B. Bhatia. Standard reference in the field provides a clear, systematically organized introductory review of fundamental concepts for advanced graduate students, research workers. Numerous diagrams. Bibliography. 440pp. 5⅜ x 8½.   64917-2

QUANTUM THEORY, David Bohm. This advanced undergraduate-level text presents the quantum theory in terms of qualitative and imaginative concepts, followed by specific applications worked out in mathematical detail. Preface. Index. 655pp. 5⅜ x 8½.   65969-0

ATOMIC PHYSICS (8th edition), Max Born. Nobel laureate's lucid treatment of kinetic theory of gases, elementary particles, nuclear atom, wave-corpuscles, atomic structure and spectral lines, much more. Over 40 appendices, bibliography. 495pp. 5⅜ x 8½.   65984-4

AN INTRODUCTION TO HAMILTONIAN OPTICS, H. A. Buchdahl. Detailed account of the Hamiltonian treatment of aberration theory in geometrical optics. Many classes of optical systems defined in terms of the symmetries they possess. Problems with detailed solutions. 1970 edition. xv + 360pp. 5⅜ x 8½.   67597-1

THIRTY YEARS THAT SHOOK PHYSICS: The Story of Quantum Theory, George Gamow. Lucid, accessible introduction to influential theory of energy and matter. Careful explanations of Dirac's anti-particles, Bohr's model of the atom, much more. 12 plates. Numerous drawings. 240pp. 5⅜ x 8½.   24895-X

ELECTRONIC STRUCTURE AND THE PROPERTIES OF SOLIDS: The Physics of the Chemical Bond, Walter A. Harrison. Innovative text offers basic understanding of the electronic structure of covalent and ionic solids, simple metals, transition metals and their compounds. Problems. 1980 edition. 582pp. 6⅛ x 9¼.   66021-4

HYDRODYNAMIC AND HYDROMAGNETIC STABILITY, S. Chandrasekhar. Lucid examination of the Rayleigh-Benard problem; clear coverage of the theory of instabilities causing convection. 704pp. 5⅜ x 8¼.   64071-X

INVESTIGATIONS ON THE THEORY OF THE BROWNIAN MOVEMENT, Albert Einstein. Five papers (1905–8) investigating dynamics of Brownian motion and evolving elementary theory. Notes by R. Fürth. 122pp. 5⅜ x 8½.   60304-0

THE PHYSICS OF WAVES, William C. Elmore and Mark A. Heald. Unique overview of classical wave theory. Acoustics, optics, electromagnetic radiation, more. Ideal as classroom text or for self-study. Problems. 477pp. 5⅜ x 8½.   64926-1